MICROWAVE IMAGING TECHNIQUES

WILEY SERIES IN REMOTE SENSING

Jin Au Kong, Editor

Tsang, Kong, Shin · THEORY OF MICROWAVE REMOTE SENSING
Hord · REMOTE SENSING: METHODS AND APPLICATIONS
Elachi · INTRODUCTION TO THE PHYSICS AND TECHNIQUES OF REMOTE SENSING
Szekielda · SATELLITE MONITORING OF THE EARTH
Maffet · TOPICS FOR A STATISTICAL DESCRIPTION OF RADAR CROSS SECTIONS
Asrar · THEORY AND APPLICATIONS OF OPTICAL REMOTE SENSING
Steinberg and Subbaram · MICROWAVE IMAGING TECHNIQUES

MICROWAVE IMAGING TECHNIQUES

Bernard D. Steinberg
University of Pennsylvania

Harish M. Subbaram
INTERSPEC, Inc.

A WILEY-INTERSCIENCE PUBLICATION

JOHN WILEY & SONS, INC.

New York · Chichester · Brisbane · Toronto · Singapore

Library of Congress Cataloging-in-Publication Data:

Steinberg, Bernard D.

Microwave imaging techniques / Bernard D. Steinberg, Harish M. Subbaram.

p. cm.—(Wiley series in remote sensing)

"A Wiley-Interscience publication."

Includes bibliographical references (p.

ISBN 0-471-50078-X

1. Microwave remote sensing. I. Subbaram, Harish M. II. Title. III. Series.

G70.4.S74 1991

621.36′78–dc20 90-12941

CIP

Printed in the United States of America

10 9 8 7 6 5 4 3 2 1

To My Loving Wife Jacqueline, My Lifetime Companion, Inspiration, and Support

To My Parents M. R. Subbaram and Sharada Subbaram

CONTENTS

PREFACE

This book is the third of a series dealing with large arrays and microwave imaging. The first book, *Principles of Aperture and Array System Design: Including Random and Adaptive Arrays* (Wiley, 1976), introduced the designers of communications and radar systems to the spatial domain. Its purpose was to bridge the gap between two disciplines, one practiced by the antenna designer and the student of electromagnetics, and the other by the student of statistical communication theory and system design. At the time that the book was written the antenna designer typically viewed the end point of his work as the single-port connector to the radio or radar set. A similarly narrow point of view was taken by the set designer, who considered his job to begin at that same connection. In short, the antenna designer dealt with the spatial problems and the set designer dealt with the temporal problems. The function of the antenna was to convert signals that were received simultaneously over space to a single time-varying waveform which the antenna delivered to the receiver. The receiver accepted the single time-varying waveform and provided the temporal processing or filtering. It was a simple division: spatial filtering, the job of the antenna designer; temporal filtering, the job of the set designer. A similar division of labor took place on transmission.

When interest in the phased array developed in the late 1960s and early 1970s, it became evident that system design should combine the spatial and temporal properties together rather than deal with them in serial fashion. A phased-array receiving antenna intrinsically has a multiport output. It is not essential that the ports be combined at RF and the output signal delivered as a single time-varying waveform to the single-port input of the receiver. The multiple output ports can be delivered directly to a multiport receiver. More powerful signal processing would then become available without the artificial constraint of spatial first, temporal next.

Treating the entire system as a combined spatial–temporal processor was strange at the time both to the electromagnetic and radar/communications communities. It was not strange, however, in the field of underwater sound, where spatio-temporal system design had been common for many years. The purpose of the book was to acquaint workers in radar and communications with the basic geometric properties of antennas and phased arrays, and the spatial properties of the radiation patterns. They were shown that their own tools—Fourier theory, convolution theory, linear systems theory, and the theory of stochastic processes—were the same basic tools that they needed to bridge the gap of understanding into the spatial domain. In short, they were adequately educated to deal with the spatial domain; they only needed exposure to the world of electric current excitations in an aperture and radiation fields and sidelobes and so forth.

The book had one more objective. The work of the Valley Forge Research Center of the University of Pennsylvania was focused on very high-resolution microwave imaging. Because of that fact the theory of the second part of the book dealt with the very large antenna arrays necessary to achieve high resolution, the problems of these arrays, and the properties of radiation fields that may result from very large antenna arrays. The thinned aperiodic array and the even more highly thinned random array were discussed and the mathematics of the random array was developed. Also, because very large arrays are intrinsically error-prone, considerable attention was given to tolerance theory for large systems.

While the first book provided the intellectual basis for the study of large antenna arrays, the second one, *Microwave Imaging with Large Antenna Arrays: Radio Camera Principles and Techniques* (Wiley, 1983), was a how-to-do-it book describing the intrinsic problems of large microwave systems and ways to overcome these problems. It demonstrated that it is possible to make very large antenna arrays that are highly distorted and highly thinned, and by using self-calibration procedures called adaptive beamforming to cause these arrays to produce microwave imagery as good as from diffraction-limited antennas. The second book amplified and extended the properties of the random array and developed the theory of diversity-combining as applied to random antenna arrays for the purpose of reducing their high sidelobe content.

The primary thrust of the second book was the achievement of high angular resolution in microwave imagery. The book recognized that high resolution was a necessary condition for high-quality imagery but that it was not a sufficient condition. It left the development of the subject "How To Make *Good* Microwave Pictures" to this book, for which that is an appropriate subtitle.

This book has two distinct parts. The first describes what microwave imagery is, how it is obtained, and what the problems are in obtaining good microwave images. This part consists of the first four chapters. The remaining eight chapters describe many techniques for enhancing image quality. Chapters 5 and 6 deal with ways for achieving diversity-combining of images. Chapter 7 introduces deconvolution for eliminating image artifacts and for the general improvement of image quality; it dwells mainly on a technique called CLEAN, which was first developed for radio astronomy. Chapters 8 and 9 discuss the self-calibration methods necessary for successful operation of very large arrays. Chapter 10 describes means for reducing the data-handling requirements of large, high-resolution phased-array systems. Chapter 11 treats the theory and practice of superresolution or non-Fourier processing of spatial data, extending it into the near field and applying it to microwave imaging for the purpose of enhancing the available resolution from a given aperture. Display techniques found useful at the Valley Forge Research Center end the book in Chapter 12.

Much but not all of the work described in this book came out of the Valley Forge Research Center. Faculty, students, and staff members who have contributed are Professors R. S. Berkowitz, C. N. Dorny, F. Haber, and S. H. Taheri (deceased); my students Drs. E. H. Attia, J. Tsao, H. Subbaram, Z. Liang, and B. Kang, and Mr. W. Lee and Ms. S. Patrick; and members of the technical staff Messers D. Carlson, W. Whistler, S. Seeleman, T. Seeleman, and W. Borders. The book itself is a joint effort with my student, now colleague, Dr. Harish Subbaram. Preparation and production of the finished manuscript was the responsibility of Mrs. Shirley Levy, to whom the authors are deeply grateful. Mr. Sam Seeleman, longtime site manager of VFRC and close friend, designed and constructed all the large, complicated antenna arrays, including the 83-m array that for the first time provided us with human-optical angular resolution at microwaves.

BERNARD D. STEINBERG

Philadelphia, Pennsylvania
April 1991

MICROWAVE IMAGING TECHNIQUES

1

INTRODUCTION TO MICROWAVE IMAGING

1-1 CATEGORIES OF MICROWAVE IMAGING SYSTEMS

Microwave imaging is picture-taking with long-wavelength microwave energy. The target or the scene to be imaged is illuminated by a microwave transmitter and the reflected or scattered energy is intercepted by the receiving antenna. The receiving system processes the microwave echoes much as the lens of a camera transforms the optical field at the lens into an image in the focal plane. A microwave source such as a radar transmitter plays the role that the sun does in optical photography.

Figure 1-1 shows the basic similarity between optical and microwave imaging: both require a source of radiation, a scattering or reflecting surface, and receivers. The images are entirely different, however, and the figure indicates one reason why this is so: the rough surface scatters optical waves while the same surface may be smooth to microwaves and produce specular reflections. This distinction and its consequences are discussed in Section 4-2.

Another difference between optical and microwave imaging is shown in Fig. 1-2. The camera images the boy in the boat from a position above the boat. The optical photograph is an angle–angle image. The distance dimension has been collapsed onto the image plane.

To achieve the same image orientation with a pulsed microwave system, the direction to the viewing equipment is rotated by 90°. The reason is that the image dimensions are range–angle. The radar transmitter launches a pulse (or a succession of pulses) toward the boat. The first echo to be received is from the stern of the boat; the last echo is from the fish. The range or distance dimension replaces one of the angular dimensions in the optical photo. The other dimension is angle or cross range. Resolution in this direction requires a

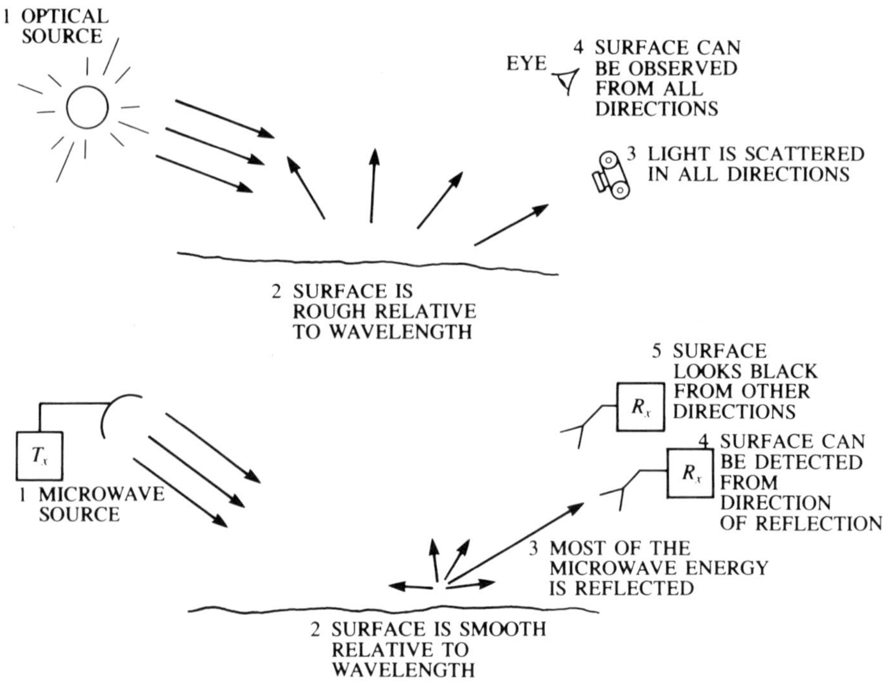

Figure 1-1 Short-wavelength light is scattered; long-wavelength microwave radiation is largely reflected. (From ref. [97])

large microwave aperture. This antenna, which usually is a phased array, and the image processor that follows it constitute the analog of the optical lens.

This book addresses terrestrial microwave imaging, which may be defined as picture-taking with microwaves at long distances on or near the surface of the earth. Another field of microwave imaging is radio astronomy, in which huge antenna arrays receive microwave radiation from cosmological sources and produce images displaying the distribution of sky brightness [117]. Still another field of microwave imaging research is work confined to the laboratory and the anechoic chamber. Here the targets are often scale models of full-size targets.

The three activities share many common theoretical tools and some experimental tools. The differences that result from the different natures of the targets and their distances, however, justify the distinction. For example, radio astronomy is a purely passive technique, whereas both terrestrial and laboratory microwave imaging require active transmitters. Each target of the radio astronomer (such as a star) is an energy source that generates and radiates energy in the microwave region. The targets or sources are not "seen" by reflected or scattered illumination. Because the sources themselves are active, the system—not having its own illuminator—is termed *passive*. Terrestrial and

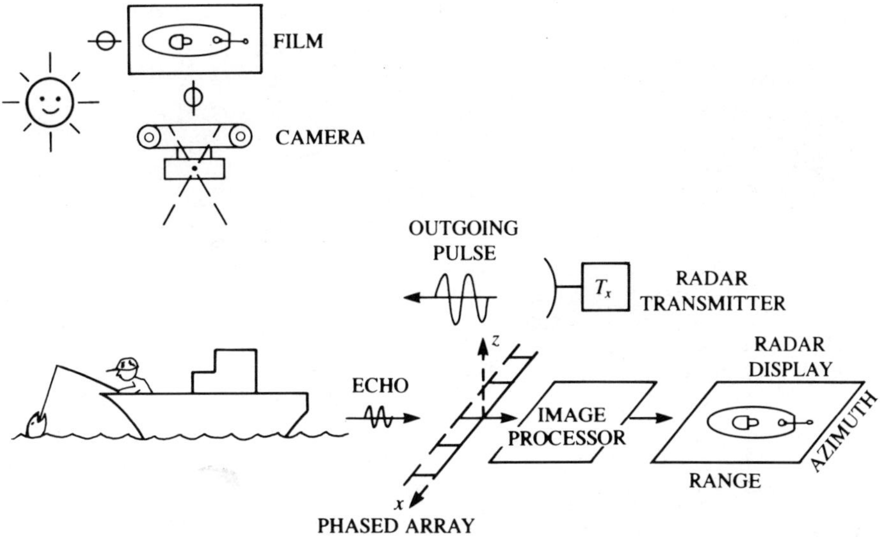

Figure 1-2 Aerial photograph and surface-to-surface microwave image provide the same view of the target. The camera's image plane is normal to the direction of the target, while the radar's image plane contains the target direction.

laboratory imaging systems, on the other hand, require transmitters and are called *active*.

Another distinction also results directly from the fact that radio astronomy targets are energy sources. Because they are physically independent of each other (stars, pulsars, quasars), their radiated waveforms are independent. As a consequence the radiation field at the receiving array is a time-varying field. By contrast, the radiation field at an array resulting from reflections from a radar transmitter is unchanged following each pulse transmission, provided only that the geometry of the target scene does not change. The radio astronomy radiation field is called *incoherent* or *noncoherent*. The microwave imaging field based upon an active transmitter is called *coherent*. This distinction affects the availability and choice of signal processing techniques used in forming the images.

Figure 1-3 illustrates the difference. On the left are shown two independent waveforms $f_1(t)$ and $f_2(t)$ from active sources. Their spherical radiation fields interfere in space and the complex sum arrives as a spatially and temporally varying wavefront, illustrated in the sketch as $e(x, t)$, at the radio astronomy antenna array. Because of the independence of the source waveforms, the spatial distribution of the energy at the antenna array varies continuously in time. The statistics of the radiation field do not vary, however. Consequently, from a time series of measurements made at each antenna, statistical averages of functions of these measurements can be estimated and these prove useful in image

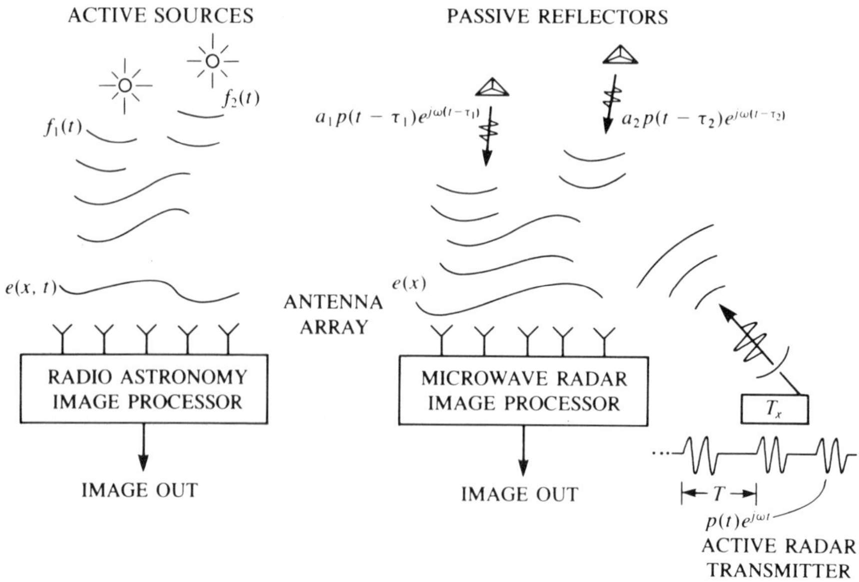

Figure 1-3 (*Left*) Radio astronomy sources are active and independent. The radiation field at the imaging array varies spatially and temporally. (*Right*) Microwave imaging targets are passive reflectors of the sequence of identical RF pulses from the radar transmitter. Successive radiation fields at the imaging array are identical.

formation. The aperture synthesis technique, used commonly at radio astronomy observatories today, is an example [113].

On the right of Fig. 1-3 is the microwave imaging case. A radar transmitter launches a wave $p(t)\exp(j\omega t)$, which consists of a sequence of RF pulses spaced by the interpulse period T. The transmitted wave arrives at the target reflectors at times τ_1 and τ_2 following transmission. Reradiation from the first reflector in the direction of the microwave imaging array is $a_1 p(t - \tau_1)\exp[j\omega(t - \tau_1)]$, and similarly for the second reflector. The coefficients a_1 and a_2 are the complex reflection coefficients of the reflectors in the direction of the receiving antenna array. The reradiated waves are not independent. They are derived from a common source and differ only by the reflection coefficients and time delays. Provided that the a_i and the τ_i are fixed, the reflected waves interfere identically from pulse to pulse. As a consequence, a succession of radiation fields across the microwave imaging array in response to a series of transmitting pulses are identical. Other than increasing the signal-to-noise ratio, no amount of averaging of functions of the measurements increases the information content derivable from the radiation field.

A third difference lies in the constancy of the radio astronomical scene and the ability to observe it for many hours as the earth rotates. Rotation moves the antenna array, thereby effectively creating a gigantic synthetic aperture

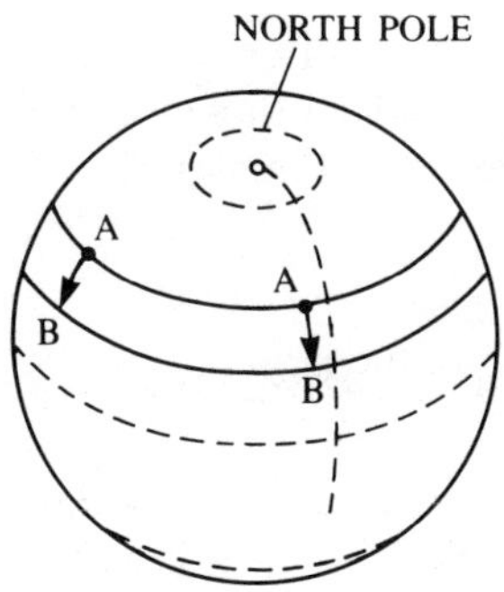

Figure 1-4 Rotation of the earth changes locations of antenna elements and their projected vector separations.

(see Fig. 1-4). Except for the very special case of inverse synthetic aperture imaging (ISAR), this relative motion of the target and the receiving antenna is not available in terrestrial imaging. (ISAR is discussed in Section 5-7.)

The distinctions between terrestrial and laboratory microwave imaging are of a different nature. Both utilize microwave transmitters and therefore the designer has control over the radiated waveform. It may be pulse or continuous wave, fixed frequency, variable frequency or step-frequency, long pulse transmission, or anything else. The difference lies instead in the target environment and is generally much more limiting for long-distance terrestrial imaging. The radiated wavelength and bandwidth are relatively unrestricted in the laboratory and operation can take place with as short a wavelength and as wide a band as modern experimental equipment permits. In practice, laboratory imaging systems can be designed for arbitrarily low frequencies up to several tens of gigahertz. Terrestrial imaging systems, on the other hand, are limited both in wavelength and bandwidth. The wavelength is limited by atmospheric absorption and scattering to centimeter and larger waves, while the range of accessible frequencies is limited by the Federal Communications Commission; ten percent contiguous bandwidth is the upper limit as a rule of thumb. Although in certain circumstances the experimental bandwidth can be extended considerably, it is rarely worth the investment in developing design techniques for systems whose bandwidths exceed FCC regulations, for such systems are not capable of satisfactory testing and hence are unlikely candidates for procurement. Conversely, a severe size limitation pertains to systems designed to operate in a laboratory whereas outdoor long-distance systems are relatively unconstrained in the land available for the implementation of large antenna arrays.

Another factor that distinguishes indoor and outdoor systems is the relatively high degree of environmental control available to the designer of a laboratory imaging system operating within an anechoic chamber. By contrast, the outdoor imager always encounters multipath and scattering phenomena. Lastly, the indoor discipline is confined to imaging small models of targets while the other performs imaging experiments on the targets themselves.

These distinctions are neither good nor bad per se, but they influence design procedures sufficiently to render them three different areas of imaging research. This book deals only with terrestrial imaging.

1-2 THE OBJECTIVE OF TERRESTRIAL MICROWAVE IMAGING

The purpose is *not* to take beautiful pictures; microwave imaging is not an art form. The objective is to obtain images of sufficient quality that targets may be counted, their deployment or relative positions observed, and their natures determined. It is desired to measure the size of a target, its shape, and its orientation. Is it a school bus or an Army tank? Is it a military or a civilian airliner? Better still, what aircraft (or ship or automobile) is it? Given a troubled airplane approaching an airport, are its wheels down?

None of these questions intrinsically demands an image of the target to obtain the answer. Experience has shown, however, that without such images answers are often not forthcoming. Thus microwave picture-taking is an aid to target recognition and classification.

1-3 TERRESTRIAL IMAGING SYSTEMS

1-3-1 Radio Camera [97]

The term *radio camera* pertains to a phased array radar system in which the array is sufficiently large to obtain very high angular resolution. It operates upon a snapshot of data, that is, the echoes from a single pulse transmission received by each of the antennas in the system. Generally the aperture is so large as to be mechanically or electrically unstable, thereby requiring dynamic self-calibration. The information used to achieve self-calibration is based upon measurements of the radiation field. A procedure called *adaptive beamforming* performs this function. Signal processing is entirely in the spatial domain. Many of the microwave images presented in this book are radio camera images.

1-3-2 Synthetic Aperture Radar (SAR) [6]

Instead of requiring the multiplicity of antenna elements employed in a phased array, SAR time-shares a single antenna element to form a large aperture. One pulse is transmitted for each successive position of the receiver and the received echo sequence is stored. Because of the time-shared nature of the operation, the data are acquired in time and the signal processing is entirely temporal. The antenna element is typically transported in an airplane; hence SAR is usually identified with airborne radar.

1-3-3 Inverse Synthetic Aperture (ISAR) [19]

The motion of the target replaces the motion of the aircraft that SAR requires for successful operation. The operation is again time-shared and the processing is entirely temporal. Adaptive beamforming is used to compensate for irregularities in the flight path of the airplane. The method is described in Chapter 5 and several aircraft images based upon ISAR plus adaptive beamforming are shown there and elsewhere in the book.

1-3-4 Inverse Synthetic Radio Camera (ISRC)

It is possible in principle to combine the radio camera and ISAR to form gigantic arrays capable of telescopic imaging resolution. The elements of the radio camera are placed hundreds or thousands of wavelengths apart. The grating lobes or the aliasing images that would normally result are eliminated by ISAR processing. The signal processing in this case is both spatial and temporal. Again, adaptive beamforming is central to the success of the procedure. An experiment demonstrating the validity of the concept is presented in Section 5-7.

1-4 IMAGE QUALITY

There is no known theory of image quality from which a metric can be derived and applied to an arbitrary image *independent of the source distribution that produced it*. Many measures can be designed, however, if the constraint of the italicized phrase in the sentence above is removed. For example, comparison of the true source distribution and its image in a normalized mean square sense is one measure. Another is the normalized correlation between the source distribution and its image. Both require knowledge of the source distribution, which, however, is denied the user in practical long-range imaging systems. Hence a useful theory of image quality would provide an absolute numerical measure of the quality of an image without requiring knowledge of the source distribution. Such a measure has not appeared in the literature.

A source-free measure of image quality would be valuable because it would permit the use of the concept called *image feedback control* (IFC) in the design of imaging systems [97, section 4.5]. With IFC design the coefficients of the image signal processor are adjusted by feedback instructions from the image-quality-measuring circuit or algorithm so as to maximize the image quality (see Fig. 1-5). Notwithstanding the absence of such a metric, certain attributes of images are known to be important factors in the subjective judgment of image quality. Four are discussed below.

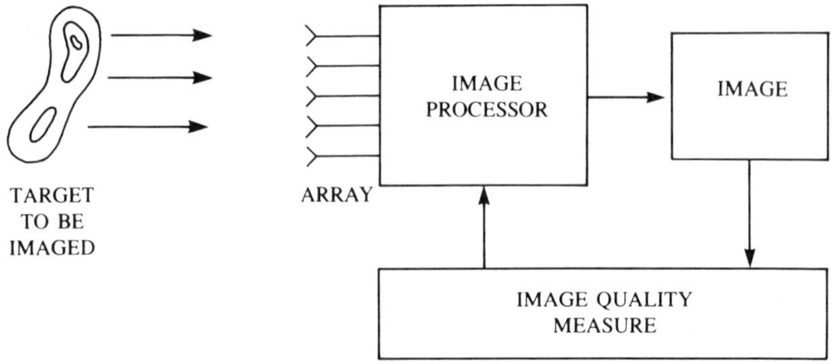

Figure 1-5 Image feedback control. The coefficients of the image processor are automatically adjusted so as to maximize image quality.

1-4-1 Resolution

High resolution is a necessary attribute of image quality. Resolution pertains to the cell size in the target or scene to be imaged and not to the pixel size in the image. The cell size shown in Fig. 1-6 is $R\,\Delta\theta$ transverse to the direction of the beam and ΔR in the longitudinal or range direction. R is the range or distance to the target, $\Delta\theta$ is the beamwidth of the imaging system, and ΔR is the extent of the range cell. $\Delta\theta$ is approximately λ/L for any antenna, where λ is the wavelength, L is the size of the aperture (Section 2-5), and ΔR is approximately $c/2B$, where c is the speed of light and B is the signal bandwidth. More precisely, resolution in the direction r is the minimum spacing in that direction at which two targets of nominally equal strength can be resolved. As will be shown later, resolution cells the order of 1 m are useful in terrestrial imagery. This is approximately the transverse resolution of the human eye at

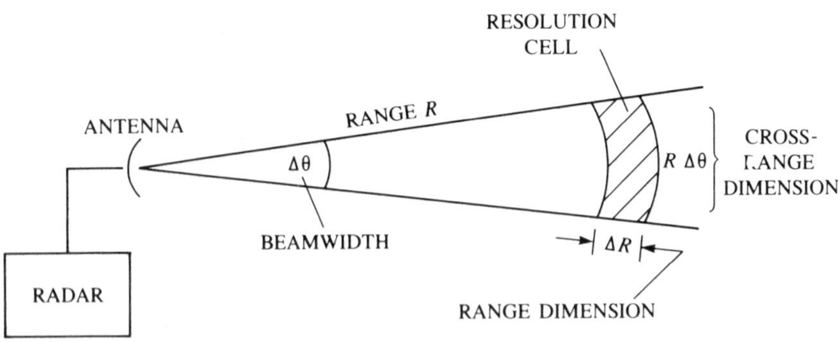

Figure 1-6 The resolution cell is defined by the range dimension ΔR of the radiated waveform and the cross-range dimension $R\Delta\theta$ of the beam. The beamwidth $\Delta\theta \approx \lambda/L$, where λ is the wavelength of the radiated energy and L is the size of the antenna.

a distance of 3 km and of a camera in bright sunlight at 10 km. A qualitative sense of human optical resolution is obtained by noting that, when flying at an altitude of 30 000 to 40 000 feet, one can distinguish between automobiles on a road but cannot discern or count the wheels.

1-4-2 Target Dynamic Range

The dynamic range of microwave scattering centers within a manmade target is very large, as much as 50 to 90 dB [68]. The large echoes are returns from flat plates or corners, such as the broadside mirrorlike reflection from a fuselage of an airplane or the corner formed by the junction of the vertical and horizontal stabilizers. Weak scattering from corners or edges of surfaces is easily obscured by the specular highlights. Yet it is the edge and corner scattering that defines the shape, size, and orientation of a target, whereas the large specular reflections are primarily useful for detection. Thus the easy-to-image bright spots in an image tend to mask the shape-defining weaker scattering.

The fundamental limitation to target dynamic range comes from the sidelobe structure of the radiation pattern of the antenna. The radiation pattern of a uniformly illuminated aperture has the form $(\sin x)/x$, where x is proportional to the sine of the angle from the normal to the antenna. The magnitude of the first sidelobe is $2/(3\pi)$, which is approximately 13 dB below the main lobe. Successive sidelobes fall off at the rate of 6 dB per octave (in x), which is very slow. Thus targets 10 to 15 dB smaller than the largest target are very difficult to distinguish from the sidelobe responses of the system to the larger target. Figure 1-7 illustrates the confusion that results from the high sidelobes of a random array of antenna elements, which is a common aperture distribution for the very large arrays required for high-angular-resolution imaging.

In principle the side radiation pattern can be reduced as much as desired through design of the antenna weight function or aperture taper, and one can thereby eliminate the masking effect of high sidelobes. Most antenna practice, however, leads to sidelobe levels no lower than 30 or 40 dB, although special procedures can produce antenna arrays with the entire side radiation 60 dB or more below the main lobe [45]. These methods require severe mechanical and electrical tolerances that are unrealistic in large imaging systems. In fact, very high-resolution systems often must be designed to perform satisfactorily even if, during operation, the properties of the side radiation pattern decay to those of the random array [94]. The large-aperture effects upon design practice are discussed in Section 4-3.

Practical factors limit target dynamic range. One is saturation of the front end of the receiver by strong echoes and the other is the limited dynamic range in the display. Prevention of loss of target detail requires careful receiver design to avoid saturation. Proper choice of the nonlinear transformation from the voltage out of the signal processor to intensity in the display can be used to enhance the dynamic range in the display.

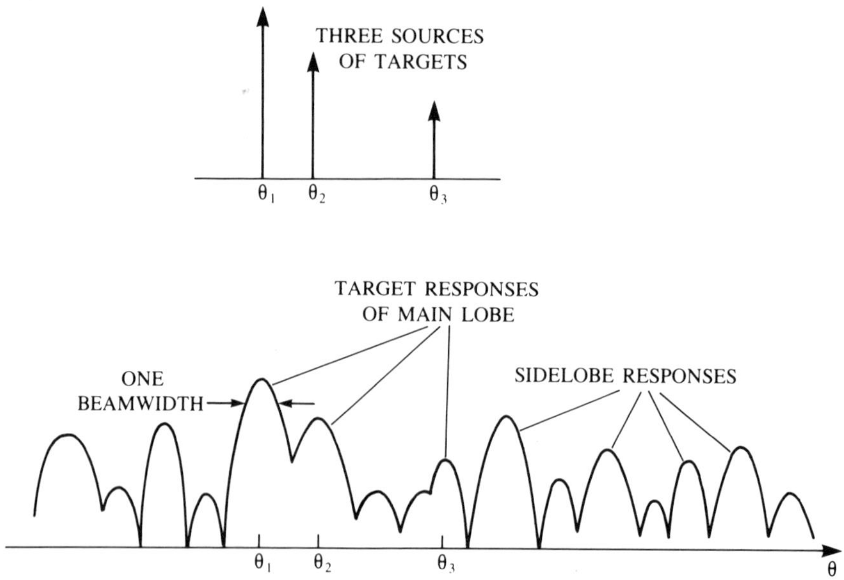

Figure 1-7 Image (in one dimension) obtained by scanning a three-target scene with a random array. (From ref. [97]).

1-4-3 Image Contrast

Image contrast is defined as the ratio of the brightest intensity in the image to the average background or gray level. Large image contrast is required for the same reason that large target dynamic range is required, namely, to prevent the suppression of the weakly scattered signals.

1-4-4 Edge and Boundary Continuity

There is no difficulty in determining the nature of objects in optical photographs even when the objects are partially obscured. Shape, color, texture, and shading provide innumerable clues. By and large these factors are missing in microwave images. Furthermore, the information content per unit area is huge in an optical photograph as compared to a microwave image, due both to the wavelength ratio as well as the disparity in resolution cell size. As a consequence, continuity of boundaries and the preservation of edges, while important in optical images, are vital qualities in microwave images (Section 4-2).

Edges often disappear because the weak scattering from them fails to compete with the strong specular echoes, as discussed earlier. Boundary continuity is often affected for the same reason. In both cases, a large target dynamic range is valuable.

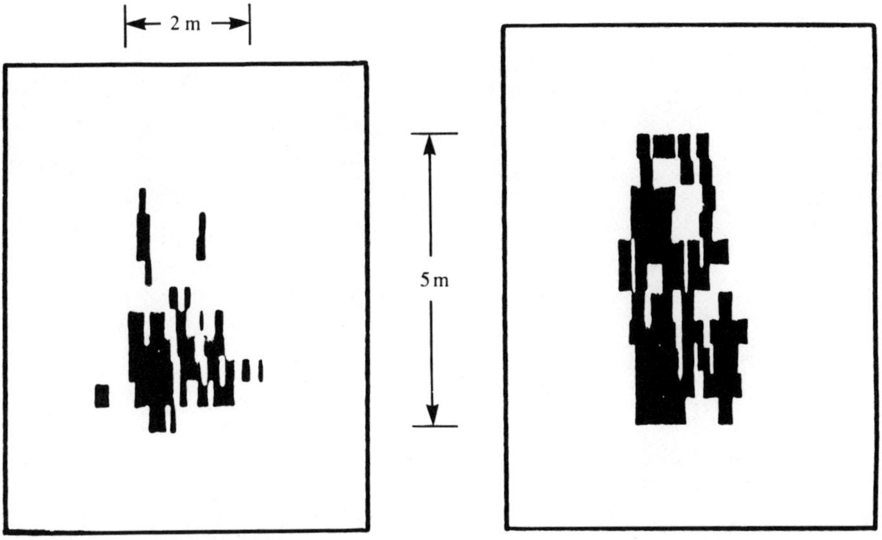

Figure 1-8 Two microwave images of a small truck made from the same data. (*Left*) Original image, illustrating decomposition of vehicle by speckle into several smaller targets. (*Right*) After deconvolution by CLEAN technique, plus diversity. The wavelength is 3 cm. The range resolution is 1 m. The target distance from the imaging array is 120 m. The array size is 30 m.

Another factor that breaks boundaries and causes large targets to appear as a multiplicity of smaller ones is the coherent interference phenomenon called *speckle* (Section 4-2). An example is seen in the left image of Fig. 1-8. The image is of a small truck about 2 × 5 m. Diversity (Chapter 5) and deconvolution procedures (Chapter 7) that directly attack the speckle problem are helpful in preserving target boundary continuity. The right image, which is made from the same microwave echoes, illustrates the improvement in image quality available from these techniques: here the shape, size, and orientation of the vehicle are apparent.

2

THE MATHEMATICS OF MICROWAVE IMAGING

2-1 WAVELENGTH

The long wavelength required for terrestrial imaging is responsible for many of the problems encountered in imaging system design. Figure 2-1 shows atmospheric attenuation of radio waves resulting from gaseous absorption as a function of wavelength and frequency. The attenuation coefficient rises at a modest rate from low frequencies until the 10 to 30 GHz region, beyond which it grows very rapidly. This chart shows that wavelengths no shorter than 1 to 3 cm are appropriate. Such wavelengths are useful for target distances of a few tens of kilometers; 10 and 25 cm wavelengths are more common for the greater distances required by search radars. Still larger wavelengths are used when especially low propagation loss is needed.

It is interesting to compare the wavelength regime useful for microwave imaging, which is nominally 1 cm to 1 m, to the optical/infrared wavelength regime, which extends from 5×10^{-7} to nominally 10^{-5} meters. The wavelengths employed in microwave imaging are 10^3 to 10^6 times longer. This huge difference dramatically alters the scattering characteristics of targets and induces a corresponding effect upon their images. Secondly, the large wavelength ratio plays a crucial role in the physical characteristics of microwave imaging systems.

2-2 VERY LARGE APERTURE REQUIRED

The long wavelength imposes a requirement for very large apertures. To achieve the resolving power of an optical instrument, the microwave aperture must be larger by the wavelength ratio. This is because the resolution of an aperture is

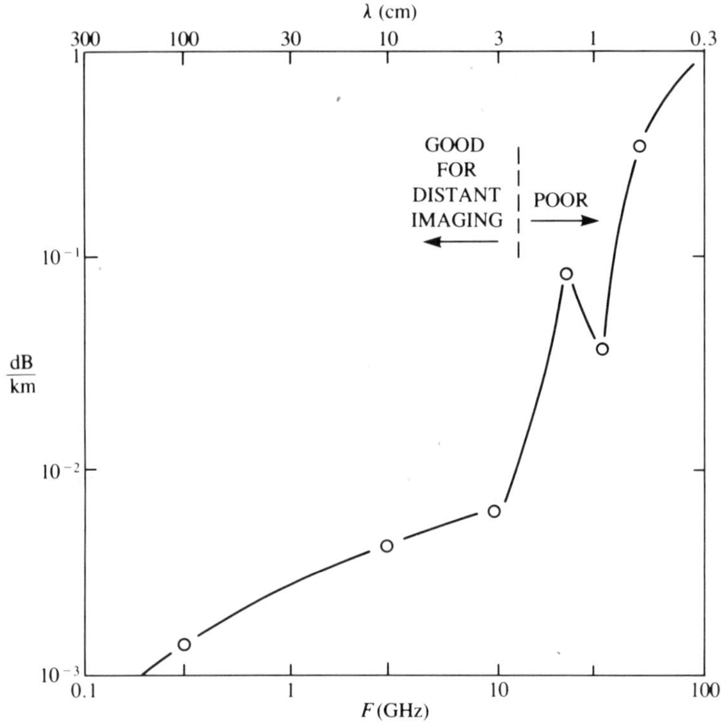

Figure 2-1 Gaseous absorption (dB/km) vs. frequency and wavelength at 5000 ft altitude. (Data from ref. 92).

determined not by its physical size but by the number of wavelengths across it, a fact that follows from scalar diffraction theory. Assuming a one-dimensional aperture defined by its current density $i(x)$, the field strength E at point r, θ is given by

$$E(r, \theta) = K \int \frac{i(x) \exp[-jk\rho(x, r, \theta)]}{\rho(x, r, \theta)} dx \tag{2-1}$$

where K is a constant, $k = 2\pi/\lambda$ is the wave number of the radiated energy, and $\rho(x, r, \theta)$ is the distance from a point in the aperture to the field point. The notation is indicated in Fig. 2-2. The distance ρ can be expressed as a polynomial:

$$\rho^2 = r^2 + x^2 - 2rx \sin \theta$$

$$\rho = r\left[1 - \frac{2xu}{r} + \frac{x^2}{r^2}\right]^{1/2}$$

$$= r - xu + \frac{x^2(1 - u^2)}{2r} + \text{higher-order terms} \tag{2-2}$$

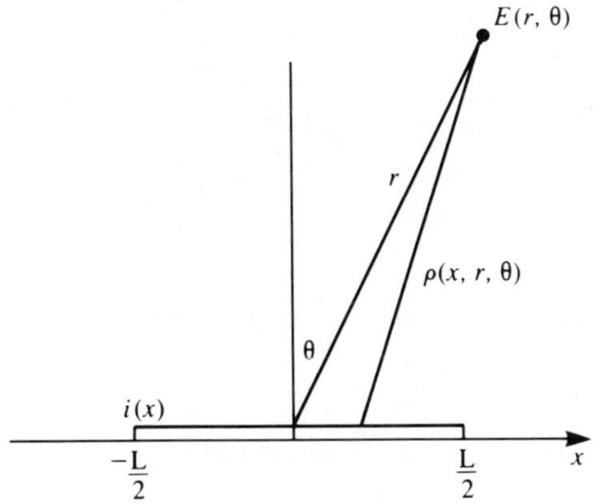

Figure 2-2 $E(r, \theta)$ is the complex field strength due to radiation from current $i(x)$ in the aperture.

where the angular variable θ has been replaced by a reduced angular variable $u = \sin\theta$.

The phase delay of the wave traveling between point x in the aperture and r, θ is $k\rho$. At a distance $r > L^2/\lambda$, ρ can be approximated by the first two terms of Eq. 2-2, meaning that the phase error at the extremes of an aperture of length L is acceptably small and the magnitude of the integral in Eq. 2-1 is in error by less than 0.25 dB. Under this condition, too, the denominator hardly varies with x and can be approximated by r. Hence the far-field approximation of Eq. 2.1, in which the distance is characterized by

$$\rho = r - xu \tag{2-3}$$

simplifies to

$$E(u) = K' \int_{\text{array}} i(x)\, e^{jkxu}\, dx \tag{2-4}$$

where $K' = Kr^{-1} \exp(-jkr)$.

Equation 2-4 is a simple Fourier integral. It is well known that the product of the widths (durations, extents) of a function and its Fourier pair is approximately constant. Consider the Gaussian Fourier pair shown in Fig. 2-3. σ_t and σ_ω may be identified with the half-widths in the two domains. σ_t and σ_ω bear a reciprocal relation and their product is unity. A very good approximation to the 3 dB beamwidth of an aperture is given by [94]

$$\Delta\Theta_{3\text{dB}} = \frac{\lambda}{4\sigma_x} \tag{2-5}$$

where σ_x is the square root of the second central moment of the aperture excitation

$$\sigma_x^2 = \frac{\int x^2 i(x)\, dx}{\int i(x)\, dx} \tag{2-6}$$

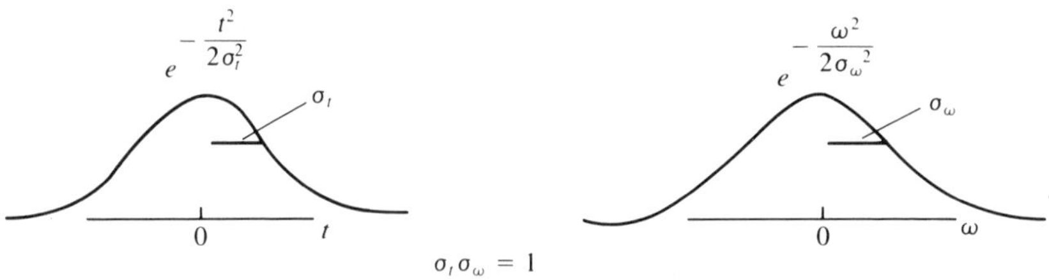

Figure 2-3 The product of the standard deviations of a Gaussian Fourier pair is unity.

where the origin of the x-axis has been chosen to make the first moment

$$\int xi(x)\, dx = 0 \tag{2-7}$$

The variance of a uniformly illuminated one-dimensional aperture of length L is $\sigma_x = L^2/12$ and the beamwidth calculated from Eq. 2-5 is $0.875\lambda/L$ radians. Other excitation functions yield other coefficients, and all are of the order of unity. Table 2-1 illustrates the relationship between 3 dB beamwidth and aperture size for various aperture weighting functions.

It is evident that to maintain a given beamwidth and therefore a given angular resolution, aperture size must scale with wavelength. Because the wavelength ratio between optics and microwaves is so huge, microwave imaging antennas themselves are huge. Table 2-2 lists the microwave apertures required

TABLE 2-1 3 dB Beamwidths

Aperture		*Beamwidth*
Uniform	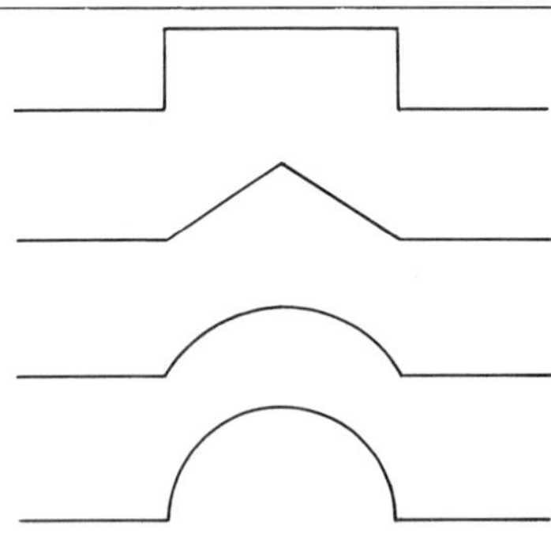	$0.88\,\frac{\lambda}{L}$
Triangular		1.27
Circular		1.02
Parabolic		1.16
	$\vert\leftarrow L \rightarrow\vert$	

TABLE 2-2 Size of antenna (dish, phased array) required to provide resolution comparable to eye, camera, and small (2 in.) telescope[a] (5 cm radiation assumed).

Eye	150 m (Nearly two football fields)
Camera	500 m
Small telescope	5000 m (The size of an airport)

[a] Aperture sizes are in meters.

to achieve resolving powers comparable to those of common optical instruments. A wavelength of 5 cm is used for the examples.

2-3 THE PHASED ARRAY

The huge antenna size needed to achieve useful resolving power precludes use of the most basic microwave antenna, the parabolic dish. The largest practical dish is in the order of 30 m, which is many times too small. Even if it were large enough, another factor would limit its use. Because of its large mass, the speed by which it can be scanned is slow. Electronic steering is preferable; for all practical purposes it can be accomplished nearly instantaneously. Both factors lead to the use of phased arrays for microwave imaging.

The phased array is a collection of small, broad-beamwidth apertures called *elements* or *antenna elements* that are electrically connected in such a fashion that a very narrow radiation pattern results (Fig. 2-4). The beam can be scanned typically $\pm\frac{1}{2}$ to ± 1 radian from the array normal. The antenna elements are shown as dipoles. They may also be horns, spirals, slots in waveguides or any convenient electromagnetic transducers. The arrows in the figure between the dipoles and the beamforming and scanning circuitry, which indicate signal-flow direction, depict the system as a receiving array. Reversing the arrows converts the array to transmission. An electromagnetic array is a linear, reciprocal device; its geometric radiation properties are the same for transmission and reception [89]. Thus Eq. 2-1 is the field strength of r, θ for a transmitting array that is excited by a current density distribution $i(x)$ delivered to the antenna elements by the electrical beamforming and scanning circuits. The same equation defines the complex gain of a receiving array to a source at r, θ where $i(x)$ is the current induced in the antenna elements by the radiation field from the source.

The functional equivalents to the lens, dish, and phased array are indicated in Fig. 2-5. By virtue of the shape of the lens, a spatially varying time delay from end to end is introduced to a wave passing through it, such that rays diverging from a light impulse originating at a distance arrive simultaneously at a common point in the image plane. The parabolic reflector of the reflecting telescope and the microwave dish provide the same time-alignment function.

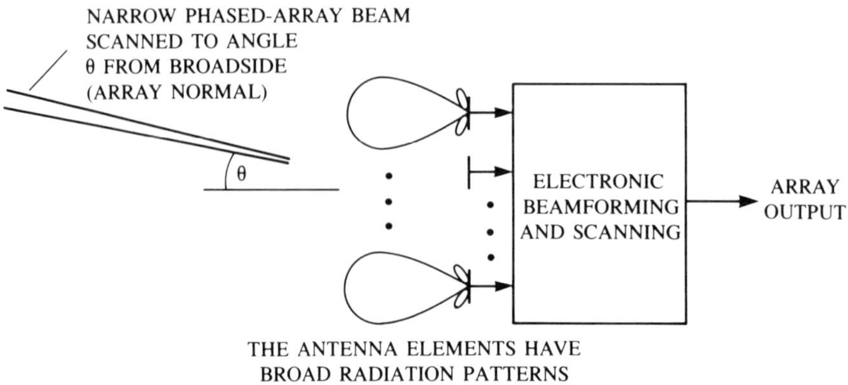

Figure 2-4 A phased array forms and scans a narrow beam by interconnecting antenna elements that have broad, fixed radiation patterns.

When the required size of the lens or dish becomes excessive, the structure must be decomposed into subunits and the received signals combined with time delays or phase shifters such that the output waveform is close to the desired function. This is the phased array.

The primary functional components of the electrical beamforming and scanning circuits are the bank of phase shifters and the summer shown in the lower part of Fig. 2-5. A point source at a great distance radiates energy at wavelength λ to the antenna array. The rays arriving at the antenna elements are assumed to be parallel. This is the far-field assumption and it implies planarity of the wavefront at the antenna array. Let the source be at angle θ to the array normal and let the array consist of antenna elements on a line spaced by a constant distance d. This array is called a linear, periodic array. As noted earlier, the antenna elements may be any convenient design. They are modeled mathematically as isotropic point receivers and are treated as sampling points in space of the radiation field. The antenna elements are followed by phase shifters ϕ_1 to ϕ_N. Their outputs are summed to provide the array output.

From the geometry, it is evident that the wave arriving at the nth element travels an extra distance

$$l = -d \sin \theta \tag{2-8}$$

relative to the preceding element, which is consistent with the far-field approximation given in (2-3). This extra travel distance represents an added phase delay

$$\Delta\phi = -\frac{2\pi l}{\lambda} = \frac{2\pi d \sin \theta}{\lambda} = kdu \tag{2-9}$$

where, on the right, $2\pi/\lambda$ is replaced by the wave number k and the sine of the angle is replaced by the reduced angular variable u.

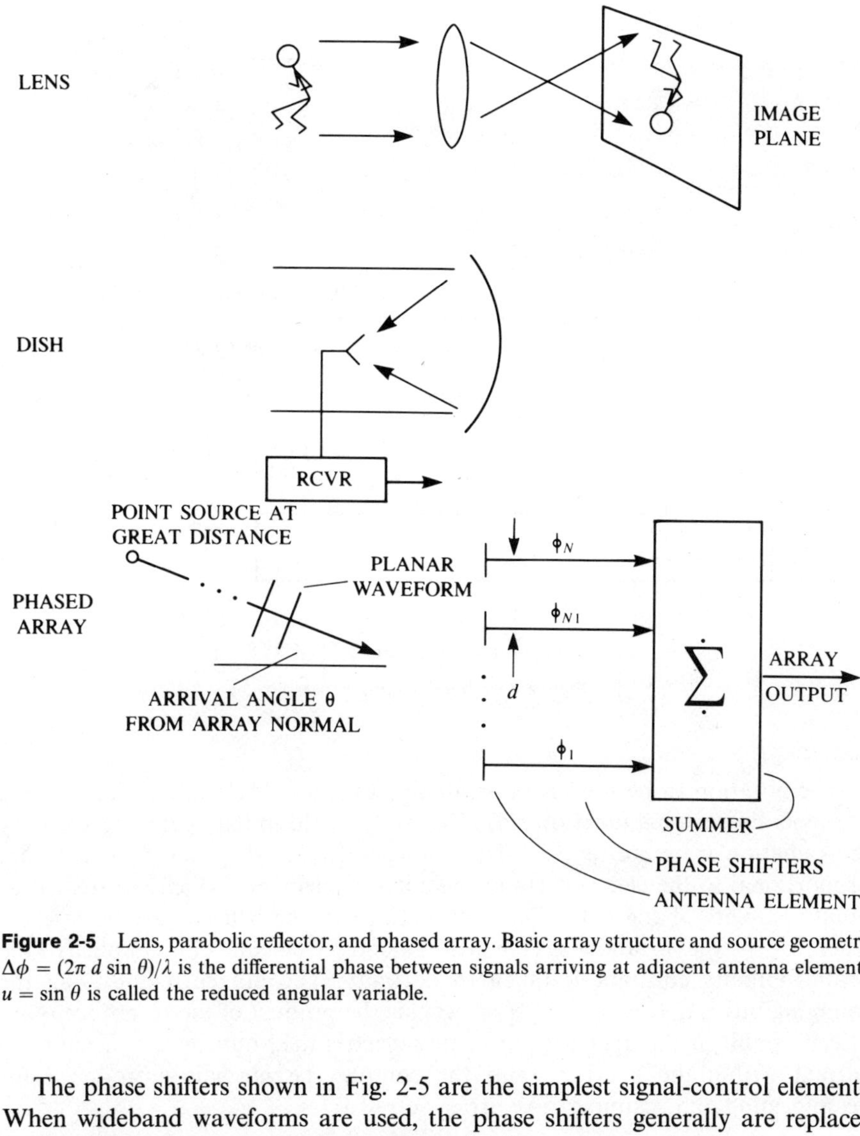

Figure 2-5 Lens, parabolic reflector, and phased array. Basic array structure and source geometry. $\Delta\phi = (2\pi\, d \sin\theta)/\lambda$ is the differential phase between signals arriving at adjacent antenna elements. $u = \sin\theta$ is called the reduced angular variable.

The phase shifters shown in Fig. 2-5 are the simplest signal-control elements. When wideband waveforms are used, the phase shifters generally are replaced by time delay elements to properly reflect the differential travel times from source to antenna elements (see Chapter 3). An amplitude variation a_n from element to element often is introduced for the purpose of reducing the sidelobe level of the radiation pattern [45, 94]. The complex gain in the nth channel between antenna element and summer is called the nth weight. For narrowband waveforms

$$w_n = a_n\, e^{j\phi_n} \tag{2-10}$$

where the ϕ_n determine the beam-steering direction $\theta = \sin^{-1} u$. The ϕ_n are evaluated in the discussion that follows.

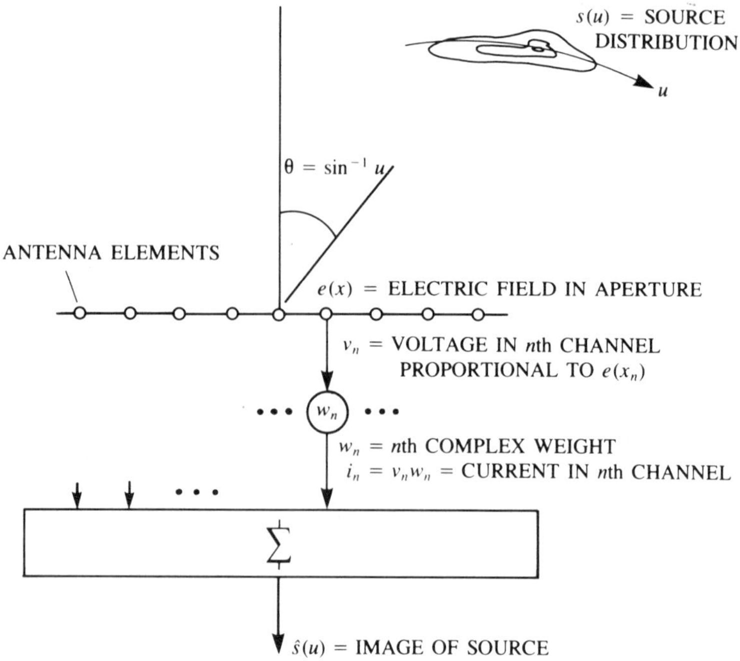

Figure 2-6 The imaging process.

The notation to be used is illustrated in Fig. 2-6. The source is considered as a one-dimensional function $s(u)$. The electric field in the aperture caused by the radiation from $s(u)$ is $e(x)$. The voltages driving the imaging system are proportional to the electric field at the antenna elements, which are treated as point samplers of the field. Thus the voltage in the nth channel $v_n \propto e(x_n)$, where x_n is the location of the nth antenna element. For simplicity, the proportionality constant is taken to be unity. The nth current driving the summing bus is $i_n = w_n v_n = w_n\, e(x_n)$. $w(x)$ is the product of the transducer gain at each point in the aperture converting electric field outside the aperture to current within the aperture, and the complex circuit gain introduced for beamforming and scanning. The array output is

$$\hat{s} = \sum_{n=1}^{N} w_n v_n \tag{2-11}$$

where the "hat" over the s indicates that Eq. 2-11 is the estimate of the source strength in the beam-steering direction. In matrix notation, the array may be defined by the weight vector

$$\mathbf{W}^T = [w_1\ w_2 \ldots w_n \ldots w_N] \tag{2-12}$$

where the superscript T denotes the transpose. Similarly,

$$\mathbf{V}^T = [v_1\ v_2 \dots v_n \dots v_N] \tag{2-13}$$

represents the signal vector. The array output becomes

$$\begin{aligned}\hat{s} &= [w_1 \dots w_N]\begin{bmatrix} v_1 \\ \vdots \\ v_N \end{bmatrix} \\ &= \mathbf{W}^T\mathbf{V} = \mathbf{V}^T\mathbf{W}\end{aligned} \tag{2-14}$$

Let the distance from the source in Fig. 2-5 to the first element be r. The signal arrives with a phase delay kr and is characterized by the complex phasor e^{-jkr}. The signal at the second element is $e^{-jk(r-du)}$, and so forth. Thus the signal vector (2-13) received by the array can be characterized by

$$\mathbf{V}^T(u) = [e^{-jkr}\ e^{-jk(r-du)} \dots e^{-jk(r-ndu)} \dots e^{-jk(r-Ndu)}] \tag{2-15}$$

The common factor e^{-jkr} introduces a constant, common phase shift across the array that has no influence upon the geometric properties of the array and may be dropped. It is only the internal phase relations between the signals at the antenna elements that are relevant. Consequently, Eq. 2-15 may be rewritten

$$\mathbf{V}^T(u) = [1\ e^{jkdu} \dots e^{jkndu} \dots e^{jkNdu}] \tag{2-16}$$

The beamsteering phase shifts are the arguments of the exponential components of (2-16), or $\theta_n = kndu$.

The objective of a receiving array is to adjust the phase shifts behind each element so that the array responds maximally to the signal from direction θ. When this is done, a beam is said to have been formed. Varying the direction of that beam is called *scanning*. It is accomplished by varying the weights according to the angle to which it is desired to steer the beam. By observing the output sum as a function of the scan angle θ (or u), one measures the angular distribution of arrival energy. Thus

$$w_n = w_n(u) \tag{2-17}$$

and

$$\hat{s} = \hat{s}(u) \tag{2-18}$$

This angular spectrum is called the image.

For the simple problem illustrated in Fig. 2-5, the weight vector $\mathbf{W}$ that forms a beam point in direction θ is the complex conjugate of $\mathbf{V}$:

$$\mathbf{W}^T(u) = [1\ e^{-jkdu} \dots e^{-jkndu} \dots e^{-jkNdu}] \tag{2-19}$$

The products $w_n v_n = 1$, all n, and the array output exactly equals N.

Let the source radiate a time-varying function. The received signal (current or voltage) is designated $i(t, u, n)$. It is a function of time, arrival angle and element position. The array output is

$$\hat{s}(t, u) = \sum_{n=1}^{N} i(t, u, n) \exp(-jkndu) \tag{2-20}$$

When the antenna is a continuous aperture, the position variable is no longer discrete and nd is replaced by the continuous variable x. The output is then the integral over the array of the received waveform:

$$\hat{s}(t, u) = \int_{\text{array}} i(t, u, x) \exp(-jkxu)\, dx \tag{2-21}$$

Equation 2-20 is a discrete Fourier transform and Eq. 2-21 is a continuous Fourier transform.

The conjugate Fourier variables are $u = \sin\theta$ and nd/λ or x/λ. The transforms are spatial transforms and both are functions of time because the source function is assumed to be a time-variable. In terrestrial microwave imaging the typical source is a pulsed radar transmitter. Its waveform can be modeled as a constant-amplitude sinusoid during the pulse duration, zero outside, and identical in successive pulse repetitions (see Fig. 1-2). Such a constant and repetitive pulse sequence discloses only target distance; consequently, the way it varies in time may be dropped from the mathematical development.

Equations 2-20 and 2-21 illustrate that an imaging system responsive to a distant source provides a Fourier transformation to the signals received across it. In practice, the transformation can be affected in a phased array with RF hardware, as in a Butler matrix [17], or in a video signal processor following coherent quadrature demodulation on each receiver and antenna element channel, or in a computer. The images shown throughout this book were obtained using a computer.

It is easy to see that the signal vector **V** also bears a Fourier relationship to the source function. We begin with a point source at angle $\theta_0 = \sin^{-1} u_0$. The source may be characterized by

$$s(u) = \delta(u - u_0) \tag{2-22}$$

where δ is the Dirac delta function. Its radiation field in the plane of the array is (to within a constant)

$$e(x) = \exp(-jkr) \exp(jkxu_0) \tag{2-23}$$

As in Eq. 2-15 the first term is a constant across the array. It is a fixed phase shift proportional to the target distance. It does not contribute to the spatial properties of the radiation field and may be ignored. The second term contains the spatial information. This term is a linear phase shift across the array with slope proportional to the source angle u_0. This term and the source function (2-22) are a Fourier transform pair

$$\delta(u - u_0) \Leftrightarrow \exp(jkxu_0) \tag{2-24}$$

A single point source such as (2-22) is not a common or realistic target. Often microwave targets can be modeled as collections of point sources. Man-made targets such as vehicles and buildings can be realistically represented in this fashion. Natural terrain sometimes also images as a collection of individual spots, most often in ground-to-ground imaging at low grazing angles. When viewed from the air at steep angles, however, the same terrain appears more continuous.

We will model the source as a collection of point targets of relative amplitudes A_t, phases ψ_t and located at angles $\theta_t = \sin^{-1} u_t$ from the normal to the array, as shown in Fig. 2-7. Its spatial description is given by

$$s(u) = \sum_t A_t \exp(j\psi_t)\delta(u - u_t) \tag{2-25}$$

The electric field in the aperture consists of a sum of terms of the form (2-23). Again the first term may be dropped because it is constant across the array. Thus the electric field in the aperture due to a source distribution of the form (2-25) located at a great distance from the array is

$$e(x) = \sum_t A_t \exp[j(\psi_t + kxu_t)] \tag{2-26}$$

which is the inverse Fourier transform of Eq. 2-25.

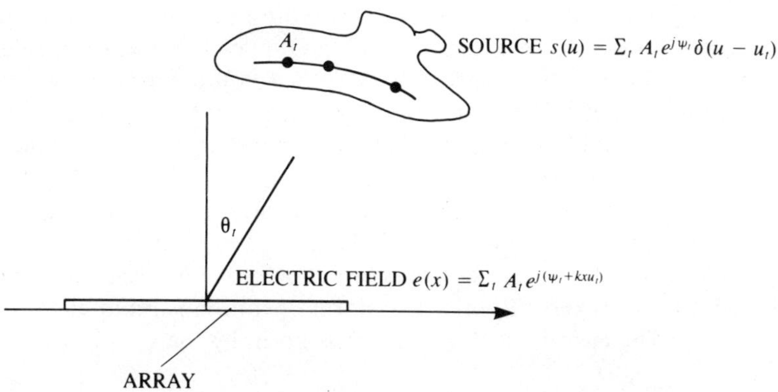

Figure 2-7 The source is modeled as a one-dimensional collection of point isotropic radiators located at $u = u_t$, $t = 1, \ldots, T$. The electric field at the array is a weighted sum of exponentials, each representing the plane wave contribution from one component of the source.

The current $i(x)$ flowing in the aperture is proportional to the product of the electric field (2-26) and the aperture weighting function $w(x)$. A discrete antenna defined by weights w_n at locations x_n is characterized by

$$w(x) = \sum_n w_n \delta(x - x_n) \tag{2-27}$$

The locations x_n of the elements in a periodic aperture with element spacing d are

$$x_n = nd \tag{2-28}$$

which is the simplest form of phased array. It is described by the weight function

$$w(x) = \sum_n w_n \delta(x - nd) \tag{2-29}$$

The current excitation in the aperture

$$i(x) = e(x)w(x) \tag{2-30}$$

is the product of the inverse Fourier transform of the source distribution $s(u)$ and the weight function $w(x)$. The complex image out of the array is the Fourier transform of Eq. 2-30, which is given by

$$\hat{s}(u) = \int i(x)\, e^{-jkxu}\, dx \tag{2-31}$$

2-4 OPERATOR NOTATION

The imaging operation defined above is easily represented in operator notation. It consists of a Fourier process followed by a multiplication operation, followed by a Fourier inversion. The process as described is too simple to warrant a change of notation. Later, however, complexities arise that are very cumbersome when treated as manipulations of integrals and summations. Operator notation is then helpful. It is introduced in this section.

The source function in Fig. 2-8 is designated s, meaning $s(u)$ as given by Eq. 2-25. The electric field in the aperture is designated e, meaning $e(x)$, as given by Eq. 2-26. The direct Fourier transform operation is designated by the operator $\mathscr{F}$ and the inverse Fourier transform operation is represented by the operator $\mathscr{F}^{-1}$. The electric field in Fig. 2-8 is given by the operator equation

$$e = \mathscr{F}^{-1} s \tag{2-32}$$

The antenna weight vector $w(x)$ is characterized by the operator $\mathscr{W}$. The

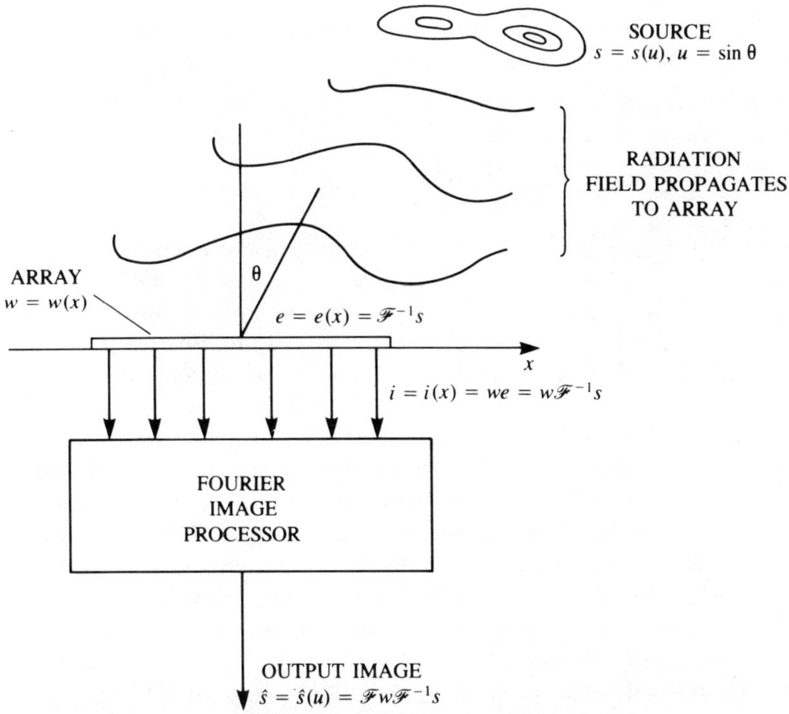

Figure 2-8 Fourier imaging expressed in operator notation.

current $i(x)$ is written

$$i = \mathscr{W} e = \mathscr{W} \mathscr{F}^{-1} s \qquad (2\text{-}33)$$

The image $\hat{s}(u)$ is the Fourier transform of (2-33). Thus the basic imaging operation may be written in operator notation as

$$\hat{s} = \mathscr{F} \mathscr{W} \mathscr{F}^{-1} s \qquad (2\text{-}34)$$

The order of application of the transformations is from right to left. That is, $\mathscr{F}^{-1}$ first operates on s. Its product with $\mathscr{W}$ is then operated upon by $\mathscr{F}$. This order is made explicit in Eq. 2-35 by the addition of brackets to the equation above:

$$\hat{s} = \mathscr{F} [\mathscr{W} \mathscr{F}^{-1} s] \qquad (2\text{-}35)$$

Substituting Eq. 2-32

$$\hat{s} = \mathscr{F} \mathscr{W} e \qquad (2\text{-}36)$$

which by the multiplication-convolution property of the Fourier transform leads to

$$\begin{aligned}\hat{s} &= \mathscr{F}\mathscr{W} * \mathscr{F}e \\ &= f * s \qquad (2\text{-}37)\end{aligned}$$

where $f \triangleq \mathscr{F}w$ is the radiation pattern of the antenna and $*$ denotes convolution.

Equation 2-37 shows the intrinsic problem in achieving high-quality, high-resolution imaging. Convolution is a smoothing or smearing process. In linear system theory terms, it is a low-pass filtering operation. For $s(u)$ to be a good representation of $s(u)$, $f(u)$ should approximate a δ-function. That is, the radiation pattern should have a narrow main lobe and substantially no sidelobes. Because of the Fourier relation between $f(u)$ and $w(x)$, the requirement of a narrow lobe implies that a huge aperture must be provided. However, because the Fourier kernel is a product of geometric terms, the geometric or mechanical tolerances applied to the antenna structure are severe, a situation which becomes more difficult to accommodate as the antenna grows large. The tolerance problem (in manufacture, installation, and operation) therefore limits how large a phased array may be. Thus an intrinsic physical problem besets the imaging system designer. Chapter 4 discusses the problem further and Chapters 8 and 9 describe methods that compensate for the difficulty.

For example, in a very large aperture there will be errors of several kinds in the complex gains of each of the antenna element channels. Mechanical instability will cause antenna elements to be misplaced from the expected positions. The result is a set of signal phases different from what it would be otherwise. The design of the weight vector, on the other hand, is based on the assumed element locations. Thus an error develops in the array that can be characterized by a complex gain operator $\mathscr{G} = g(x)$ for which compensation must be introduced. Furthermore, large systems can experience multipath distortions in the signal that results from reflections and scattering of energy into each receiver by inadvertent targets. Multipath shifts the signal phase and alters its amplitude. When the array is very large, the physical sources of multipath differ from antenna element to antenna element. Hence the signal gain variation becomes a random variable across the array and can be included in the operator $\mathscr{G}$. Also, when the aperture is sufficiently large, the weather across the array may vary enough to influence the phase shifts through the radomes of the antennas, assuming that radomes are included as part of the system design. This too has the effect of introducing a random phase variable across the array. Another factor is the electromagnetic coupling that always exists between an antenna element and its local environment. Such coupling influences the driving point impedance of the antenna element as seen by the electrical network behind it. The local electromagnetic coupling may vary from antenna element to element in a very large system. Another source of system error is uncompensated gain errors and phase mistunings in the antenna element channels. These phenomena are local to the array and can be combined into a simple gain-distortion term

$$\mathscr{G} = \prod_{j=1}^{J} \mathscr{G}_j, \qquad j = 1, 2, \ldots, J \tag{2-38}$$

where J is the number of gain-error-inducing functions or sources and $\prod$ denotes the product. The image now is

$$\hat{s} = \mathscr{F}\mathscr{G}\mathscr{W}\mathscr{F}^{-1}s \tag{2-39}$$

A well-designed system would introduce a correction or compensating function $c(x)$ represented by the operator $\mathscr{C}$. Ideally, it is the inverse operator to $\mathscr{G}$. Introducing a compensating network to the imaging equation leads to

$$\hat{s} = \mathscr{F}\mathscr{C}\mathscr{G}\mathscr{W}\mathscr{F}^{-1}s \tag{2-40}$$

Distortion may arise from sources that are neither in nor close to the aperture. Turbulence in the propagation medium is an example. Turbulence may induce sufficient variations in the refractive index of the medium across the array to induce random phase (and sometimes amplitude) perturbations in the signal received at different points in the aperture. Medium turbulence is a well-known phenomenon in optical telescopy. Temporal and spatial variations in the refractive index limit the quality of optical imagery in astronomy. Image distortion appears when the aperture exceeds the distance-correlation length of the dielectric constant of the medium, which is as little as 10 cm in a city atmosphere. Medium distortion may be represented by the operator $\mathscr{M}$. In general, $\mathscr{M}$ means $m(x, \theta)$ for it is also a function of the arrival angle θ. This is because waves arriving from different directions pass through different portions of the medium and experience different perturbations in the refractive index. The coupling of the variables x and θ destroys the neat Fourier relationship for which the operator formulation is most useful. However, in many important cases, and specifically in narrow field-of-view imaging systems, the medium distortion function is dominated by position in the aperture and is only weakly influenced by direction of arrival. For such systems, we can consider θ a parameter designating the general direction in which imaging is taking place and treat the medium operator only as a function of position x. We identify such an operator as $\mathscr{M}_\theta$, meaning $m(x; \theta)$, that is, $m(x)$ for a given value of bearing angle θ. To the extent that this is an accurate description of a physical process, the Fourier relationship is preserved. The electric field is modified from Eq. 2-32 to

$$e = \mathscr{M}_\theta\mathscr{F}^{-1}s \tag{2-41}$$

The current in the aperture, including the effect of gain errors, is

$$i = \mathscr{G}\mathscr{W}\mathscr{M}_\theta\mathscr{F}^{-1}s \tag{2-42}$$

It is evident that the gain error $\mathscr{G}$ and the medium error $\mathscr{M}_\theta$ commute and therefore a common compensating network $\mathscr{C}$ can correct both functions simultaneously. The imaging equation now becomes

$$\hat{s} = \mathscr{F}\mathscr{C}\mathscr{G}\mathscr{W}\mathscr{M}_\theta\mathscr{F}^{-1}s \tag{2-43}$$

To some extent, the position and angle variables in $\mathscr{M}$ do decouple in long-distance terrestrial imaging. The energy source, as indicated for example in Fig. 1-2, is a radar transmitter. Its beamwidth $\Delta\theta$ is a few degrees at most. At any one time the imaging system operates upon echoes arriving from only those targets illuminated by the transmitting beam, which is a narrow angular sector a few degrees wide oriented towards the direction of the transmitting beam. Imaging, therefore, is a narrow field-of-view operation. The imaging *system*, however, is not confined to a small angular sector. By scanning the radar transmitter beam through one or more radians, a high-resolution wide field-of-view map can be obtained. Within each narrow sector of width $\Delta\theta$, pointed to some direction θ_i, the medium operator is $\mathscr{M}_{\theta_i}$, meaning $m(x; \theta_i)$. The compensation operator must be a function of transmitter pointing-direction θ_i, that is, $\mathscr{C}$ in Eq. 2-43 becomes $\mathscr{C}_i$, meaning $c(x; \theta_i)$.

The design of high-quality imaging systems is, to a large extent, the design of the compensator $\mathscr{C}$. Ideally, it is the inverse to the product $\mathscr{G}\mathscr{M}_\theta$, provided that the medium turbulence phenomenon can be described as relatively invariant with direction of arrival over a narrow angular interval.

2-5 THE NEAR FIELD

The near field is characterized by a distance R_{NF} from the array such that radiation from a point radiator located closer to the array than that range cannot be represented by a planar wavefront. The plane wave approximation to the actual spherical wavefront is appropriate when the distance is sufficiently large and the aperture sufficiently small that phase measurements across the aperture do not disclose the curvature of the phasefront, or, if they do, that they are sufficiently small that they may be ignored.

The near-field to far-field transition distance R_{NF} may be defined in more than one way, which implies that it is determined less by theory than by subjectively imposed tolerances by the user of the theory. Figure 2-9 illustrates a curved wavefront from a point source that induces a phase variation across the array given by

$$\phi(x) = -k(r^2 + x^2 - 2rx \sin\theta)^{1/2} \tag{2-44}$$

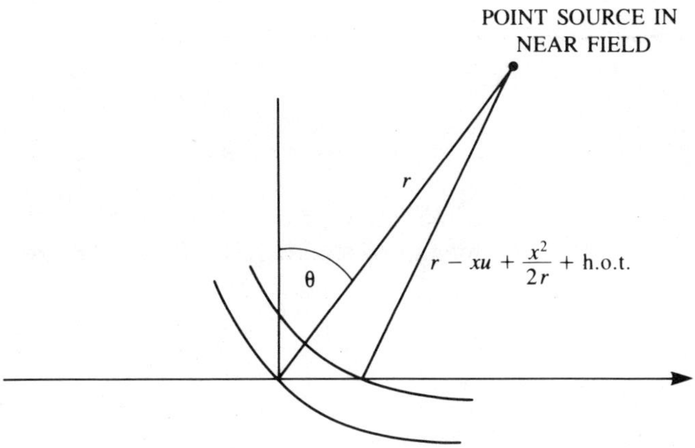

Figure 2-9 The phase variation in the array due to a point source in the near field is no longer linear with position. The first-order or Fresnel correction is quadratic. The phase is $\phi(x) = -k(r - xu + \frac{x^2}{2r} + \text{h.o.t}),$ where $u = \sin\theta$.

which may be expanded as the polynomial

$$\phi(x) = -k\left[r - xu + \frac{x^2}{2r} + \text{higher-order terms}\right] \tag{2-45}$$

where the term $kx^2u^2/2r$ has been dropped for this analysis because this term is negligible for most practical radar systems in which the transmitting beamwidth is at most a few degrees. The first term is a constant across the array determined by the distance from the source to the center of the aperture. The second term is the linear phase variation with position in the array associated with the direction of arrival of the energy from the source relative to the normal to the array at its center. The third term is a quadratic variation with position in the array. It gives the first-order effect of the curvature of the radiation field. It is this term that is ignored in Fourier image processing. The maximum phase error so incurred occurs at the maximum value of x, which is half the aperture size. Letting L be the aperture length, the maximum phase error is

$$\delta\theta_{\max} = \frac{kL^2}{8r} \tag{2-46}$$

How large (2-46) may be without inducing undesirable distortion into the imaging process is a matter of choice by the system designer. Typically a maximum phase error of $\pi/4$ to $\pi/2$ is tolerable. Substituting R_{NF} into the equation above and equating it to $\pi/4$ or $\pi/2$ leads to the near-field to far-field transition distance:

$$R_{NF} = \frac{L^2}{\lambda} \quad \text{or} \quad \frac{L^2}{2\lambda} \tag{2.47}$$

The former value leads to the interesting and simple interpretation shown in Fig. 2-10. Consider an antenna of length L. Its Fourier beamwidth is the

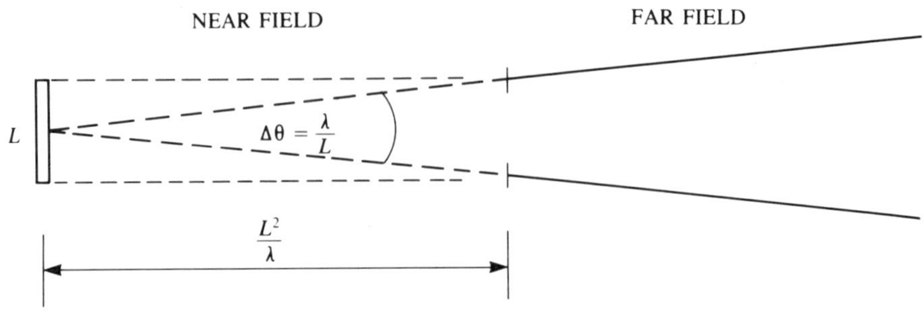

Figure 2-10 Near-field transition geometry. Solid lines indicate the beam profile in the far field, where the beamwidth is approximately λ/L. Dashed lines crudely indicate the beam profile in the near field. The near-field to far-field transition is $R_{NF} = L^2/\lambda$.

order of λ/L, as shown, meaning that in the far field the radiation pattern of the antenna has a beamwidth of that angular cross section. It is known from a study of diffraction theory that the radiation in the neighborhood of the array is confined to a cross section of roughly aperture size, as indicated by the dashed lines in the figure. Thus the radiation pattern changes from a form roughly described by the dashed lines to a beam of linearly increasing width. The transition takes place at a distance from the aperture equal to L^2/λ, which may be identified as R_{NF}. This distance is taken as the nominal near-field to far-field transition distance throughout the book. Equation 2-46 shows that the maximum phase error at the edge of the aperture for a target at this distance that is treated as a far-field target is $\pi/4$. While this is a large error, the error inside the aperture decreases very rapidly because of the quadratic phase variation, and it is easy to show that the loss in antenna array gain for a target at this distance is only 0.25 dB. Furthermore, the beam broadening is trivial (2 percent). The deficiency is in the side radiation pattern. Given a uniformly illuminated aperture, the highest sidelobe rises from -13.5 dB to -12.2 dB. Many designers consider this loss in gain and rise in sidelobe level too severe a penalty and treat $2L^2/\lambda$ as R_{NF} instead.

In a larger sense, however, the distinction is immaterial because large antenna systems designed for high-resolution imaging perform the imaging operation in a digital computer. There it is no more difficult to calculate the exact distances from targets to the various positions in the antenna array as it is to use approximations. The main advantage of the far-field (plane wave) and near-field (parabolic curvature) approximations is the ease with which the problem can be thought about and discussed.

The weight vector $w(x)$ describes the antenna array in terms of its length L and the amplitude taper introduced for sidelobe control, and the number of antenna elements and their spacings. When considering near-field imaging problems, it is useful to include the phase curvature due to target distance as part of the weight vector. By so doing, one effectively transforms the range or distance coordinate of the target from its near-field distance to infinity, and the image processor remains a Fourier transformer. This is shown below.

We will treat the deviation of the electric field from a planar wavefront as a curvature-induced distortion. Let h represent the phasefront deviation, meaning

$$h(x, r) = \exp\left(\frac{-jkx^2}{2r}\right) \tag{2-48}$$

Its conjugate is built into the weight vector to compensate for the phase curvature at the target distance r. Let $w(x, r)$ be the weight vector when the target is at distance r. Then assuming quadratic phase curvature,

$$w(x, r) = w(x, \infty)h^*(x, r) \tag{2-49}$$

where $w(x, \infty)$ may be identified with $w(x)$ of the earlier sections. The one-dimensional far-field radiation pattern $f(u)$ associated with the weighting function $w(x)$ was given in Eq. 2-37. To obtain the two-dimensional pattern $f(u, r)$ we need to return to the full kernel in the diffraction integral (2-1) or at least to the Fresnel approximation (2-45). Using Eq. 2-45 minus the lead term (which is a constant across the array), the two-dimensional radiation pattern is

$$f(u, r) = \int w(x, r)h(x, r) \exp(-jkxu)\, dx \tag{2-50}$$

Now let the aperture be focused at a distance r_0, meaning that $r = r_0$ in Eq. 2-49. Substituting Eq. 2-49 with $r = r_0$ gives

$$f(u, r; r_0) = \int w(x; \infty) \exp\left[-jk\left(xu + \frac{x^2}{2r} - \frac{x^2}{2r_0}\right)\right] dx \tag{2-51}$$

which goes to $f(u; \infty)$ in Eq. 2-50 as r goes to r_0. The distance r_0 is called the focal distance. At $r = r_0$ the beam pattern is the same as the beam pattern at infinity caused by the original weighting function. In short, quadratic phase curvature across the array can compensate for phasefront curvature from a source at finite range.

The near-field phase correction could be included instead in the compensation operator $\mathscr{C}$. In practice, it is useful to apply known, deterministic compensation to the weight vector; a priori unknown distortion-inducing phenomena are corrected by self-calibration loops or calculations that are associated with the compensation operator.

Modifying the weight vector as in Eq. 2-49 alters the range profile of the beam from that crudely drawn in Fig. 2-10 to that shown in Fig. 2-11. At the

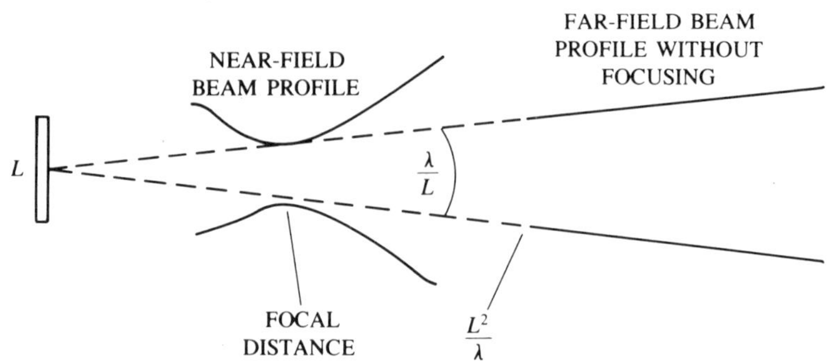

Figure 2-11 Near-field beam profile when array is focused at a distance $< L^2/\lambda$. The beamwidth is λ/L at the focal distance and is larger at lesser and greater distances.

focal range, the beam cross section pinches down to an angular width of λ/L and has a transverse linear dimension of approximately $\lambda r/L$. This beam dimension is smaller than aperture length. A penalty is paid in beam broadening at near and far ranges. Focusing at a distance $L^2/3.5\lambda$ preserves antenna gain (within 3 dB) from a near range of $L^2/7\lambda$ to infinity. The focal range $L^2/3.5\lambda$ is called the hyperfocal distance [94].

Whereas the radio astronomer has no concern with near-field focusing, the terrestrial and laboratory microwave imagers do. At the Valley Forge laboratory, an antenna 83 m long operates at 3 cm wavelength. R_{NF} is 200 km. The equipment is used to image a nearby town at distances from 4 to 15 km. Figure 2-12 shows houses and street patterns in the town with a cross-range resolution-cell dimension of 1.8 m, which is about 1/45 of the aperture size. This constitutes very near-field imaging. The propagation path from transmitter and array to the target site is along the surface of the earth. The grazing angle is essentially zero. Consequently only the upper stories of those two-storeyed houses on elevated streets are visible to the radar transmitter and the large receiving array. Nevertheless, street patterns and back yards are visible and houses can be counted.

Figure 2-13 is an image of a Boeing 727 airplane flying at a distance of 3.2 km from the laboratory. The antenna was 32 m long and operated at 3 cm wavelength. The cross-range dimension is approximately 3 m, which is about 1/10 the aperture size and R_{NF} is 34 km. The airplane was flying at an elevation

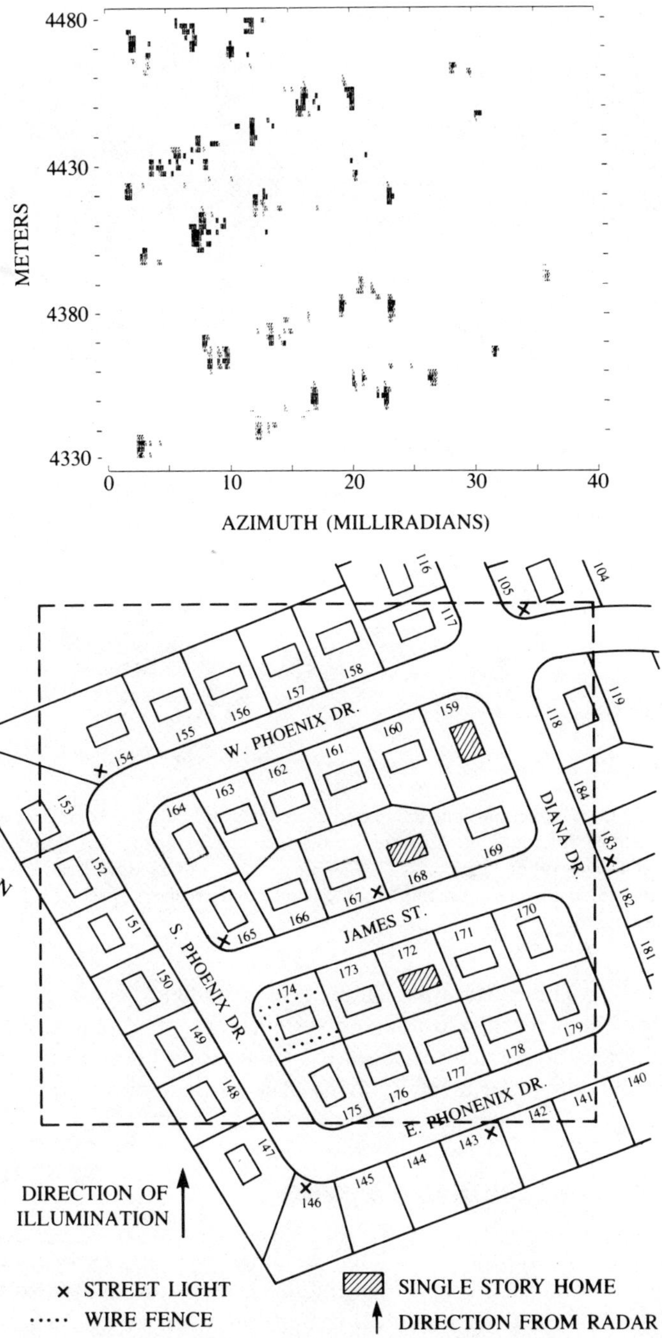

Figure 2-12 Diffraction-limited 3-cm radar map of a suburban housing development in Phoenixville, Pa., showing streets and houses. Each pixel cluster is a house. The houses are about 20 m apart. Resolution is comparable to human vision. The array length is 83 m.

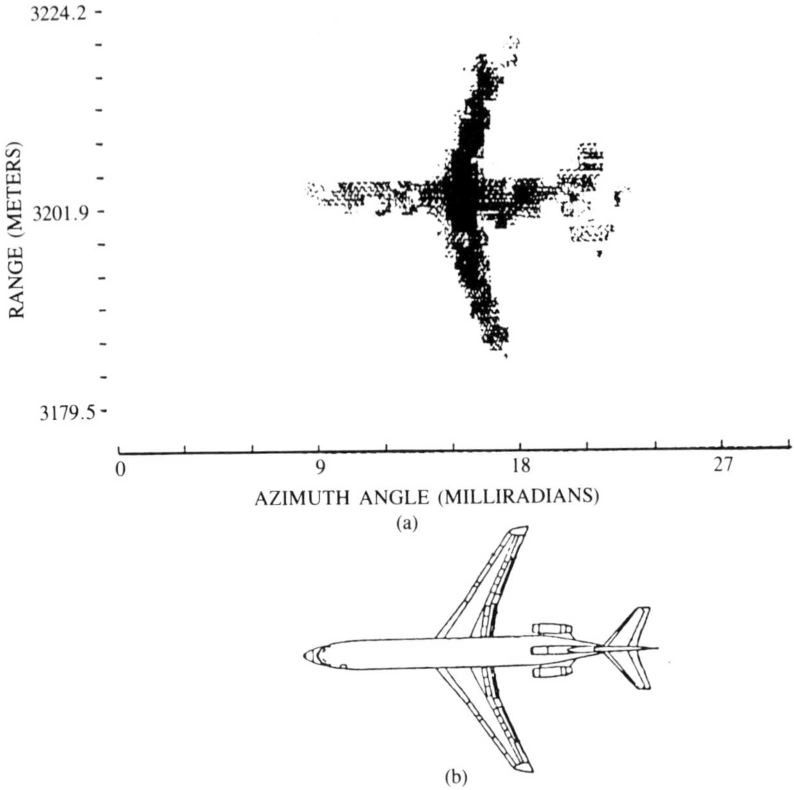

Figure 2-13 (a) Diffraction-limited 3-cm radar image of a Boeing 727 commercial airliner flying into Philadelphia International Airport. The distance is 3.2 km from the Valley Forge Research Center and the elevation angle is 32°. The image was obtained with a 32-m University of Pennsylvania radio camera consisting of 32 X-band radar receivers spaced by 1 m. (b) Plan view drawing of aircraft.

angle of 32°. These examples show that near-field focusing is an essential property of high-resolution imaging systems, except of course if the system is designed for radio astronomy. Figure 2-14 shows a high-resolution image of a small (20 m) commuter airplane flying at a distance of 5.5 km.

2-6 TWO-DIMENSIONAL IMAGING

The imaging operation as defined in Eqs. 2-14 or 2-31 or 2-43 is one-dimensional. Resolution in u, the angular dimension transverse to the array normal, is obtained from the linear dimension of the array. As we have seen in Table 2-1, the product of the transverse angular dimension of the beam and the aperture length is close to unity. How is resolution in a second dimension to be attained?

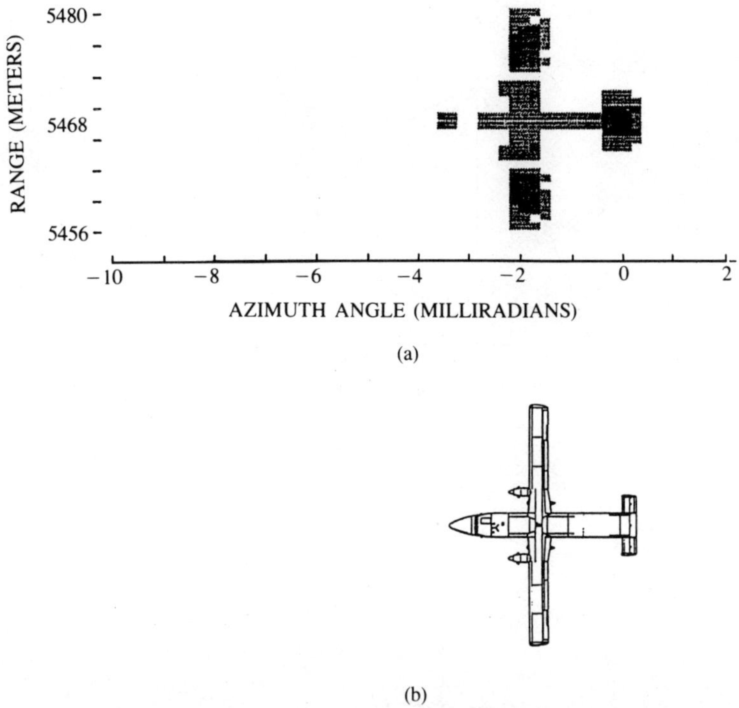

Figure 2-14 (a) Microwave image of small commuter aircraft, the Shorts-300, flying into Philadelphia International Airport at a distance of 5.5 km from Valley Forge Research Center. The picture was made by the University of Pennsylvania high-resolution radio camera. (b) Plan view drawing of the aircraft.

There are three procedures that lead to two-dimensional imaging. Conceptually simplest is the two-dimensional array, which is the spatially sampled analog of the optical lens. Antenna elements are deployed in a plane or on some surface as indicated in Fig. 2-15. This is the procedure used in radio astronomy. It is also a useful method for imaging targets that are overhead.

Figure 1-2 shows the second and more common procedure. There, a one-dimensional phased array obtains angular resolution in the transverse direction and the time of flight of the radar pulse from transmitter to target to receiver carries the longitudinal information. The longitudinal or range resolution is unrelated to aperture size. It is determined by the pulse duration of the transmitter. It is well known in radar that if the pulse duration is δ and the velocity of light is c, the resolution in range is [82]

$$\Delta R = \frac{c\delta}{2} \tag{2-52}$$

It is easy to see that ΔR is the interval in range over which scatterers produce

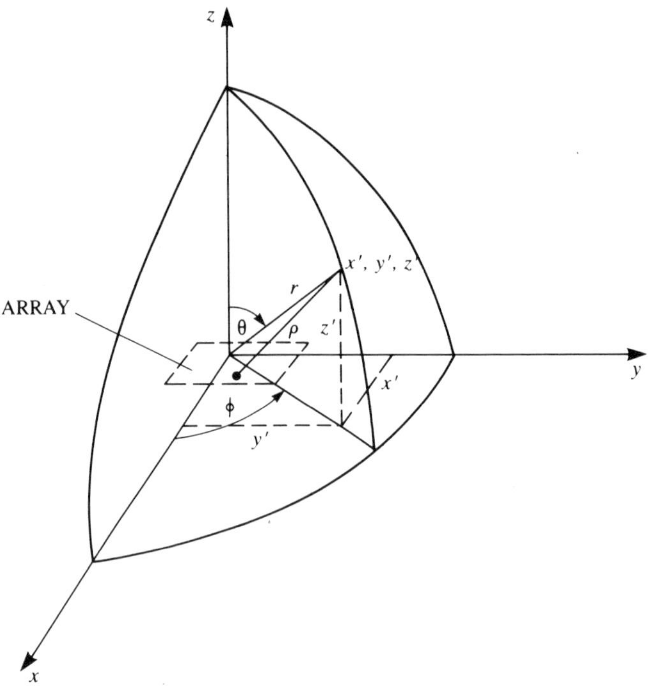

Figure 2-15 Planar aperture.

echoes that arrive simultaneously at the receiver. More fundamental than the pulse duration is the bandwidth of the radiated signal. A nominal bandwidth

$$B = \delta^{-1} \tag{2-53}$$

is required to achieve the resolution given by Eq. 2-52, although the waveform needs not be a pulse. Pulse compression techniques have been developed during the last thirty years in which long pulses (duration T) of wide bandwidth (such that $BT \gg 1$) yield a resolution in the range dimension [58]

$$\Delta R \approx \frac{c}{2B} \tag{2-54}$$

This procedure is the most practical for microwave imaging. It is useless, of course, for radio astronomy.

The third procedure for two-dimensional imaging is evident from the near-field equation of the radiation pattern given by Eq. 2-51. That equation contains the range-focus parameter r_0. A change in r_0 results in a change of the focal distance of the antenna. In principle, an aperture can be successfully focused from short range to long range by a variation of r_0. While an available tool in principle, range focusing usually is ineffective. Clearly it is useless in radio astronomy because all targets are in the far field. In near-field imaging, the range resolution or 3 dB range beamwidth is identified with the depth of field of the aperture, which is approximately [94]

$$\Delta R = 7\lambda(r_0/L)^2 \tag{2-55}$$

Unfortunately the depth of field grows quite rapidly with range. The depth of field in Fig. 2-12 is 700 m; in Fig. 2-13 it is 2000 m. In neither case would range focusing be useful.

The two-dimensional images of Figs. 2-12 and 2-13 were obtained with the second procedure. One-meter and 3-m range resolution were obtained with short transmitted pulses. One-dimensional images were formed at 1-m (or 3-m) intervals and the set of such images were made into the composites shown. While generally this is the most practical means of two-dimensional imaging, its use requires that the target be extensive in range. An airplane flying overhead is not; its extensiveness is in both cross-range dimensions. To successfully image under this condition, the antenna array must be deployed in two dimensions, as pictured in Fig. 2-15. The two-dimensional radiation pattern is expressed by the two-dimensional integral

$$\begin{aligned} f(\theta, \phi) &= \int_{\text{array}} w(x, y)\, e^{-jk\rho}\, dx\, dy \\ &\doteq \int_{\text{array}} w(x, y)\, e^{jk(x \sin\theta \cos\phi + y \sin\theta \sin\phi)}\, dx\, dy \end{aligned} \tag{2-56}$$

in which the fixed phase associated with the distance r to the center of the array has been dropped in the second expression, and the weighting function $w(x, y)$ that defines the array is expressed in two dimensions.

Equation 2-56 is easily derived from the geometry of Fig. 2-15. The solid dot is a point in an aperture located in the xy plane. The origin is taken to be the center of the array. The normal to the array is the z axis. A point on the sphere at polar angle θ and azimuth angle ϕ has coordinates x', y', z'. Expressed in polar coordinates,

$$\begin{aligned} x' &= r \sin\theta \cos\phi \\ y' &= r \sin\theta \sin\phi \\ z' &= r \cos\theta \end{aligned} \tag{2-57}$$

The distance ρ from the array point $x, y, 0$ to the field point x', y', z' is given by

$$\begin{aligned} \rho^2 &= (x' - x)^2 + (y' - y)^2 + z'^2 \\ &= r^2 \sin^2\theta \cos^2\phi - 2xr \sin\theta \cos\phi + x^2 \\ &\quad + r^2 \sin^2\theta \sin^2\phi - 2yr \sin\theta \sin\phi + y^2 + r^2 \cos^2\theta \\ &= r^2 - 2r \sin\theta(x \cos\phi + y \sin\phi) + x^2 + y^2 \end{aligned} \tag{2-58}$$

When the square root of Eq. 2-58 is expanded in a binomial series and only the constant and linear terms are preserved (far-field approximation), the distance is

$$\begin{aligned}\rho &\approx r\left[1 - \frac{\sin\theta(x\cos\phi + y\sin\phi)}{r}\right] \\ &= r - x\sin\theta\cos\phi - y\sin\theta\sin\phi \qquad (2\text{-}59)\end{aligned}$$

Equation 2-59 is the far-field approximation to the distance ρ for an unfocused aperture. It also is the near-field approximation for targets at the focal range of a focused aperture. The sum of all the contributions from the current elements in the array, each rotated in phase by an amount corresponding to the distance ρ, is the relative field strength at θ, ϕ. It is expressed by the two-dimensional Fourier integral (2-56).

2-7 POLAR PROJECTION

Equation 2-56 does not generally factor into the product of two one-dimensional Fourier transforms. Nevertheless, one-dimensional transform theory may be applied for targets at the focal distance of the aperture. The complex radiation pattern along any longitude line on the polar sphere can be calculated as the one-dimensional Fourier transform of a function of the two-dimensional aperture weighting function $w(x, y)$. The method described below is an example of the powerful projection slice theorem commonly used in computer tomography [24]. Consider a longitude line on the polar sphere, as for example the one containing point x', y', z' in Fig. 2-15. The evaluation of the two-dimensional Fourier transform along such a longitude is called a *polar cut* of the two-dimensional radiation pattern.

A polar cut is the intersection of the two-dimensional radiation pattern $f(\theta, \phi)$ with a plane containing the polar axis of the array (the z-axis). A polar plane may be oriented at an arbitrary angle with respect to the x- and y-axes, although it is convenient to rotate the axes so that the x-axis falls in the polar plane. When this is done, the azimuth angle $\phi = 0$ and Eq. 2-59 becomes

$$\rho = r - xu \qquad (2\text{-}60)$$

which is the conventional one-dimensional far-field approximation.

Equation 2-60 is not a function of y. It contains the mean distance r, which serves as an unimportant phase factor common across the radiation pattern, and a linear term in x. Equation 2-60 is the appropriate expression for the calculation of the phase shift for all current elements in the array sharing a common x coordinate. All such current elements may be added to form an equivalent current element on the x-axis. Thus the two-dimensional array may

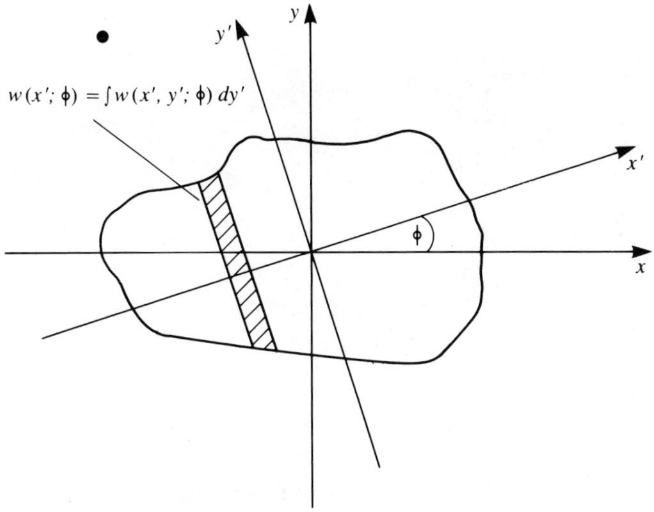

Figure 2-16 The aperture weight function is projected on to the x'-axis, forming a one-dimensional aperture from which the radiation pattern at azimuth ϕ is calculated. (From ref. [94])

be collapsed to a one-dimensional line array and the radiation pattern in the polar cut is the one-dimensional Fourier transform of the collapsed current density across the array.

Figure 2-16 shows a two-dimensional aperture of arbitrary shape. The line at azimuth ϕ becomes the new x-axis. All elements of the weighting function $w(x, y)$ are projected onto the x'-axis to form the one-dimensional weighting function

$$w(x'; \phi) = \int w(x', y'; \phi)\, dy' \tag{2-61}$$

The radiation pattern at azimuth or longitude ϕ is the Fourier transform of Eq. 2-61:

$$f(u; \phi) = F\{w(x'; \phi)\} \tag{2-62}$$

Equation 2-62 gives a one-dimensional cut through the radiation pattern of a two-dimensional aperture at an arbitrary longitude on the polar sphere. By varying the azimuth angle ϕ through 360°, the entire two-dimensional far-field radiation pattern can be calculated.

An example of the application of the procedure is illustrated in Fig. 2-17. The aperture is rectangular of size a (in x) and b (in y). The current density is a nonzero constant I_0 within the aperture and is zero outside. The projection of the excitation function upon the x'-axis is the equivalent weighting function $w(x'; \phi)$ along x', an axis at azimuth ϕ from x. This projection is trapezoidal.

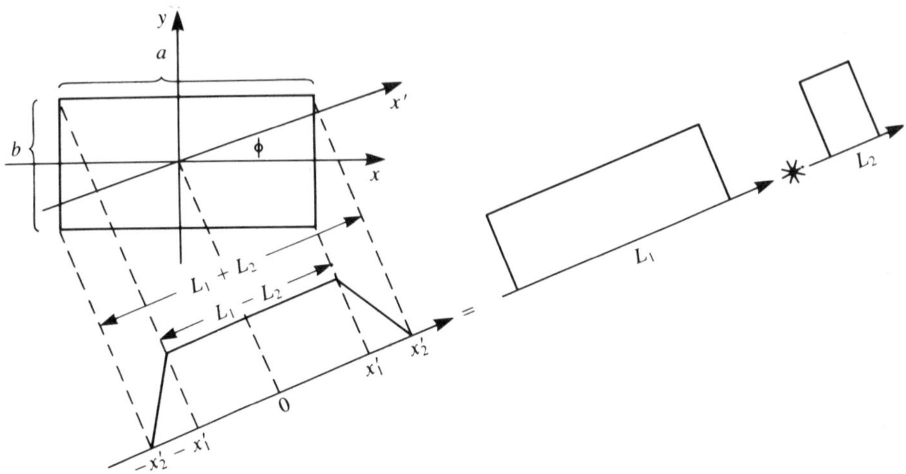

Figure 2-17 The aperture excitation $i(x, y) = I_0$ within the rectangular aperture and is zero outside. The equivalent weighting function at azimuth ϕ is a trapezoid, which is the convolution of two rectangles. The radiation pattern is the product of two sync functions.

Its breakpoints are

$$\pm x_{1'} = \pm \tfrac{1}{2}(a \cos \phi - b \sin \phi)$$

and its endpoints are

$$\pm x_{2'} = \pm \tfrac{1}{2}(a \cos \phi + b \sin \phi)$$

The far-field radiation pattern along azimuth ϕ is, by Eq. 2-62,

$$f(u; \phi) = \int w(x'; \phi) e^{jkx'u} \, dx' \tag{2-63}$$

which can be calculated by direct integration of the weighting function sketched in Fig. 2-17.

Evaluation of Eq. 2-63 can be considerably simplified for this particular problem by recognizing that a trapezoid is the convolution of two rectangles: if the lengths of the rectangles are L_1 and $L_2 < L_1$, the base of the trapezoid is $L_1 + L_2$ and the length of its top side is $L_1 - L_2$. Hence $L_1 = a \cos \phi$ and $L_2 = b \sin \phi$. Identifying the two rectangles as $w_1 = w_1(x'; \phi)$ and $w_2 = w_2(x'; \phi)$, we can write $w = w_1 * w_2$. Next, by virtue of the convolution-multiplication property of Fourier theory [74], Eq. 2-62 can be rewritten

$$f(u; \phi) = Fw = F\{w_1 * w_2\} = (Fw_1)(Fw_2) \tag{2-64}$$

which is the product of the Fourier transforms of two rectangles. The Fourier transform of a rectangle of length L is easily calculated:

$$f(u) = \frac{1}{L}\int_{-L/2}^{L/2} e^{jku}\, dx = \frac{\sin \dfrac{\pi l u}{\lambda}}{\dfrac{\pi l u}{\lambda}} \tag{2-65}$$

Hence Eq. 2-64 becomes

$$f(u; \phi) = \frac{\sin \dfrac{\pi a u \cos \phi}{\lambda}}{\dfrac{\pi a u \cos \phi}{\lambda}} \cdot \frac{\sin \dfrac{\pi b u \sin \phi}{\lambda}}{\dfrac{\pi b u \sin \phi}{\lambda}} \tag{2-66}$$

2-8 WIDE BANDWIDTH

Using modern sweep- and step-frequency oscillators, laboratory imaging experiments can be conducted with bandwidths comparable to the band center frequency. The reciprocal fractional bandwidth, Q, can be as low as the order of unity. Similarly, in undersea acoustics and in medial biological ultrasonic imaging the transducers can operate with Q values of this order. Not so in long distance microwave imaging, for the spectral occupancy of outdoor systems in the United States is severely limited by the FCC. In this context, a 10 or 20 percent bandwidth is considered wide.

Wide bandwidth affects imaging in three ways. First, time delays generally replace phase shifters in the phased array. Second, the beamwidth of the main lobe is modified. Third, the sidelobe level may be reduced. The first factor has an impact upon the hardware and software design of the imaging system. The second factor has a negligible effect. The third factor may or may not be significant, depending upon the nature of the array and its size. These topics are discussed in the next chapter.

3

WIDEBAND EFFECTS

3-1 THE NEED FOR TIME-DELAY STEERING

High resolution in the range dimension requires a wide-bandwidth signal. A pulse waveform of duration δ provides a range cell of length $c\delta/2$. A bandwidth $B = \delta^{-1}$ is required to produce such a pulse. A waveform with longer duration ($T > \delta$) modulated with bandwidth B can also provide the same resolving power in range through the use of matched filtering in the receiver. The technique is commonly called pulse compression [58].

A resolution cell of 1 meter is useful for long-distance imaging of manmade scenes. Bandwidth of 150 MHz is required to achieve such resolution in the range dimension. A resolution of 1 foot is desirable when the target is small, such as a small vehicle like an automobile. Bandwidth of 500 MHz is required in this case. Beamsteering phase shifters cease to be appropriate over so wide a band and time delays must be used instead. Time-delay steering *is* phase-shift steering in which the phase at each frequency in the signal band is shifted in proportion to the frequency. This is a necessary condition for all frequency components of the signal to form constituent components of the radiation pattern oriented in the same direction.

This fact is evident from the following argument. The phase shift in an array at an element a distance x from the center when the system is steered in the direction θ from the array normal should be

$$\phi(x) = -kx \sin\theta = -\frac{\omega x u}{c} \tag{3-1}$$

The required phase is linearly proportional to angular frequency ω. The transfer function of a delay line of length τ is $\exp(-j\omega\tau)$. Its phase $\phi(\omega) = -\omega\tau$ is linearly proportional to frequency. Thus the time delay element is the correct device for beamsteering.

Narrowband systems operate satisfactorily with phase shifters. Wideband systems require time delays. Several questions need be asked. What is the penalty for the use of phase shifters instead of time delays? A continuously variable time delay is the ideal component. Practicality demands the use of a tapped delay line instead. What is the effect of discrete rather than continuous time delay? Is there a practical compromise that combines discrete delay lines with vernier phase shifts?

The nonideal steering component (the phase shifter) reduces the mainlobe gain and increases the sidelobe level. The former reduces the detection sensitivity of the system and reduces the margin between the brightest target in an image and the noise level. Loss of mainlobe gain is equivalent to reducing the transmitter power. A nominal tolerable loss of 1/2 to 1 dB will be assumed in the analysis that follows. The lost energy in the main lobe caused by the phase errors that result from imperfect beamsteering goes into the side radiation pattern. When the errors are approximately random and uncorrelated across the array, the effect is an increase in the average background level that is statistically homogeneous across the field of view of the system. Sometimes errors in phase have components that are periodic across the array. Periodic distortions lead to grating lobes. Figure 3-1, taken from [64], is a simulation of the radiation pattern of a wideband ultrasonic imaging system that shows both phenomena.

It is possible to destroy the grating lobes by intentionally adding a known set of random phases or random time delays at each antenna element. These quantities are accounted for in the beamsteering instructions given to the steering components. Because they are random additive terms to the residual phase or time-delay errors that result from quantization, the residual periodicities across the aperture are destroyed and so are the grating lobes.

The energy in the grating lobes is not destroyed, of course; it adds to the uniform, random sidelobe background.

In general, it is preferable to redistribute the grating energy across the angular spectrum. Grating lobes represent the coherent combining of the energies of the error components. These lobes can produce artifacts such as false targets. Spreading their energies over the entire angular spectrum has a minor effect upon the average background level and eliminates these lobes as sources of false targets.

3-2 THE LIMIT TO PHASESHIFT STEERING

The phase variation across the array at midband ($\omega = \omega_0 = h_0 c$) from a wave arriving from direction $u = u_0$ is $k_0 x u_0$. If the source is in the near field, we shall assume that the array is focused at the proper distance so that the phase

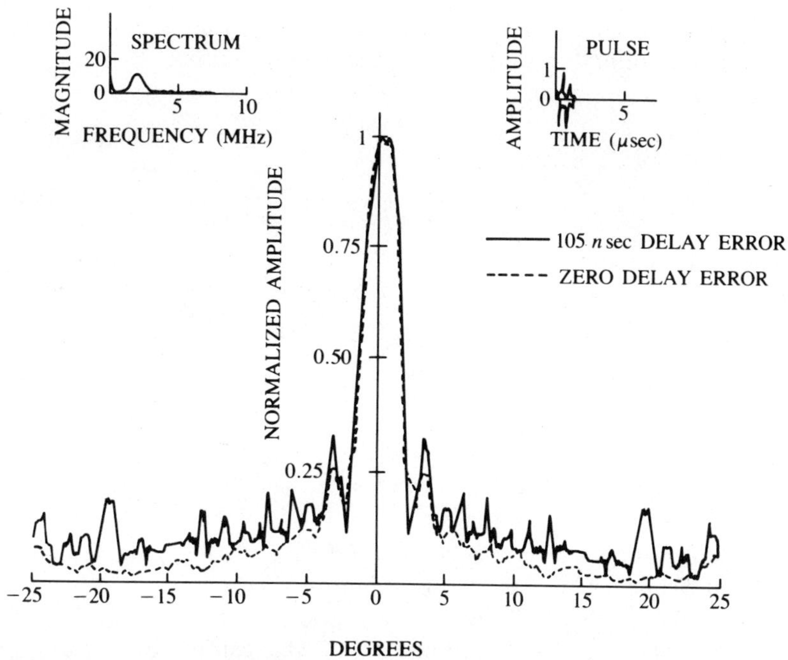

Figure 3-1 Solid lines show simulated azimuthal impulse response for a 19-mm aperture emitting a five-cycles burst of 2.0-mHz ultrasound. Drawn in dashed lines is zero-delay error response when an exponential-weighted three-cycle burst of 2.0-mHz ultrasound is transmitted. Note large error grating lobes at $\pm 19.5°$. (From ref. [64]).

variation after focusing is the same as given above. With phaseshift steering, the beam is steered at midband to the direction of arrival of the wave by phase shifters set with the complementary phase

$$\phi(x) = -k_0 x u_0 \tag{3-2}$$

At a slightly different frequency, say $\omega = \omega_0 + \delta\omega$, the array steers the beam to a slightly different direction. This phenomenon is known as *squint*. Consider the wideband waveform having Fourier spectrum $g(\omega)$ to consist of many frequency components. The density of components is proportional to $g(\omega)$. Each component is received by a radiation pattern displaced from u_0 by δu. The weighted sum of these constituent patterns is the net or effective radiation pattern of the antenna. It can be written in the form

$$f(u) = \int_{\omega} g(\omega) f(u, \omega)\, d\omega \tag{3-3}$$

which is the integral over all frequencies of the radiation pattern of the array at frequency ω, denoted $f(u, \omega)$, weighted by the spectral amplitude $g(\omega)$. Using midband phase shifting to steer the array towards u_0 according to Eq. 3-2 and an aperture weighting function $w(x)$, $f(u, \omega)$ can be written

$$\begin{aligned} f(u, \omega) &= \int w(x) \exp(-jkxu_0) \exp(jkxu)\, dx \\ &= \int w(x) \exp\left[j \frac{x}{c} (\omega u - \omega_0 u_0) \right] dx \\ &= \int w(x) \exp\left[j \frac{\omega_0 x}{c} \left(\frac{\omega u}{\omega_0} - u_0 \right) \right] dx \\ &= f\left(\frac{u\omega}{\omega_0} - u_0, \omega_0 \right) \\ &= f_0\left(\frac{u\omega}{\omega_0} - u_0 \right) \end{aligned} \tag{3-4}$$

where the integration is over the aperture and the radiation pattern of the array at midband frequency ω_0, $f(u, \omega_0)$, is abbreviated as $f_0(u)$.

The objective is to determine the beamwidth of the wideband radiation function $f(u)$ when Eq. 3-4 is used in Eq. 3-3. This can be done by examining the behavior of Eq. 3-3 at look directions u close to u_0. Let $u = u_0 + \delta u$ and $\omega = \omega_0 + \delta\omega$. Then

$$\begin{aligned} \frac{u\omega}{\omega_0} - u_0 &= u_0 \frac{\delta\omega}{\omega_0} + \delta u + \frac{\delta u \delta\omega}{\omega_0} \\ &\approx \frac{u_0 \delta\omega}{\omega_0} + \delta u \\ &= \frac{u_0 \delta\omega}{\omega_0} + u - u_0 \end{aligned} \tag{3-5}$$

where we have assumed that $\delta\omega$ (and therefore the bandwidth of the signal) is much smaller than the central frequency ω_0 and that the product of the δu with $\delta\omega/\omega_0$ may be neglected. The assumption that $\delta\omega \ll \omega_0$ holds for most terrestrial microwave imaging situations; hence the approximation in Eq. 3-5 is valid especially for look directions u close to u_0. Since only this region is of interest in determining the beamwidth of $f(u)$, we may approximate $f(u, \omega)$ in Eq. 3-4 using Eq. 3-5 as follows:

$$\begin{aligned} f(u, \omega) &= f_0\left(\frac{u\omega}{\omega_0} - u_0 \right) \\ &\approx f_0\left[u - \left(u_0 - \frac{u_0 \delta\omega}{\omega_0} \right) \right] \end{aligned} \tag{3-6}$$

Thus the use of midband phase shifters to steer the array at frequency $\omega \neq \omega_0$ results in a shift in the radiation pattern of the array. The amount of shift, denoted $\bar{u}$, is

$$\bar{u} = u_0 \frac{\delta\omega}{\omega_0} \tag{3-7}$$

Note that Eq. 3-7 holds for large values of $\omega_0/\delta\omega$. The wideband radiation pattern of the array can be determined by substituting Eq. 3-6 into Eq. 3-3 as follows:

$$f(u) = \int g(\omega) f\left(u - u_0 + u_0 \frac{\delta\omega}{\omega_0}\right) d\omega \tag{3-8}$$

which, using $\delta\omega = \omega - \omega_0$, leads to

$$f(u) = \int g(\omega) f_0\left(u - 2u_0 + \frac{u_0}{\omega_0}\,\omega\right) d\omega \tag{3-9}$$

Observe that $f(u)$ has the form of the correlation between the two functions $g(\cdot)$ and $h(\cdot)$ except that the argument of $h(\cdot)$ contains ω scaled by a factor u_0/ω_0.

We are now in a position to estimate the width Δu of $f(u)$. The width of a function is proportional to its standard deviation (or the square root of its variance). Let σ_u^2 be the variance of $f(u)$; then $\Delta u \approx 2\sigma_u$. Variances add under correlation or convolution (see [94], appendix); that is, if $a(x) = b(x) * c(x)$, where $*$ denotes convolution, and b and c are not causally related, then $\sigma_a^2 = \sigma_b^2 + \sigma_c^2$, where these terms are the variances of a, b, and c. However, Eq. 3-8 is of the form $a(x) = b(x) * c(xu_0/\omega_0)$. For such a correlation, the variance $\sigma_a^2 = (u_0\sigma_b/\omega_0)^2 + \sigma_c^2$, as can be ascertained from Fourier theory. Hence the variance of Eq. 3-8 is the variance of the monochromatic radiation pattern $f_0(u)$ plus the variance of the signal spectrum multiplied by $(u_0/\omega_0)^2$. The monochromatic variance may be taken as $(\Delta u_0/2)^2$ where $\Delta u_0 \approx \lambda_0/L$ is the midband beamwidth of the array. Similarly, the variance of $g(\omega) = (\Delta\omega/2)^2$ where $\Delta\omega$ is the 3 dB bandwidth of the signal. Thus

$$\left(\frac{\Delta u}{2}\right)^2 = \frac{1}{4}\left[(\Delta u_0)^2 + \left(\frac{u_0 \Delta\omega}{\omega_0}\right)^2\right] \tag{3-10}$$

Let the reciprocal of the normalized bandwidth of the signal be represented by Q, that is

$$Q \triangleq \frac{\omega_0}{\Delta\omega} \tag{3-11}$$

Using Eq. 3-11 in Eq. 3-10 yields

$$\Delta u = \sqrt{(\Delta u_0)^2 + \frac{u_0^2}{Q^2}} \tag{3-12}$$

Note that Eq. 3-12 was derived assuming $\Delta\omega \ll \omega_0$ or $Q \gg 1$. This condition is normally met in long-distance terrestrial imaging. The factor by which the beamwidth is increased is

$$\frac{\Delta u}{\Delta u_0} = \sqrt{1 + \left(\frac{u_0}{\Delta u_0 Q}\right)^2}$$

$$\approx 1 + \frac{1}{2}\left(\frac{u_0}{\Delta u_0 Q}\right)^2, \qquad u_0 \ll \Delta u_0 Q \tag{3-13a}$$

$$\approx \frac{u_0}{\Delta u_0 Q}, \qquad u_0 \gg \Delta u_0 Q \tag{3-13b}$$

Thus the beam broadening factor is a function of the steering angle as well as of the size of the aperture and the bandwidth.

The first expression (Eq. 3-13a) shows that when the Q is very large, scanning is virtually unlimited and beam broadening is negligible. This is the conventional narrowband case. On the other hand, Eq. 3-13b shows that when the Q is small the maximum scan angle must be small relative to $\Delta u_0 Q$ for serious beam-broadening to be avoided. Evaluation of $\Delta u_0 Q$ shows the impossibility of satisfying this condition and hence the futility of wideband imaging with phase shifters. The dimensions of the resolution cell at midband are $\Delta S = R\,\Delta u_0$ (lateral or cross-range) and $\Delta R = c/2B = \lambda Q/2$ (range). Thus $\Delta u_0 Q = 2\Delta S \Delta R / R\lambda$. Inserting values that prove useful in high-resolution imaging such as $\Delta S = \Delta R = 1$ m, $R = 10^4$ m and $\lambda = 0.03$ m, $\Delta u_0 Q$ evaluates to 7×10^{-3} radians which, by Eq. 3-13, must be large compared to u_0 to prevent significant degradation in the beamwidth. Thus scanning with phase shifters cannot be achieved under realistic conditions for long-distance, high-resolution microwave imaging.

3-3 TIME DELAYS REPLACE PHASE SHIFTERS FOR NEAR-FIELD IMAGING

The phase shifters in Fig. 2-5 that form and steer the beam are appropriate only when the signal waveform is narrowband. Equation 2-9 is the phase difference between the target echoes received simultaneously on adjacent antenna elements: $\Delta\phi = kdu$. It is a function of element spacing d, which is a constant of the array, wave number $k = 2\pi/\lambda = \omega/c$, and direction $\theta = \sin^{-1} u$, which is the parameter of the source to be measured by the imaging system.

When the bandwidth is narrow, k can be treated as a constant of the system and $\Delta\phi$ directly measures source angle. When the bandwidth is wide, however, the variation of $\Delta\phi$ with frequency over the signal band cannot be distinguished from target direction, in which case the quality of the angle measurement suffers. The severe limitation imposed upon scanning angle was discussed in the preceding section.

From the geometry shown in Fig. 2-5, it is evident that the proper component is a time delay rather than a phase shifter. The wave arriving at element n travels a distance $d \sin \theta$ further than to the $(n + 1)$th element. The differential arrival time is $(d \sin \theta)/c$, where c is the propagation speed in the medium. Errors in beam formation inevitably result when phase shifters are used instead of time delays. For example, assume that phase shifters tuned to midband are employed. The differential phase across an array of length L at the midband angular frequency ω_0 due to a wave arriving from direction $\theta = \sin^{-1} u$ is

$$\Delta\phi = \frac{\omega_0 L u}{c} \tag{3-14}$$

which is compensated by the phase shifters in the phased array. At the edges of a band of width B (Hz) the differential phases are

$$\Delta\phi_{\pm B/2} = \frac{(\omega_0 \pm \pi B)Lu}{c} \tag{3-15}$$

The difference between the differential phases (3-15) causes the problem. Phase shifters tuned for band center are misadjusted elsewhere. The maximum phase error is the difference between the values in Eq. 3-15:

$$\delta\phi_B = \Delta\phi_{+B/2} - \Delta\phi_{-B/2} = \frac{2\pi BLu}{c} \tag{3-16}$$

which must be held below some nominal value determined by the tolerable loss in gain in the main lobe of the antenna. This loss directly reduces the detection sensitivity of the array. It may be calculated by first integrating the signal voltage over frequency at each point in the aperture and then integrating the result over the aperture. The calculations are performed in Section 3-6 and are shown in Table 3-1 later. There it is seen that the BL product must be held to

$$BL \leqslant \frac{0.9c}{u} \tag{3-17}$$

to limit the loss in main beam gain to about 1 dB. Defining $\theta_{max} = \sin^{-1} u_{max}$ as the maximum scan angle from array normal at which the loss does not exceed the stated value, Eq. 3-17 can be interpreted as

$$u_{\max} \approx \frac{0.9c}{BL} = \sin \theta_{\max} \tag{3-18}$$

A few examples show that phase shifters are unduly restrictive. The largest antenna array at the Valley Forge Research Center is 83 m long. It is usually operated at 3-cm wavelength and has 3-m range resolution. The bandwidth B required to achieve 3-m range resolution is

$$\frac{c}{2\Delta r} = \frac{3 \times 10^8 \text{ m s}^{-1}}{2 \times 3 \text{ m}} = 50 \text{ MHz}$$

The BL product is 41.5×10^8 m s^{-1}. This system produced the image of the housing development in Fig. 2-12. Another experimental system, 32 m long, was operated with a 1-m pulse to obtain the airplane image of Fig. 2-13. Its BL product is 48×10^8 m s^{-1}. In both examples, the maximum scan angle from the normal of the array is about 0.06 radians which is less than 3.5°. An antenna with so small a scanning capacity is not useful for general imaging purposes.

When imaging is confined to the far field, however, Eq. 3-18 proves less restrictive. In fact, the limitation on the use of phase shifters in wide-angle-scanning imaging systems is almost entirely a near-field phenomenon. A wide-angle system scans up to about 1 radian to either side of the array normal. The limit to the size of the scanning sector is generally set by the acceptable loss in azimuthal or cross-range resolution of the array at the edges of the scanning sector. This loss in resolution is due to the reduction in effective aperture size in directions away from the array normal. The effective aperture length of an array of length L to a wavefront arriving from θ is the aperture-extent projected upon the wavefront, that is,

$$L_{\text{eff}} = L \cos \theta_0 \tag{3-19}$$

as shown in Fig. 3-2. The azimuthal dimension of the beam becomes

$$\Delta\theta(\theta_0) = \frac{\lambda}{L_{\text{eff}}} = \frac{\lambda}{L \cos \theta_0} \tag{3-20}$$

At $\theta = 60°$, the effective aperture is reduced by a factor of 2 and the beamwidth (3-20) doubles. Scanning beyond $\pm 60°$ results in such rapid deterioration of the cross-range resolution that, for the calculations in this section, 60° is taken as the maximum nominal scan angle. Letting $u_{\max} = \sin 60° = 0.866$, Eqs. 3-17 or 3-18 yield

$$BL \leqslant c \tag{3-21}$$

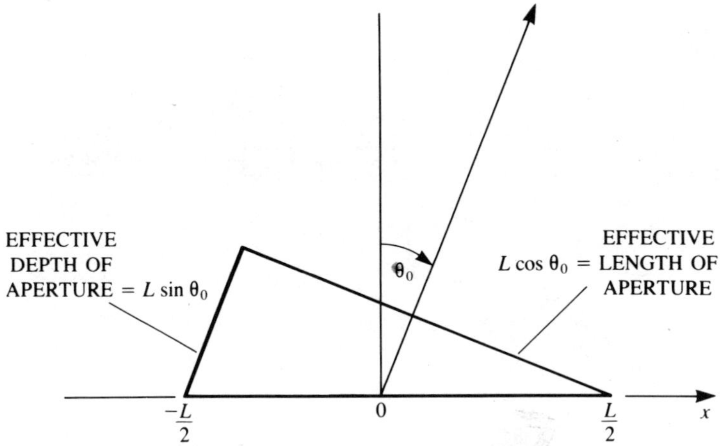

Figure 3-2 Effective aperture length reduced to $L \cos \theta_0$.

where the inequality indicates that BL may be as small as desired but may not exceed the value calculated from Eq. 3-18.

A second relevant design consideration is that it is good imaging practice to make the longitudinal and transverse (range and cross-range) dimensions of the resolution cell approximately equal. Equating $\Delta R \approx c/2B$ and $\Delta S \approx \lambda R/L$ gives

$$B \approx \frac{cL}{2\lambda R}$$

and

$$BL \approx \frac{cL^2}{2\lambda R} = \frac{cR_{NF}}{2R} \tag{3-22}$$

where the near-field to far-field transition range R_{NF} is substituted for L^2/λ (2-47). Comparing Eqs. 3-21 and 3-22 we see that they are consistent only when $R \geqslant R_{NF}$, that is, in the far field. Hence we conclude that near-field imaging requires time delays in the antenna element channels whereas far-field imaging does not.

3-4 TIME DELAYS ARE DISCRETE

A continuously variable time delay from zero to $t_{max} = L/c$ would solve all time-delay problems. Unfortunately, time delay is exceedingly difficult to achieve.* Wideband cables at RF and IF can provide delay but, because the

* See [117] for a comprehensive discussion of time delay techniques.

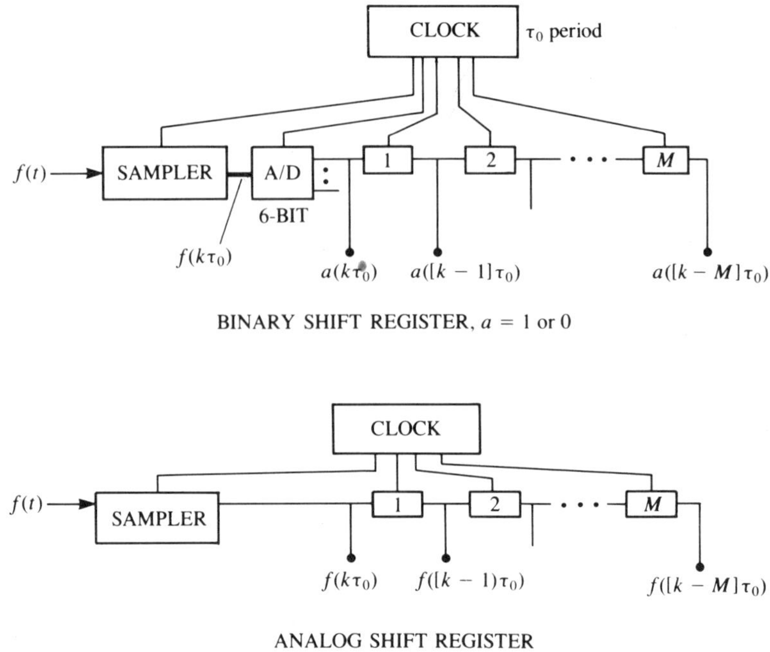

Figure 3-3 *N*-tap shift register delay lines.

speed of electromagnetic propagation in cable is large (a few tenths of the speed of light), the cable lengths are impractically large for large systems. The propagation speed can be enormously reduced in clocked, delay-quantized (or discretized) circuits. The binary shift register is the classic example. An M-bit shift register clocked by an oscillator having period τ_0 constitutes an M-tap binary delay line with intertap delay equal to τ_0. Figure 3-3 (upper) shows an input waveform $f(t)$ sampled at intervals of τ_0 and quantized to b bits in an A/D converter. Each of the b output bits is applied to a delay register of the b M-element chains. At the instants that each new sample of $f(t)$ is taken, each stored sample is transferred to the register to its right and the oldest stored sample is discarded. At $t = k\tau_0$ the signal sample at the mth tap of each of the b shift registers is $a([k-m]\tau_0)$, $m = 0, 1, \ldots, M$, where $a = 1$ or 0. Analog samples of $f(t)$ can be stored in a similar circuit. The storage element must be an analog circuit component such as the charge-coupled device (CCD). Clocked analog samples in a CCD-tapped delay line transfer analog samples from one register to another, as illustrated in the lower part of Fig. 3-3.

In Fig. 3-3 $f(t)$ is one of the coherently demodulated in-phase or quadrature video signals delivered by the microwave receiver that follows each antenna element. These waveforms, defined in Fig. 3-4, carry the target information used in the imaging process in their amplitudes and phases. Figure 3-5 shows schematically how the tap signals are selected and combined. The N antenna

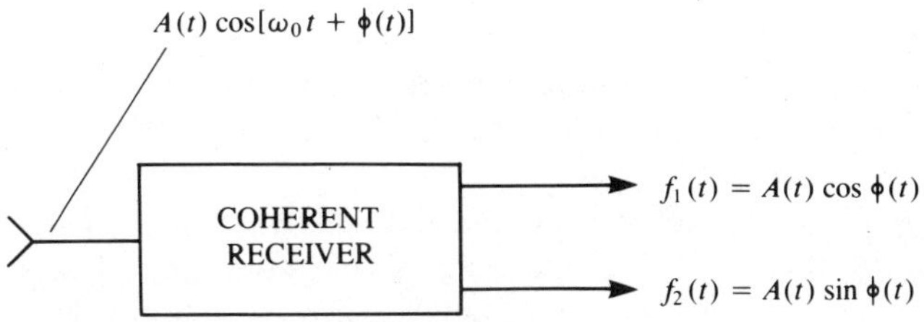

Figure 3-4 Coherent quadrature demodulation of the signal received at each antenna element.

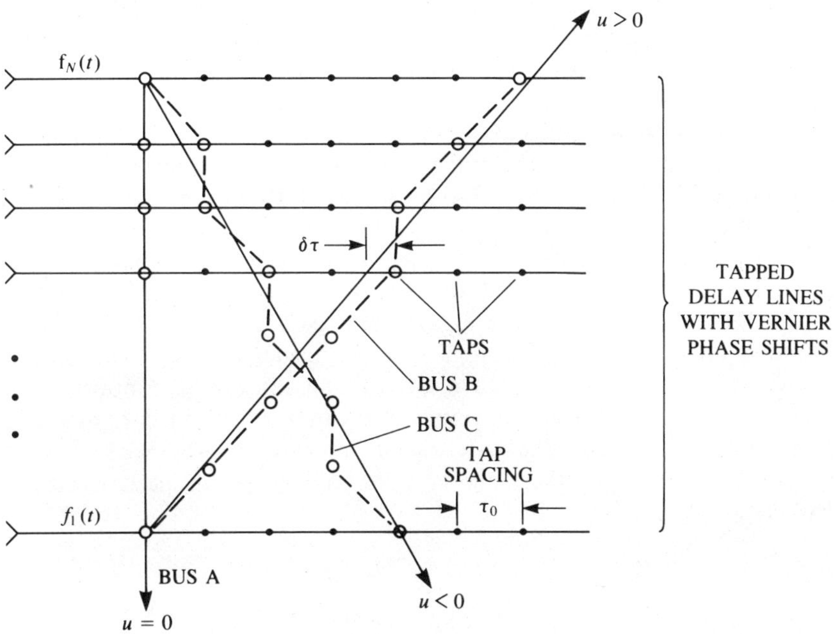

Figure 3-5 Multiple simultaneous beamforming.

elements drive tapped delay lines. The tap spacing is τ_0. A wave from broadside ($u = 0$) arrives cophased at all antenna elements and requires no differential time-delay compensation before summation. Adding the signals at the inputs to the delay lines provides the array response to such a wave. The vertical line connecting the input taps represents a summing bus 'tuned' to a broadside wave. In short, summing the inputs to the delay lines, as on bus A, constitutes beamforming in the direction of $u = 0$.

Beamsteering at other angles suffers, however, because a linearly increasing (or decreasing) time delay with element position is demanded, as shown by the solid lines for $u > 0$ and $u < 0$. The procedure eliminates only the primary error caused by wide bandwidth. A residue error remains because the time delays are discrete or quantized. The delays associated with the taps connected by the output buses differ slightly from the proper delays. Figure 3-5 illustrates the errors; they are the distances along the delay lines between the taps with open circles and the intersections with the delay-slope lines. One such error $\delta\tau$ is indicated in the figure.

Figure 3-6 shows the relative orientation of the physical aperture and the virtual aperture after time-delay steering. Continuous time delays orient the virtual aperture normal to the arrival angle. Discrete delays orient the virtual aperture in the same direction but the residual timing errors displace the virtual antenna element positions toward (or away from) the source. These errors can be corrected by a continuously adjustable vernier time delay or a discrete vernier delay having tap spacing very small compared to τ_0. Conceptually, a vernier delay is placed at each tap output. In practice one per delay line per output beam suffices. An alternate procedure using vernier phase shifters is easier to accomplish and therefore is superior provided that the residual errors are sufficiently small. Each vernier delay is replaced by a vernier phase shifter, and the set partially compensates for the effect of the discrete time delays. Each phase shifter is adjusted so that the residual phase error at midband in its channel is zero. Its value is $\phi_n = -\omega_0 \delta\tau_n$, where $\delta\tau_n$ is the timing error at the nth tap. Errors remain throughout the remainder of the passband of the signal, however. These errors result in a loss in array gain and a rise in the sidelobe level. Both are calculated in Section 3-6. The former decreases the detection sensitivity of the imaging system. The latter reduces the image contrast and the target dynamic range that the imaging system can handle. Hence residual phase errors must be restricted. The procedure is indicated in Fig. 3-7. Analysis of the residual errors and their effects upon the radiation pattern also are calculated in Section 3.6.

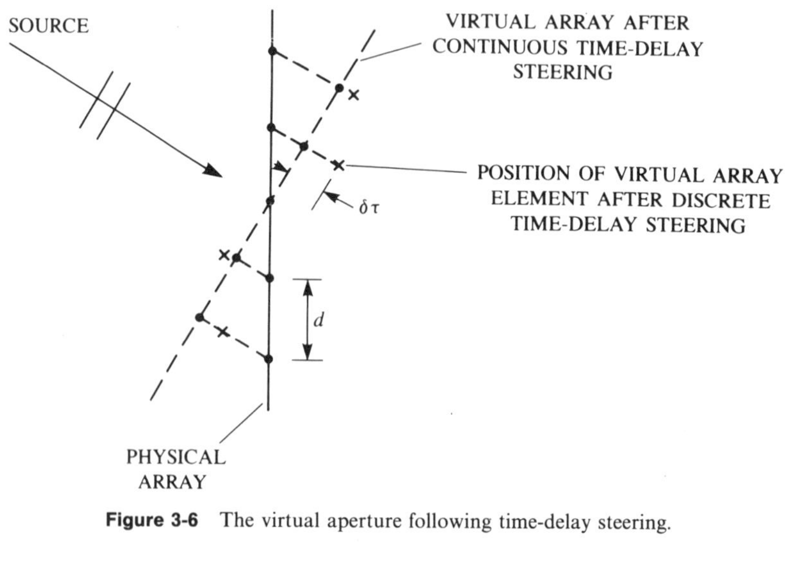

Figure 3-6 The virtual aperture following time-delay steering.

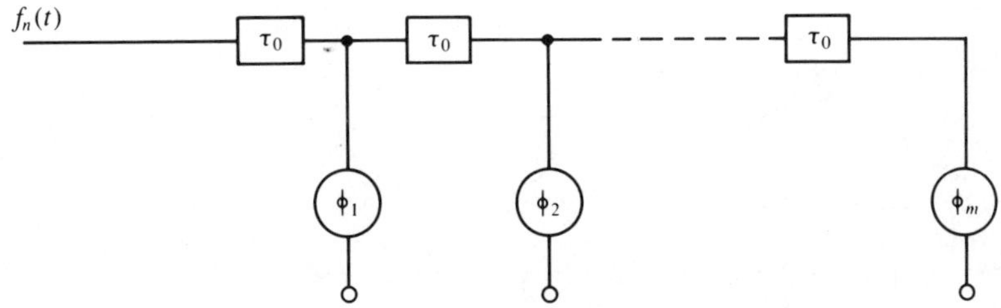

Figure 3-7 Tapped delay line with vernier phase shifters.

Returning to Fig. 3-5, assume that a wave arrives from $\theta = \sin^{-1} u > 0$, say 60°. It reaches the top element earlier than the bottom of the array. The proper time delays are the intersections of the line labeled $u > 0$ with the delay lines. The open circles are the nearest taps to these intersections. When the phase shifts of Fig. 3-7 are incorporated with the delay lines of Fig. 3-5, the sum of these tap outputs, as on bus B, gives the best available array response to this wave. The output of bus B, then, represents a beam oriented to 60°. Similarly, a wave from $u < 0$, say $\sin(-30°)$, is coaligned by time delays and vernier phase shifts at the intersections of the line labeled $u < 0$ and the delay lines. Bus C sums the outputs of the closest taps. It forms a beam at $-30°$.

A set of J such buses simultaneously forms J beams from the array. The process is indicated in Fig. 3-8. When the output beams are applied directly to the display, the imaging process is complete.

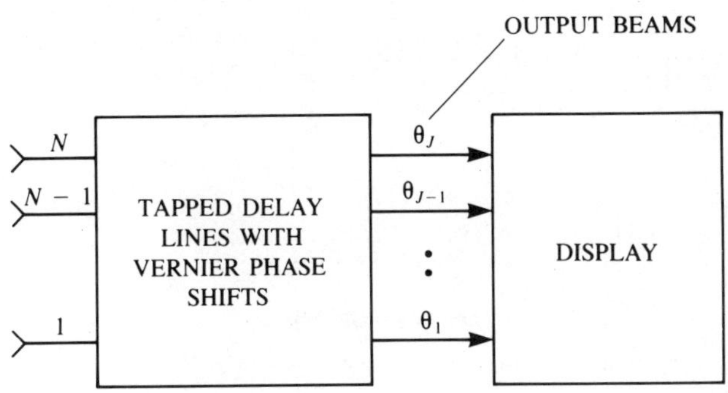

Figure 3-8 The imaging process.

3-5 IMAGE PROCESSING IN COMPUTER

The functions performed by tapped delay lines and midband phase shifters at each tap can be easily accomplished in a digital computer. The shift registers in Fig. 3-3 are replaced by memory locations. The phase shifting and summing are performed by arithmetic operations. Figure 3-9 outlines a composite of

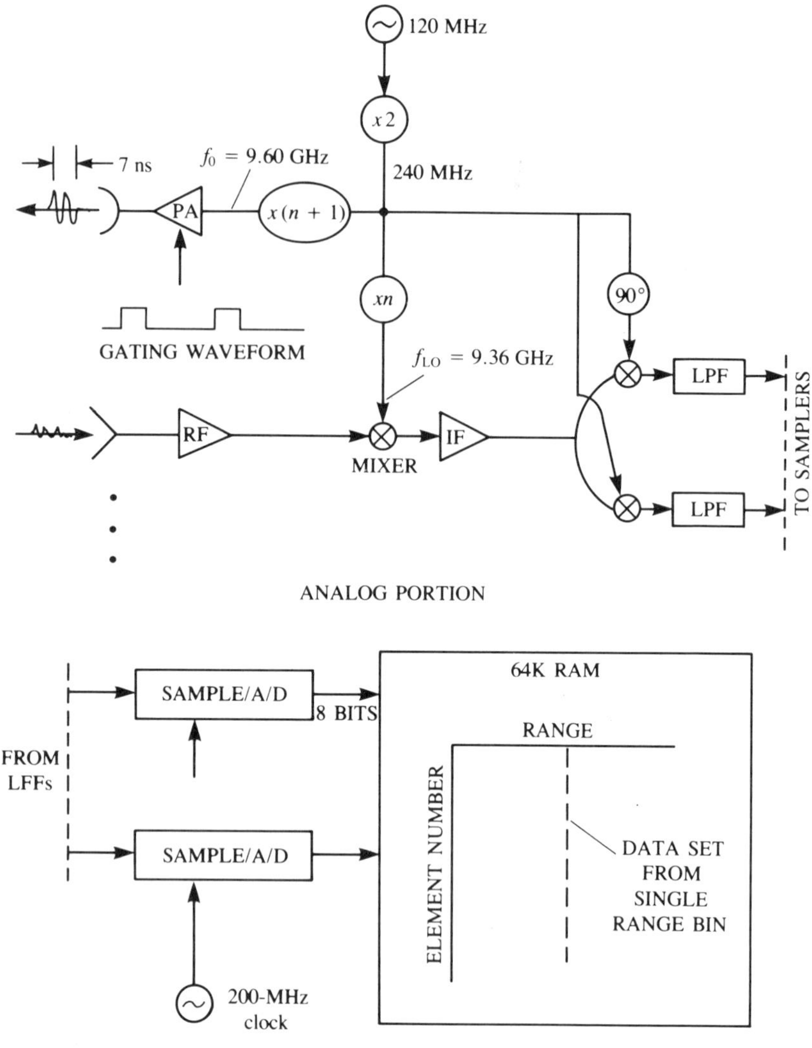

Figure 3-9 Experimental radio camera imaging system at Valley Forge Research Center.

typical transmitting–receiving procedures employed at the Valley Forge Research Center. A 120-MHz crystal oscillator provides timing for the analog (transmitter and receiver) portion of the system. Its output is doubled in frequency to 240 MHz, which is used in three places. The 240-MHz is frequency multiplied to 9.60 GHz to create the RF carrier and to 9.36 GHz to provide the local oscillator waveform in the receivers. The RF wave is gated in a power amplifier by the pulse waveform consisting of 7-ns pulses spaced typically by

200 μs. The 7-ns pulse provides 1-m range resolution. The transmitting antenna is a 1.2-m dish.

Echoes are received by horn antennas, amplified at RF in low-noise amplifiers and heterodyned to the 240-MHz intermediate frequency. The IF bandwidth is 150 MHz, sufficient to sustain the 7-ns pulse. The IF output is quadrature-demodulated at 240 MHz. The two demodulated signals in each channel are lowpass filtered, providing the outputs labeled $f_1(t)$ and $f_2(t)$ in Fig. 3-4.

These analog signals then enter the digital portion of the receiver. A 200-MHz clock paces a sampler and 8-bit A/D converter in each of the two quadrature channels. The 8-bit words are stored in a 64K RAM. The entire receiving process is duplicated for each channel in the phased array.

Each pair of 8-bit words is one complex number. It represents the instantaneous value of the radiation field at the time (radar range) and place (element number or position) of its measurement. A two-dimensional matrix of such numbers carries all the available information for imaging. The abscissa is range or time of arrival of an echo. The ordinate is element number. Each column is the data set from a single range bin. It is the current excitation $i = \mathscr{W}\mathscr{F}^{-1}s$ given in Eq. 2-36. The modulus of its transform (Eq. 2-37) is the one-dimensional image of that range bin. The set of such images is the two-dimensional picture of the airplane.

When multiple receiving antennas and receivers are employed, as illustrated above, the imaging system is called a radio camera. A single snapshot of data suffices for image formation. An alternate form of the system is possible when imaging is performed on an isolated target that is moving, such as an airplane. Although the imaging process is identical to radio camera imaging, the procedure traditionally is called *inverse synthetic aperture radar* or ISAR.

The functional diagram of the ISAR transmitter and receiver is the same as in Fig. 3-9 with two exceptions. (A more detailed description of ISAR than the one that follows is to be found in Section 5-7.) A common antenna is used both for transmitting and receiving and a transmit–receive (T/R) switch separates the outgoing and incoming signals. This is conventional radar practice. Second, the string of dots under the receiving antenna in Fig. 3-9 is missing because only a single receiving antenna is employed.

The data sequence is obtained temporally rather than spatially in ISAR. A set of echoes from the moving target in response to N successive transmitted pulses is received, demodulated, and stored in the RAM. The data set has the same form as in Fig. 3-9. The abscissa is range or time of arrival of an echo. The ordinate, however, instead of being element number is pulse or sample number.

Figure 3-10 shows a section of a data record of one such experiment. The target was a Lockheed L-1011 airplane flying at 272 mph at a distance of 2.5 km from the Valley Forge laboratory. Its elevation was 40°. Polarization was horizontal. The wavelength was 3 cm.

Recorded are the moduli of the measurements. The bins are spaced by

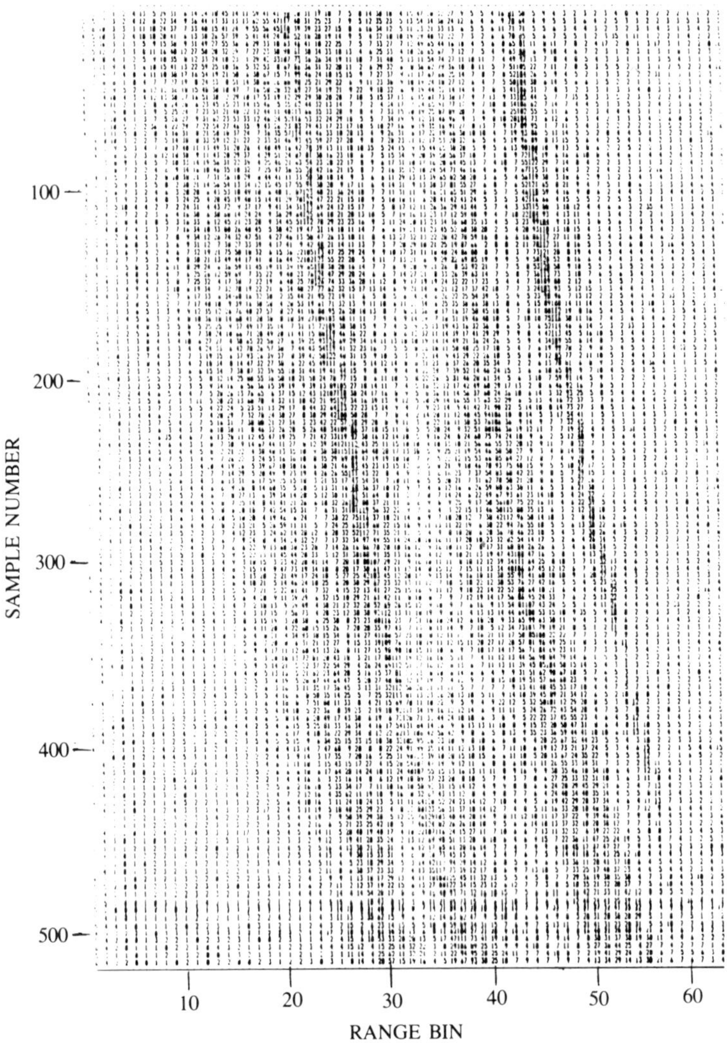

Figure 3-10 Portion of data record of ISAR experiment (run 53). The target is an L-1011 aircraft flying at 2.5 km from VFRC. The range bins are spaced by 5 nsec. Pulse duration is 1 m (7 nsec); aircraft speed is 272 mph; elevation is 40°; X-band; horizontal polarization. The range to bin 1 is 2494.25 m. Each row is the average of four successive echo traces. The echoes from the scatterer used for adaptive beamforming form the relatively isolated echo track at the far right.

$\tau_0 = 5$ ns or 0.7 m. Successive pulses are separated by 2.5 ms. The aircraft occupies about 45 range bins or 32 m, which is consistent with the wingspan (45 m) when observed at the 40° elevation angle ($45 \cos 40^\circ = 32$). The image made from this data set is shown in Fig. 3-11.

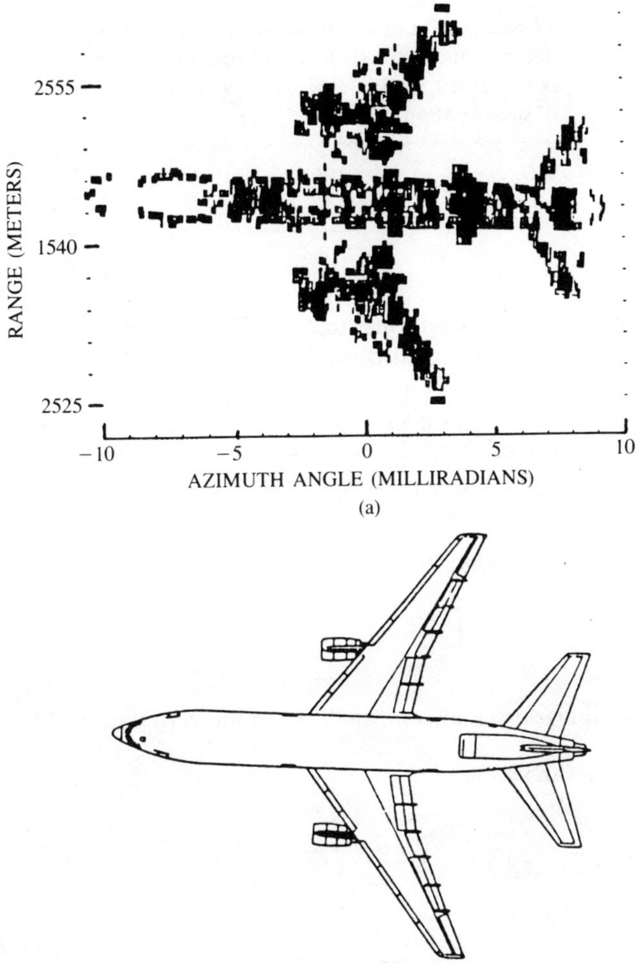

Figure 3-11 (a) High-resolution radar image of L-1011 aircraft made from the data set shown in Fig. 3-10. (b) Plan view drawing of the aircraft. The airplane was flying at a distance of 2.5 km from the Valley Forge radio camera.

3-6 LOSSES DUE TO USE OF PHASE SHIFTERS

3-6-1 Loss in Array Gain

One effect of the residual errors resulting from the use of phase shifts rather than variable time delays is a loss in main beam gain. Another is a rise in the sidelobe level. The main beam loss is calculated in two steps: the signal voltage is first integrated over the frequency band at each point in the array, and the result is then integrated over the aperture.

TABLE 3-1 Loss in phased array gain due to signal bandwidth *B*; *u* is the direction of arrival, *L* is the array length, and *c* is the speed of propagation

BuL/c	0.46	0.66	0.89
Loss in dB	0.25	0.5	1.0

For simplicity the calculations that follow are based on the far-field approximation. The results are also valid in the near field for distances beyond approximately ten aperture lengths. The signal component at point x in the array and angular frequency ω is $\exp(j\omega xu/c)$. After phase weighting at midband for beamsteering, it is $\exp[j(\omega - \omega_0)xu/c]$. The first integration, normalized to the infinite-Q (monochromatic) value, is

$$\frac{\displaystyle\int_{\omega_0 - \pi B}^{\omega_0 + \pi B} \exp[j(\omega - \omega_0)ux/c]\, d\omega}{\displaystyle\int_{\omega_0 - \pi B}^{\omega_0 + \pi B} d\omega} = \frac{\sin(\pi Bxu/c)}{\pi Bxu/c} \tag{3-23}$$

The second integration, also normalized to the zero bandwidth case, is

$$\frac{\displaystyle\int_{-L/2}^{L/2} \frac{\sin(\pi Bxu/c)}{\pi Bxu/c}\, dx}{\displaystyle\int_{-L/2}^{L/2} dx} = \frac{c}{\pi BuI} \int_{-\pi BuL/2c}^{\pi BuL/2c} \frac{\sin y}{y}\, dy$$

$$= 1 - \frac{Y^2}{3\cdot 3!} + \frac{Y^4}{5\cdot 5!} - \cdots, \qquad Y = \frac{\pi BuL}{2c} \tag{3-24}$$

The loss in decibels is 20 log (3-24). Table 3-1 lists BuL/c for three values of loss.

As shown in Section 3-3, this loss proves too restrictive on the maximum scan angle and thereby limits the available scanning sector to a few degrees. For this reason time delays must be substituted for phase shifts. Because time delays are discrete and not continuously variable, vernier phase compensation at each tap is required. Phase shifts can correctly compensate for mistiming only at a particular frequency, which is set at midband. At all other frequencies in the passband a residual error, which is now calculated, remains.

Let the signal received at the nth antenna element from direction θ be

$$v_n(t) = s\left(t - T_0 + \frac{(n-1)\, d \sin\theta}{c}\right) \tag{3-25}$$

where $s(t)$ is the radiated waveform, T_0 is the propagation time to the first element, and the third term in the argument is the differential travel time. The compensatory time delay introduced by the tapped delay line is $m_n \tau_0$, which is the closest integral multiple of τ_0 to $(n-1)\, d \sin\theta/c$. The residual delay error is

$$\delta\tau_n = \frac{(n-1)\, d \sin\theta}{c} - m_n \tau_0 \tag{3-26}$$

This error is common for all frequency components in the wave. Consider a signal component at angular frequency ω at location x_n in the array. It is given by $\exp(j\omega x_n u/c)$. The proper time delay would zero its argument. Instead a residual argument $\omega\, \delta\tau_n$ remains. We compensate at midband by a phase shift

$$\phi_n = -\omega_0\, \delta\tau_n \tag{3-27}$$

The nth delay channel is modified as shown in Fig. 3-7. A phase shifter couples each delay element to the tap output. Beamsteering constitutes selection of the proper set of delays *and* vernier phase shifts. The phasor representation of the signal component following delay compensation by $m_n \tau_0$ and phase shifting by ϕ_n is $\exp[j(\omega - \omega_0)\delta\tau_n]$, which results in a loss in main beam gain when $\delta\tau_n \neq 0$.

This loss is similar in kind to that calculated above for phase-shift steering (Eqs. 3-23 and 3-24). It is very much smaller, however, because the primary steering burden is carried by the time delays. The calculation of the loss is the same. The first step is integration of the signal component over the band of width B, normalized to the monochromatic value:

$$\frac{\displaystyle\int_{\omega_0 - \pi B}^{\omega_0 + \pi B} \exp[j(\omega - \omega_0)\delta\tau_n\, d\omega}{\displaystyle\int_{\omega_0 - \pi B}^{\omega_0 + \pi B} d\omega} = \frac{\sin \pi B\, \delta\tau_n}{\pi B\, \delta\tau_n} \tag{3-28}$$

Integration over the array is next. The formal procedure in Eq. 3-24 is no longer appropriate, however, because the independent variable, the timing error, is no longer a linear function of position in the array. As seen in Fig. 3-5, the timing error can vary erratically from channel to channel. This condition will always prevail for random arrays as well as for very large periodic arrays with considerable error in element locations. It is also the case at large scan angles for smaller periodic arrays having negligible position error, because when the scan angle is large enough the tap number changes rapidly from delay channel to channel. In these cases the residual error $\delta\tau$ can be treated as a random variable uniformly distributed in the interval $[-\tau_0/2, \tau_0/2]$. At smaller scan angles the tap-number progression in the periodic array is slow and the timing error $\delta\tau$ is correlated between delay channels. This correlation can be removed artificially by the addition (subtraction) of random time-delay shims in each channel. It is most easily accomplished by introducing randomly cut cable lengths in the RF portion of the system, a normal and common practice *unless* special care is taken to avoid it. In this case the common practice is desirable.

Assuming a uniform distribution of timing error, the second calculation becomes the evaluation of the expectation $E[\cdot]$ of Eq. 3-28 over this interval:

$$E\left[\frac{\sin \pi B\,\delta\tau}{\pi B\,\delta\tau}\right] = \frac{1}{\tau_0}\int_{-\tau_0/2}^{\tau_0/2} \frac{\sin \pi B\,\delta\tau}{\pi B\,\delta\tau}\, d(\delta\tau) \qquad (3\text{-}29)$$

Eq. 3-29 has the same form as Eq. 3-24 and has the same solution in Y when $Y = \pi B\tau_0/2$. When a short pulse of duration $\delta \cong B^{-1}$ is used, the relation between intertap delay, pulse duration and loss in array gain is given by Table 3-2. By way of comparison, in imaging the airplane of Fig. 2-13 the values of δ and τ_0 were 7 and 5 ns, respectively, which evaluates to a loss of 0.6 dB. Table 3-2 lists τ_0/δ for other values of loss in main beam gain.

TABLE 3-2 Loss in array gain due to intertap delay τ_0 relative to pulse duration δ

τ_0/δ	0.46	0.66	0.89
Loss in dB	0.25	0.50	1.00

3-6-2 Rise in Sidelobe Level

The effects of random timing errors are different for the random array and the conventionally designed periodic or quasiperiodic array. The random array is characterized by a set of element locations that can only be described statistically. For example, an array formed from antenna elements placed on roofs of buildings is one such array because the elements' positions would be determined by available space and not by design formulas. Another is an array whose elements are mounted directly into the structure or skin of a vehicle such as an airplane or large ship. Again the locations would be chosen from available space. Further, the frames of both such vehicles are nonrigid, implying that the element locations would change with time. The sidelobe properties of such arrays are known in a statistical sense [94]. The main lobe strength is $(\bar{a})^2$ where $\bar{a}$ is the mean modulus of the weighting function. The sidelobe energy is distributed uniformly in angle, its expected (or average) power level (ASL) is $\overline{a^2}N^{-1}$ and the peak sidelobe level (PSL) is about 10 dB higher. The ASL relative to the main lobe is $\overline{a^2}/N(\bar{a})^2$.

The addition of random timing errors does not alter the statistical properties of the radiation pattern in the sidelobe region. Thus the ASL remains $\overline{a^2}N^{-1}$. However the main lobe is reduced according to Eq. 3-29 and Table 3-2; therefore the ASL relative to the main lobe rises by the same amount. The increase is shown in Table 3-3.

TABLE 3-3 ΔSLL is the increase in the ratio of the average sidelobe level of a random array to the mainlobe gain. τ_0/δ is the intertap delay relative to pulse duration

τ_0/δ	0.46	0.66	0.89
ΔSLL (dB)	0.25	0.5	1.0

These values of ΔSLL are acceptably small. Hence we may conclude that the sidelobe level of the random array is not materially degraded by discrete time delays with midband phase shifters at each tap.

The evaluation of the effect of timing error in the sidelobe region for the more conventional array proceeds differently. It is again assumed that $\delta\tau$ can be regarded as a uniformly distributed random variable. It is shown in [94] that the sidelobe level at any angle rises from the design value by a random quantity due to random amplitude and phase errors. The expectation of the power pattern is

$$Eff^* = |E[\exp(j\,\delta\phi)]|^2 f_0 f_0^* + \frac{1}{N^2}(1 - |E[\exp(j\,\delta\phi)]|^2) \sum_n a_n^2 + \frac{1}{N}\sigma_a^2 \quad (3\text{-}30)$$

where $f = f(u)$ is the radiation pattern with errors, $f_0 = f_0(u) = (1/N)\sum_n a_n \times \exp(jkx_n u)$ is the design pattern, N is the number of antenna elements, $\delta\phi$ is the zero mean, random phase error, a_n is the modulus of the weight vector at the nth element, σ_a^2 is the variance of the weight moduli relative to their design values, and E is the expectation operation.

Equation 3-30 can be simplified without significant compromise to its accuracy when the phase and amplitude errors are small, as must be the case for a well-designed imaging system. In this case $|Ee^{j\delta\phi}|^2 \approx 1 - \sigma_\phi^2$, nearly independent of the distribution of the phase error and Eq. 3-30 can be approximated by

$$Eff^* \approx e^{-\sigma_\phi^2} f_0 f_0^* + \frac{\overline{a^2}\sigma_\phi^2 + \sigma_a^2}{N} \quad (3\text{-}31)$$

where $\overline{a^2}$ is the mean square modulus of the weighting function and σ_ϕ^2 is the residual phase error variance following vernier phase compensation at the delay taps. The second term in Eq. 3-31 is the rise in the sidelobe level. This approximation is satisfactory for $\sigma_\phi \leqslant \frac{1}{2}$ radian.

The problem that concerns us is the effect on the sidelobe level of the uncompensated time-delay errors. These errors result in phase errors and therefore contribute to the residual phase error variance σ_ϕ^2. Amplitude errors in the signal as it passes along a delay channel are not zero, but gain adjustments normally can be made relatively easily and accurately enough so that the amplitude variance term becomes inconsequential. Therefore, the last term in Eq. 3-31 usually can be ignored, leaving

$$Eff^* \approx e^{-\sigma_\phi^2} f_0 f_0^* + \frac{\overline{a^2}\sigma_\phi^2}{N} \tag{3-32}$$

The residual phase error at $\omega = \omega_0 + \delta\omega$ at the nth tap is $\delta\phi_n = \delta\omega\, \delta\tau_n$. The timing error $\delta\tau_n$, $n = 1, 2, \ldots, N$, is assumed to be uniformly distributed in the interval $[-\tau_0/2, \tau_0/2]$ for all n. Therefore

$$\sigma_\phi^2(\omega) = (\omega - \omega_0)^2 \sigma_\phi^2 = \frac{(\omega - \omega_0)^2 \tau_0^2}{12} \tag{3-33}$$

where σ_τ^2 is the variance of the time delay error. The contributions to the average rise in sidelobe power level from frequency components across the band of width B add. The total effective phase variance is

$$\begin{aligned} \sigma_\phi^2 &= \frac{1}{2\pi B} \int_{\omega_0 - \pi B}^{\omega_0 + \pi B} \sigma_\phi^2 \, d\omega \\ &= \left(\frac{\pi \tau_0 B}{6} \right)^2 = 0.27 (t_0 B)^2 \end{aligned} \tag{3-34}$$

which may be inserted into Eq. 3-32.

The first term in Eq. 3-32 is the design pattern attenuated by the effects of phase errors. The design pattern is the Fourier transform of the aperture weight function. It is usually tapered across the array to avoid the abrupt discontinuities in aperture excitation that otherwise occur at the edges of the antenna, which give rise to high sidelobes. Consequently the first term in Eq. 3-32 drops below the error-induced second term for $|u - u_0| \gg 0$ and can be ignored. The second term becomes a floor to the side radiation pattern. It is the main contribution to the average background level in the image. When normalized to $f_0 f_0^* = (\bar{a})^2$ at $u = 0$ it may be identified with image contrast. It is not only a function of σ_ϕ^2 but also of N and $\overline{a^2}/(\bar{a})^2$. The ratio $\overline{a^2}/(\bar{a})^2$ expresses the degree of aperture tapering for purposes of sidelobe control. In a conventionally designed array this ratio is between 1.0 and 1.5, a typical value, perhaps, being 1.3. Table 3.4 lists $10 \log[0.27\overline{a^2}(\tau_0 B)^2/(\bar{a})^2 N]$ for $\overline{a^2}/(\bar{a})^2 = 1.3$ and $\delta = B^{-1}$. For comparison purposes, the right-hand column lists the corresponding values for the random array without the timing errors. To these values may be added the ΔSLL increases due to the errors that are given in Table 3-3. It is evident that when τ_0 becomes comparable to δ the ASL rises nearly to that of a random array and the image contrast suffers accordingly.

The calculations up to this point were based on the assumption that the

TABLE 3-4 Average asymptotic background power level relative to main lobe for periodic array with tapered aperture. Values for the random array without errors are shown in the right column for comparison. All entries are in decibels

	τ_0/δ			
N	*0.46*	*0.66*	*0.89*	*Random array*
10^2	−31.3	−28.2	−25.6	−20
10^3	−41.3	−38.2	−35.6	−30
10^4	−51.3	−48.2	−45.6	−40

timing errors were random and independent from antenna element channel to channel. This assumption is generally valid in large, random arrays, as well as large periodic arrays with random position errors. As noted earlier, it is not valid in linear, periodic arrays, particularly at small scan angles, and a design correction is required. Correction is easily achieved by adding (subtracting) a random set of time delays distributed over an interval of approximately τ_0 between the antenna elements and the tapped delay lines. Without this randomization, the delay errors are correlated and performance may suffer. It is instructive to consider what happens in this case.

Figure 3-12a shows a linear, periodic, tapped delay line array receiving a

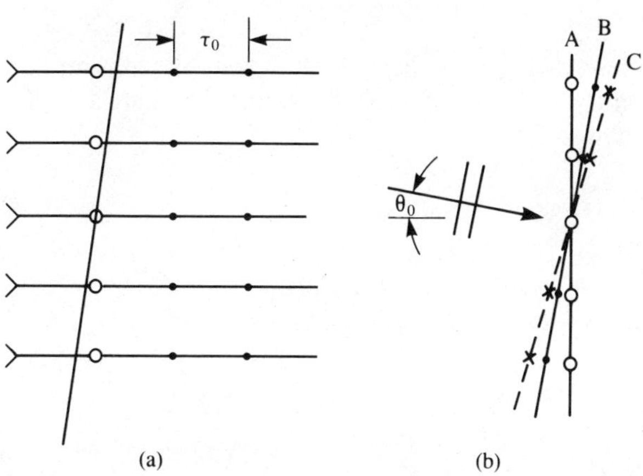

Figure 3-12 Beam-pointing error caused by quantized time delays, with and without phase compensation. (a) Open circles are closest taps to proper delays for wave arriving from direction θ_0 close to array normal. (b) Virtual arrays after discrete time-delay steering to θ_0: A, without phase shifters at each tap, beam remains steered to normal ($\theta = 0$); B, with phase shifters tuned to midband, beam is steered to θ_0 at midband frequency; C, phase shifters tuned to midband, beam is steered slightly away from θ_0 at nonmidband frequency.

wavefront from direction θ_0 that is close to the normal to the array. So long as the differential travel time of the wave between an end element and the center does not exceed $\tau_0/2$, the open circles shown are the closest taps to the proper delays. This condition is given by

$$\frac{L \sin \theta_0}{2c} < \frac{\tau_0}{2} \tag{3-35}$$

When this small steering-angle condition is met, the delays are all the same and the time-delay error is monotonic with element position. Under far-field conditions, the error is linear with position, as illustrated, and is curved when the source is in the near field. In the absence of phase shifters the beam remains fixed at 0 degrees. This situation is shown in virtual array A of Fig. 3-12b.

Phase shifters tuned to midband correct the error at the midband frequency. Virtual array B illustrates this case. The nth phase shift is

$$\phi_n = -\frac{\omega_0 x_n \sin \theta_0}{c} \tag{3-36}$$

These phase shifts properly steer the beam to θ_0 at $\omega = \omega_0$ but at no other frequency. Elsewhere in the band, say a distance $\delta\omega$ from ω_0, the phase is in error by

$$\delta\phi_n = -\frac{\delta\omega x_n \sin \theta_0}{c} \tag{3-37}$$

This linear phase variation with position, when $\omega = \omega_0$, implies that the beam is steered slightly away from θ_0. The result, as sketched in Fig. 3-13, is a broadened beam.

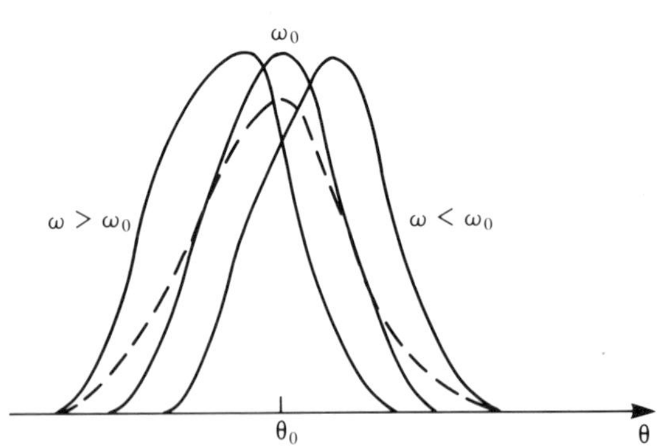

Figure 3-13 Solid curves are beam patterns at different frequencies within the band. The composite, shown wider, is the dashed curve. ω_0 is the center frequency.

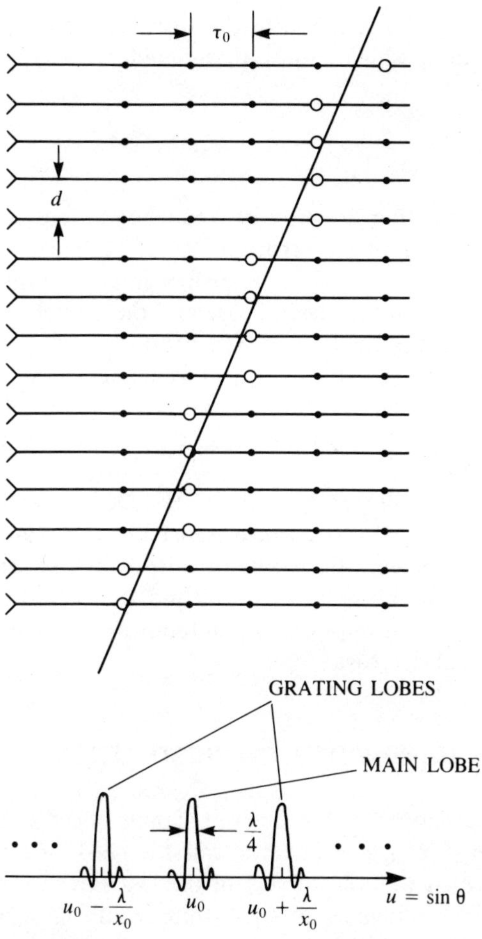

Figure 3-14 A periodic time-delay error across the aperture creates grating lobes. $L = Nd$ is the size of the aperture. $x_0 = 4d$ in the illustration.

Another phenomenon is introduced at somewhat larger steering angles. The delay and the residual phase error change from linear to periodic (Fig. 3-14); that is, the residual phase at frequency ω and position x is

$$\phi(x, \omega) = \phi(x + x_0, \omega) \tag{3-38}$$

where x_0 is the period (four interelement spacings in Fig. 3-14). Because of the Fourier relationship between the weighting function in the aperture and the radiation pattern, Eq. 3-38 leads to pattern periodicities known as grating lobes [94]. These periodicities or spurious lobes are spaced by

$$\Delta u = \frac{\lambda}{x_0} \tag{3-39}$$

It is evident that x_0 is a function of the steering angle θ_0 and therefore so is the radiation pattern, which can be designated $f(u;\theta_0)$ or f_{θ_0}. In operator notation,

$$f_{\theta_0} = \mathscr{D}_\omega^{-1}\mathscr{F}\mathscr{W}_{\omega;\theta_0} \tag{3-40}$$

$\mathscr{W}_{\omega;\theta_0}$ is the error weighting function at frequency ω and scan angle θ_0 that results from the residual phase error (3-38). $\mathscr{W}_{\omega;\theta_0}$ means $w(x,\omega;\theta_0) = e^{j\phi}$ where $\phi = \phi(x,\omega)$ in Eq. 3-38. $\mathscr{F}$ is the proper integral transform operator; it is the diffraction integral in the general case and the discrete Fourier transform for the far-field, periodic phased array. The product $\mathscr{F}\mathscr{W}_{\omega;\theta_0}$ is the component of the radiation pattern due to the energy at frequency ω. All such components are added coherently to form the radiation pattern at θ_0. $\mathscr{D}^{-1}$ means integration and the subscript ω indicates that the integration is with respect to frequency. The result is Eq. 3-40.

Although the formal solution is easily written down, the evaluation of Eq. 3-40 is tedious and the detailed pattern is usually irrelevant (because of the ease with which the problem of periodic time delay errors can be avoided). In general the solutions for both the random and the periodic array are those given earlier under the assumption of randomness and independence of the channel-to-channel delay errors.

3-7 EFFECT OF BANDWIDTH ON RESOLUTION

In Section 2-8 it was noted that wideband signals affect phased-array design and performance in three ways. The first was the need for time delays instead of phase shifters, which was the subject of the last three sections. In addition there are the possible influences of signal bandwidth upon resolution in the image (through the beamwidth of the array) and the background or gray level (through the sidelobe level). The effect on the resolution is discussed in this section. It is found that the effect is minimal because the available bandwidth in long-distance, terrestrial imaging is severely restricted. For bandwidth to play a role in image resolution the bandwidth must be of the order of the midband frequency.

Notwithstanding the limited practical effect, the theory is interesting. Let $w = w(x)$ be the weighting function of the aperture and $\mathscr{F}_\omega$ the imaging function (the subscript ω explicitly identifies it with a particular frequency component in the signal band). The contribution to the radiation pattern at ω is (in the operator notation of Section 2-4)

$$f_\omega = f_\omega(u) = \mathscr{F}_\omega w \tag{3-41}$$

All such components of the radiation pattern $f = f(u)$ must be weighted by $\mathscr{G} = g(\omega)$, the Fourier spectrum of the signal, and integrated. Letting $\mathscr{D}_\omega^{-1}$ be the integral with respect to ω, the composite radiation pattern becomes

$$f = \frac{\mathscr{D}_\omega^{-1}\mathscr{G}\mathscr{F}_\omega \mathrm{w}}{(\mathscr{D}_\omega^{-1}\mathscr{G}\mathscr{F}_\omega \mathrm{w})_{u=0}} \tag{3-42}$$

where the denominator normalizes f to unity at the origin.

For simplicity, let $\mathscr{F}_\omega$ be the far-field Fourier transform. Then (3-42) is the simple function

$$f(u) = \frac{\iint g(\omega) w(x) e^{j\omega xu/c}\, dx\, d\omega}{\iint g(\omega) w(x)\, dx\, d\omega} \tag{3-43}$$

Evaluation of Eq. 3-43 leads to the conclusion stated earlier; namely, that beamwidth is negligibly influenced by the width of $g(\omega)$.

3-8 EFFECT OF BANDWIDTH ON SIDELOBE OR BACKGROUND LEVEL

Unlike the result found in the previous section, the effect of bandwidth upon the sidelobe or background level is not trivial. Again two cases are considered, the random array and the conventionally designed, periodic array. The random array is discussed first.

The side radiation pattern of a random array looks like and has the statistics of the envelope of narrowband Gaussian noise. Figure 3-15 shows the visual similarity. Assuming that the array consists of N equal-amplitude, isotropic antenna elements, the probability density function (pdf) is Rayleigh, that is,

$$w(x) = \frac{x}{\sigma^2} \exp \frac{-x^2}{2\sigma^2}$$

The average sidelobe power level relative to the main lobe is N^{-1} and the peak or highest sidelobe is about 10 dB higher. The pdf of the highest sidelobe is compact with standard deviation slightly greater than 1 dB [94]. Thus the expected value of the PSL, multiplied by a factor somewhat greater than unity, is an appropriate measure of the largest image artifact due to sidelobes of the strongest target. Its reciprocal is proportional to the target dynamic range that can be imaged.

The effect upon the PSL, then, discloses how bandwidth affects image quality through its influence upon the background level. Figure 3-16a indicates why frequency has any effect at all. It shows the radiation patterns of an array at two different frequencies within the signal band. The array is larger, in units of wavelength, at the higher frequency than at the lower. Hence the pattern is compressed at the higher frequency relative to the lower. There is little difference

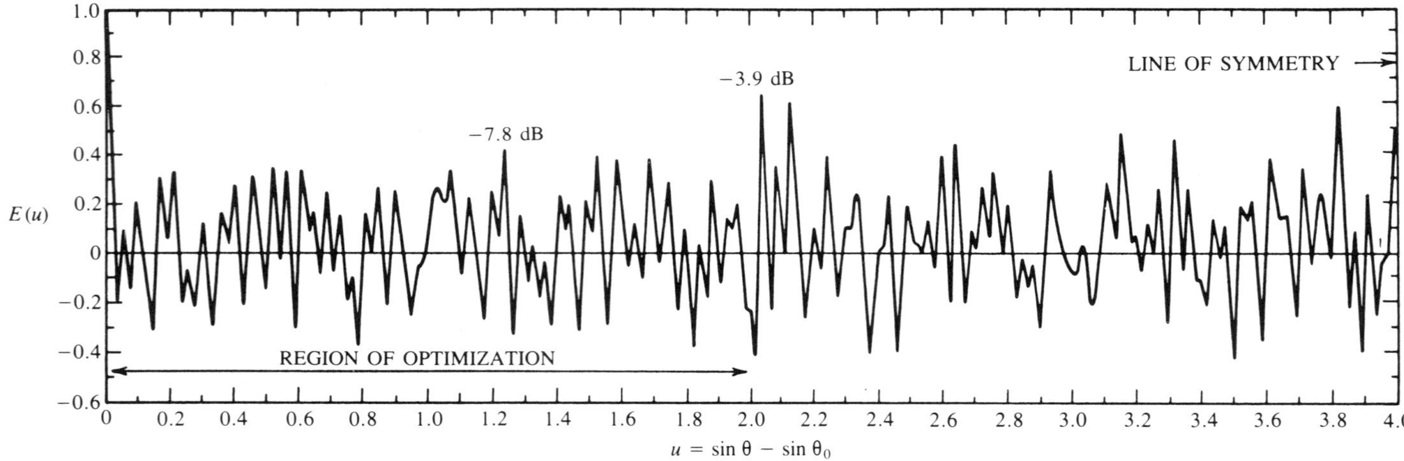

(a)

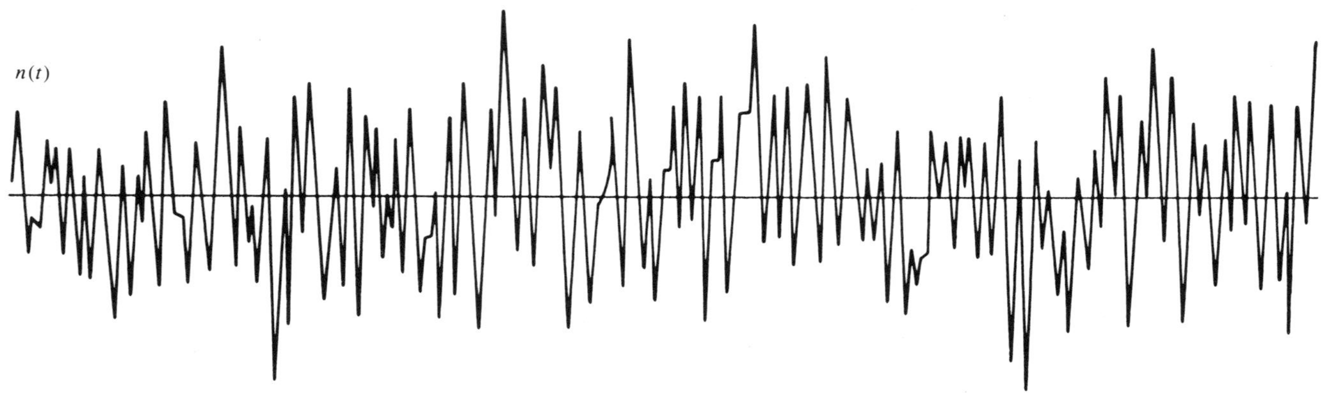

(b)

Figure 3-15 Comparison of aperiodic array pattern and bandlimited random noise. (a) Pattern of 25-element unequally spaced array in 50λ aperture. (From ref. [91]). (b) Bandlimited noise. $n = 2WT$ for bandlimited waveform $= (L/\lambda)(1 + \sin|\theta|_0)$. (From ref. [127]).

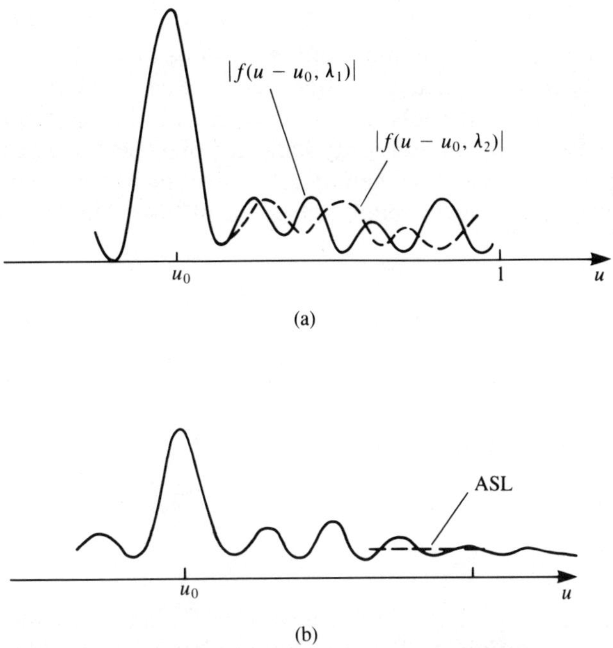

Figure 3-16 Changing frequency changes the scale of the radiation pattern, thereby decorrelating the more distant sidelobes.

between the main lobe and close-in sidelobes and therefore the composite pattern near the main lobe is relatively unaffected by bandwidth. At progressively greater distances from the main lobe, the patterns drift apart. Sidelobes at one frequency tend to fill the valleys of the pattern at another frequency. The composite pattern in the distant sidelobe region approaches the ASL. These conditions are sketched in Fig. 3-16b.

The transition region where the composite pattern ceases to be that of a monochromatic array and approximates the ASL is about Q lobe widths from the main lobe. This conclusion follows from the fact that the abscissa scales of the patterns at wavelengths λ_1 and λ_2 are in the ratio λ_1/λ_2. Let λ_1 and λ_2 be the wavelengths at the edges of the band. The number of lobes of the pattern at $\lambda = \lambda_1$ between the direction of the main lobe u_0 and angle u is $n_1 = (u - u_0)/(\lambda_1/L)$, and similarly for λ_2. Define the transition distance u_T as that value of $u - u_0$ where the numbers of lobes differ by one. Then

$$n_2 - n_1 = 1 = \frac{u_T L}{\lambda_2} - \frac{u_T L}{\lambda_1} = \frac{\Delta\lambda\, u_T L}{\lambda_1 \lambda_2}$$

$$\approx \frac{u_T L}{\lambda_0}\frac{\Delta\lambda}{\lambda_0} = \frac{u_T}{Q\Delta u} \tag{3-44}$$

where $\Delta\lambda = \lambda_1 - \lambda_2$, λ_0 is the midband wavelength, Δu is the nominal lobe width, and Q is the reciprocal of the fractional bandwidth. From Eq. 3-44, $u_T \cong Q\Delta u$, as stated, and the number of lobes in this interval is Q.

Equation 3-44 indicates that the lower the Q the smaller the portion of the sidelobe region in which the lobe crests are as high as in the monochromatic array. The theory of the random array indicates that the PSL, relative to the ASL, is approximately equal to the natural logarithm of the number of lobes in the high-lobe region plus a small additive term [94]. Thus we may expect

$$\frac{\text{PSL}}{\text{ASL}} \approx \ln Q + C \tag{3-45}$$

provided that $u_T < 1 + |u_0|$, the distance from the main lobe to the more distant horizon.

When Q is large such that $u_T > 1 + |u_0|$

$$\frac{\text{PSL}}{\text{ASL}} \approx \ln \frac{1 + |u_0|}{\Delta u} + C \tag{3-46}$$

and is independent of Q. The additive term $C \approx 1 - \ln(1 - \beta)$ where β is the probability that no sidelobe will exceed the calculated value of PSL/ASL. The expected value of the peak sidelobe is found by setting $\beta = 0.5$, which corresponds to $C = 1.7$. Using $C = 1.7$ in Eqs. 3-45 and 3-46, the ratio of these equations gives the expected reduction in the peak sidelobe of a random array. Table 3-5 lists the reduction in decibels for several antenna sizes and Q values. The beamsteering angle $u_0 = 0$.

The periodic array pattern responds differently to wide bandwidth. Consider each frequency component of the signal spectrum to have a radiation pattern associated with it. The abscissa scale of such a pattern varies directly with the wavelength of the component. These constituent patterns of the composite,

TABLE 3-5 Reduction in sidelobe level of the random array, in decibels, due to signal bandwidth. L/λ is the antenna size in units of wavelength. Q is the reciprocal of the fractional bandwidth of the signal waveform

	L/λ				
Q	10	10^2	10^3	10^4	10^5
10	0	1.8	3.3	4.4	5.2
10^2	0	0	1.4	2.4	3.4
10^3	0	0	0	1.0	1.9
10^4	0	0	0	0	0.8
10^5	0	0	0	0	0

wideband pattern progressively expand (compress) as the frequency varies across the band. Figure 3-16a illustrates the phenomenon for two frequency components. A transition distance u_T nominally separates the intervals in u in which the pattern constituents are highly correlated ($|u - u_0| < u_T$) and in which they are not ($|u - u_0| > u_T$). The value of u_T is the same as that calculated above for the random array. In the uncorrelated region ($|u - u_0| > u_T$) the pattern components add incoherently and the (voltage) sum tends to the rms level of the monochromatic pattern.

Unlike the random array, the side radiation pattern is not statistically independent of angle. Instead, $f(u)$ at an arbitrary λ tends to fall off with u. The radiation pattern of a uniformly illuminated aperture of length L and wavelength λ is

$$f(u) = \frac{\sin(\pi L u/\lambda)}{\pi L u/\lambda} \tag{3-47}$$

Beyond the main lobe region its envelope falls as u^{-1} or at a rate of 6 dB per octave. Apertures with tapered or shaped weighting functions have patterns that fall off more rapidly. The radiation pattern of any continuous aperture whose excitation goes to zero at the edges but whose derivatives are not zero there falls off as u^{-2} or 12 dB per octave. The triangular and parabolic apertures are examples. If the first derivative also is zero at the edges the fall-off rate is u^{-3}, etc. [94]. Thus in the region $|u - u_0| > u_T$ the composite wideband pattern, which is the rms sum of the constituent patterns, is not a constant. Instead, its amplitude at any angle is the rms sum (or integral over the frequency band of the signal) at that angle; hence it exhibits the fall-off rate of the monochromatic pattern.

Consider, for example, the radiation pattern (3-47) of the uniform aperture. Away from the main lobe the pattern in any small interval u is approximately sinusoidal. The rms value of a unit-amplitude sinusoid is $1/\sqrt{2}$. Hence the wideband radiation pattern in the region $|u - u_0| > u_T$ is approximately $\lambda/(\pi L|u - u_0|\sqrt{2})$.

A more significant difference between the periodic and random wideband arrays is in the peak sidelobe level. In the coherent region ($|u - u_0| < u_T$) the radiation pattern for either type of aperture is the same as the corresponding monochromatic pattern. We found earlier that because of the statistical properties of the side radiation pattern of the random array the expected value of the peak sidelobe dropped with bandwidth (Table 3-5). This reduction does not carry through to the periodic array. The highest sidelobe tends to be close to the main lobe (it often is the first sidelobe) and hence falls in the coherent region where the wideband pattern does not differ materially from the monochromatic pattern. The extent of this region is approximately Q sidelobes. Thus, unless the peak lobe is Q or more sidelobes from the main lobe (an unlikely situation in a conventionally designed, periodic array), the PSL is nearly unaffected by bandwidth.

4

PROBLEMS IN MICROWAVE IMAGING

4-1 LONG WAVELENGTH

The long wavelength required for terrestrial microwave imaging is responsible for many of its problems. The problems broadly fall into two categories, those induced by the scattering properties of long waves and those resulting from the large size of the aperture needed to achieve satisfactory resolving power. The required wavelength regime was discussed in Section 2-1 and the apertures needed to compete with common optical instruments were listed in Table 2-2. Several of the problems are described in this chapter; first, those due to scattering and second, those attributable to large antenna size.

In addition to the smooth-surface problems, which are discussed in the next four subsections, there arises a critical problem called (or due to) *speckle.* Coherent interference between components of a radiation field mottles an otherwise smooth surface and breaks the continuity in its edges. This target break-up phenomenon causes large targets to look like a multiplicity of smaller ones, denying target count, shape, and size information to the viewer.

Two categories of techniques address both types of problems: diversity is discussed in Chapters 5 and 6 and deconvolution (CLEAN) in Chapter 7.

4-2 SCATTERING EFFECTS

Figure 1-1 was introduced to show the intrinsic similarity between optical and microwave imaging systems. In each case an energy source illuminates the scene to be imaged and a receiver forms an image from the scattered or reflected radiation field. More significant than the similarities are the differences, however. Most surfaces that we think of as smooth are actually sufficiently rough, at a scale of the order of optical wavelengths, to scatter light. Thus walls, floors, tables, and most common "smooth" objects can be seen from all directions (e.g., positions 3 and 4 in the upper part of Fig. 1-1) as if they were rough surfaces. Very few objects in a typical room produce mirror-like reflections. The rare exceptions are mirrors and highly polished metal surfaces. The classical measure of roughness, due to Rayleigh, is $(h \sin \gamma)/\lambda$, where h is the rms deviation of the surface from its mean and γ is the angle that the illuminating wave makes with the surface [12]. Scattering largely occurs when the measure exceeds unity; otherwise, specular reflection dominates. The small wavelength of optical radiation ensures that most surfaces scatter, whereas the same surfaces are smooth to microwaves and do not scatter. The result is that most microwave surfaces appear black (position 5 in Fig. 1-1) except when viewed from the direction of reflection (position 4).

4-2-1 Low Target Mass

There are many consequences of the apparent smoothness of most surfaces. When a surface does not scatter, the target mass associated with the surface in the image is low; that is, the image is hardly visible. At best the weak scattering from the edges and corners defines the surface. There is little texture and no color; these two important optical clues to the nature of a target are missing in a microwave image.

4-2-2 Large Dynamic Range

Edge and corner scattering carry target shape information. Size, shape, orientation, and number of targets can often be determined by silhouettes or outlines alone. Such information-bearing scattered energy is easily obscured, however, by strong specular reflections. Figure 4-1 shows a plane wave incident upon two surfaces. The rough surface scatters the energy broadly over a large solid angle. As a consequence the intensity variation in the image is not great, being typically no more than 20 to 30 dB in optics [38]. The smooth surface concentrates the energy in a small angular interval, resulting in a very much larger target dynamic range: 50 to 90 dB variation is observed in radar [68]. This huge dynamic range has two effects upon the image. First, every imaging system is limited in the range of target strengths that can be handled by its receivers, its signal processors, and its displays. Bright specular reflections raise the threshold below which the weaker scattered signals are ignored. Hence the target shape detail carried by the scattered energy from the edges and corners will often be swamped in the imaging system by the specular reflections.

4-2-3 Specular Reflections Dominate

The second effect of large dynamic range is that the specular reflections tend to dominate the image. The images of scenes containing large, smooth surfaces

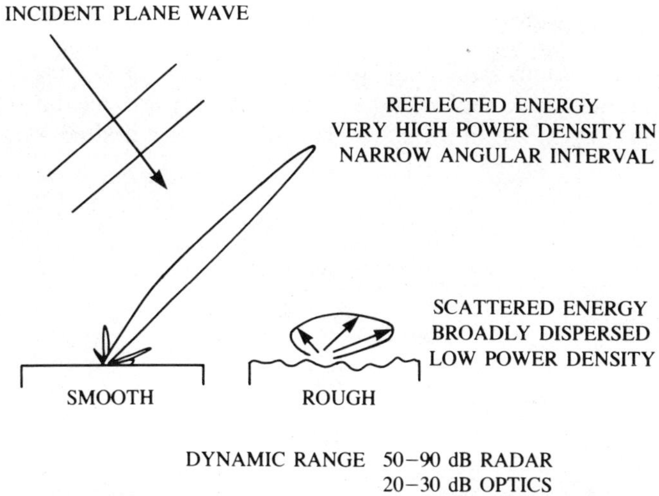

Figure 4-1 The high intensity of reflected energy from a smooth surface relative to scattered energy from a rough surface results in a very large range of signal strengths.

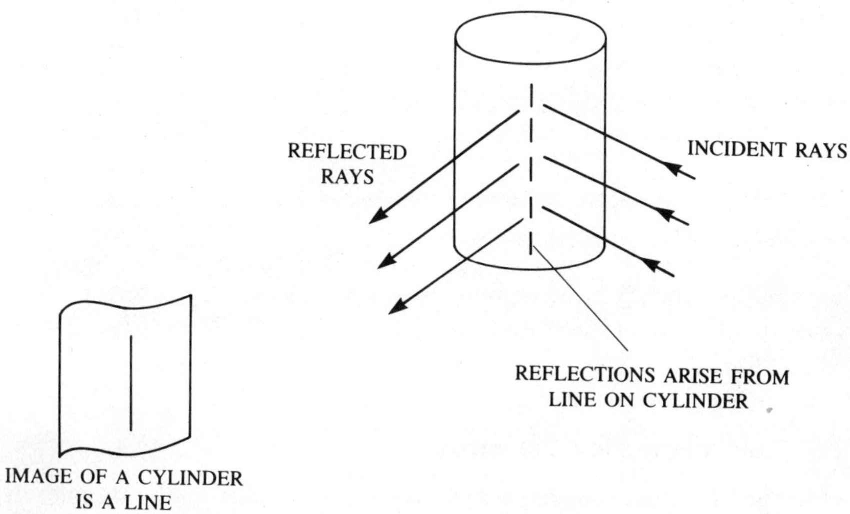

Figure 4-2 A cylinder illuminated at a wavelength at which the surface is smooth appears as a line in the image.

are largely characterized by the reflections from those surfaces. An ordinary room would appear as a hall of mirrors to microwave "eyes." Smooth, curved surfaces lose their character as well. Figure 4-2 illustrates a cylinder illuminated at a wavelength at which the surface is smooth. Reflections appear to arise from a line on the cylinder and, as a consequence, the cylinder images as a line. Similarly, a smooth sphere does not image as a sphere; it images instead as a point. In neither case would size information be retained, let alone shape.

4-2-4 Image Artifacts

Other strong reflections can dominate a scene. When these reflections arise from objects unrelated to the target being imaged, they are called *artifacts*. The desk-scene in Fig. 4-3 is illustrative. The desk is the desired target to be imaged.

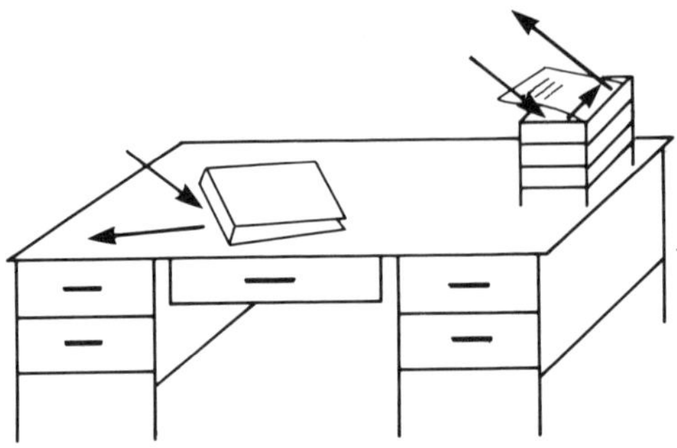

Figure 4-3 Edges and corners of the desk define its shape but scattering from these regions is often below the level of strong reflections from objects that do not convey "desk."

Its top surface is black at microwave frequencies because it does not scatter. Edges and corners define its shape. However, reflections from objects on the desk that are movable and not permanent can dominate the image. A microwave picture taken at another time might look very different simply because of a change of objects on the desk or alterations of their locations or orientations.

4-2-5 Low Information Content

A radiation field can probe or sense a target with no finer resolution than the order of the wavelength of the radiation; in effect, wavelength becomes the sampling interval. Thus the number of sample points on a two-dimensional target is inversely proportional to the square of the wavelength. The information content (regardless of how it is defined) is proportional to the number of sample points. As noted in Section 2-1 the wavelength ratio between microwaves and optics is 10^3 to 10^6. Hence the information content in microwave images is 6 to 12 orders of magnitude lower.

Consider the photograph in Fig. 4-4 of a desk largely covered by papers. While very little of the surface is exposed, it is unquestionably a desk. A few square centimeters of its exposed surface is $10^9\lambda^2$ to the eye or to an optical camera. The same area is only a few λ^2 to a microwave camera.

Figure 4-4 This cluttered surface is unquestionably a desk top to the eye or an optical camera.

4-2-6 Speckle

Coherent interference of components of a radiation field can impart a speckled appearance to an otherwise smooth and untextured surface [33]. This phenomenon is common to all imaging based on coherent radiation fields. The term *speckle* is ascribed to any physical phenomenon that induces a mottled appearance.

Speckle can be found within the image of a target or on its boundary. The latter is the most troublesome in microwave imaging. Edge and boundary continuity are very important to preserve for the reasons discussed earlier in this section (as well as in Section 1-4).

Figure 4-5 shows how coherent destructive interference causes target breakup. The upper portion is the reflection profile of a one-dimensional slice (or scan) through a target. The scattered intensity is fairly constant between the left and right edges of the target except for two strong reflectors or highlights. The profiles of these reflectors, convolved with the radiation pattern of the antenna, are their complex images (middle). The insert is a phasor diagram at a particular location P between the highlights. The three components are the target echo from that location and the two sidelobe responses from the strong reflectors. The phasor sum is nearly zero because the highlight sidelobe responses largely cancel the proper return from the target. As a consequence a void appears in the target image at that location. The one-dimensional image,

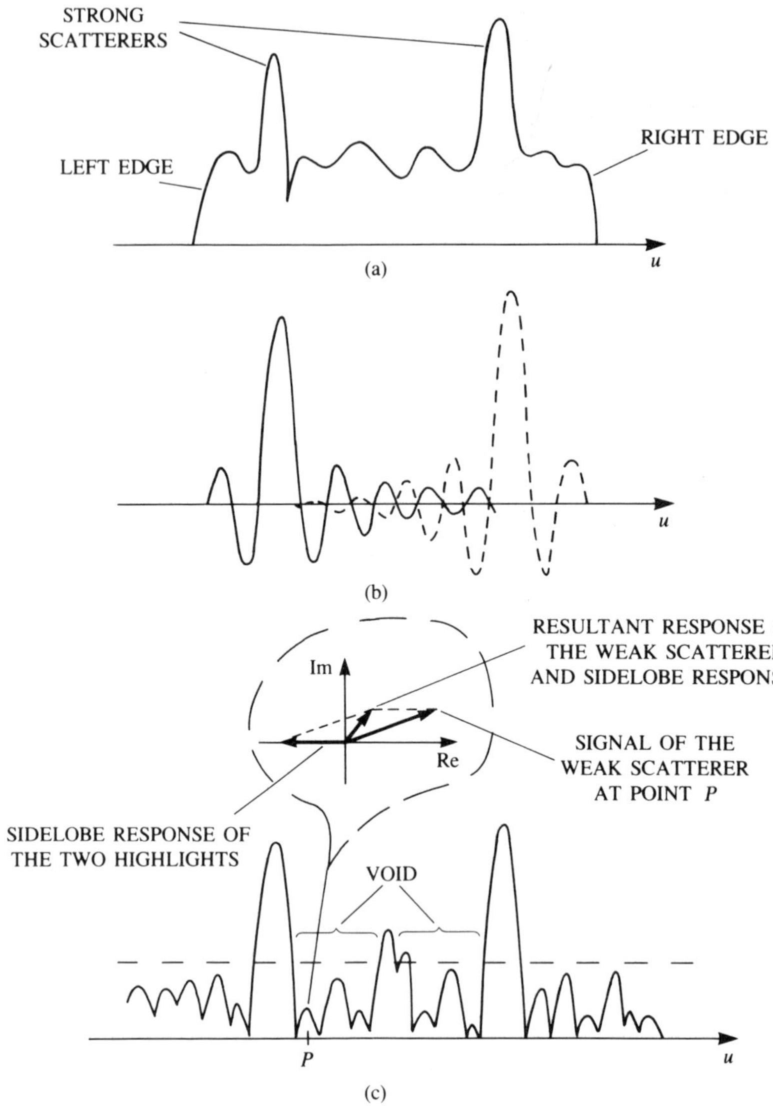

Figure 4-5 Destructive interference between sidelobes of target highlights can cause target to appear as two or more targets. (a) Profile of target with two strong scatterers. (b) Main lobe and sidelobes of the two strong scatterers. (c) Image of (a) showing target cancellation at point P in detail. (From ref. [121])

shown in the bottom portion of Fig. 4-5, looks like three targets, each smaller than the real target. Two of the "targets" contain one highlight each.

Figure 4-6 shows the result of a high-resolution experiment illustrating how speckle can decompose targets into a collection of smaller objects. A small truck of dimensions 2×5.5 m was oriented with its front toward a large receiving

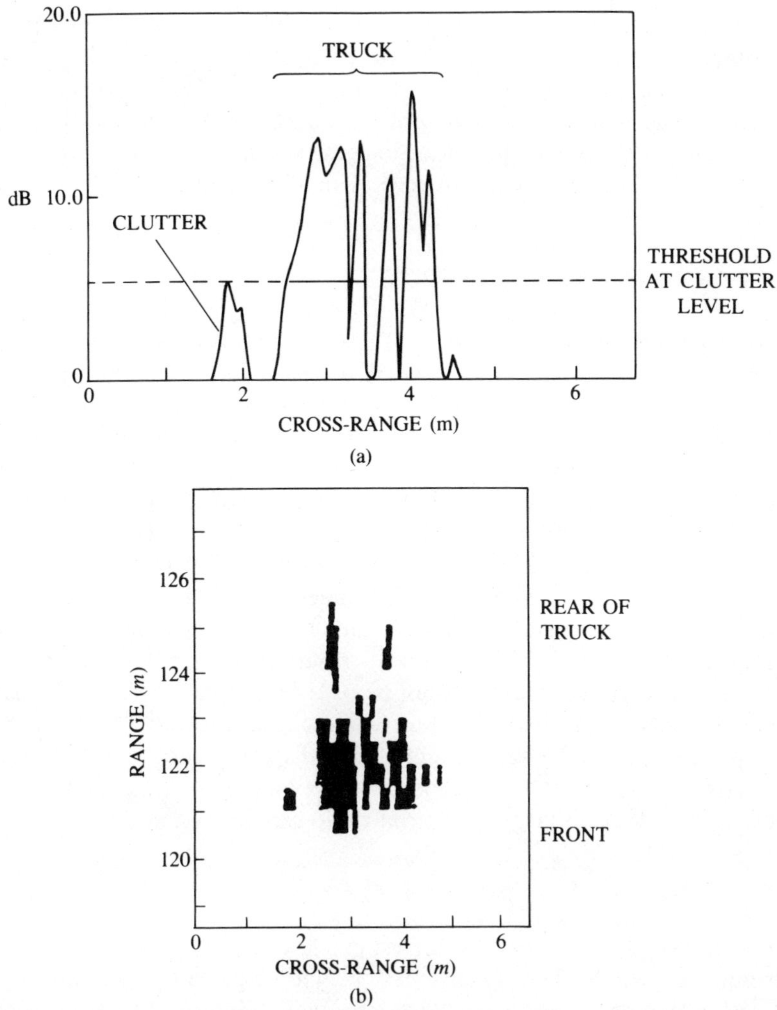

Figure 4-6 Illustration of target breakup caused by speckle. (a) One-dimensional and (b) two-dimensional images of a 2 × 5.5-m truck viewed from high-resolution array operating at 3 cm. Range dimension of resolution cell is 1 m; cross-range dimension is 15 cm. Front of truck is toward the array. (From ref. [121])

array operated at X-band ($\lambda = 3$ cm). The receiving array was a synthetically-formed 200-element periodic array 30 m long. The illuminator was a pulsed radar transmitter radiating 7-ns (1-m) pulses.

The upper figure is a one-dimensional image of a slice through the cab of the truck. The highlights and voids predicted above are evident. If a threshold is placed a few decibels above the level of the clutter so as to eliminate clutter from the display, the target strength drops below the threshold in 3 or 4 locations and the target decomposes into 4 or 5 targets. A raster of such one-dimensional images laid down as intensity-modulated lines yields the two-dimensional picture shown in the lower figure. While the resolution is sufficiently fine (1 m × 15 cm) to obtain a recognizable image of the truck, speckle destroyed the potential of the imaging system to produce a recognizable image by breaking the target into several smaller objects.

4-3 LARGE-APERTURE EFFECTS

Figure 2-1 shows why the wavelength used in long-distance imaging must be orders of magnitude larger than optical waves. Table 2-2 shows the aperture size required to achieve the resolving power common to optical instruments when using long, microwave radiation. Antennas of such size are employed only in radio astronomy. The logistics and economics in that field are very different from terrestrial imaging, however. Because of the rotation of the earth and the relatively unchanging source distribution of the sky, a small number of suitably located radio telescopes can provide extensive astronomical surveys. Not so in terrestrial imaging. If a radio camera is to aid in air traffic safety in the neighborhood of a given airport, the camera must be installed there. Military use may demand mobility. The source distribution or target scene is continually changing. The earth's rotation cannot be used for scanning; electronic scanning is required instead. And so on. Thus, while a great deal can be learned from the sophisticated hardware, computer and display techniques developed for radio astronomy during the last decade and a half, there are additional special concerns attendant on large, terrestrial radio camera systems.

4-3-1 Siting

A radio camera at an airport will consist of a few hundred or thousand receivers distributed about the field. Their locations will be partially determined by the designer of the system and partially by the availability of suitable locations for the receiver modules. The design procedure must be sufficiently robust to accommodate considerable *a priori* uncertainty. A radio camera mounted into a nonrigid body such as an aircraft or a large ship will have its parts continuously in motion relative to each other. A radio camera the size of a city will have its parts located on roofs of buildings. The locations, while fixed, are essentially selected at random. More difficult to build is the radio camera distributed among independently moving vehicles. A distributed orbiting array is an example. Another is a floating buoy field carrying hydroacoustic transducers. A third is an imaging system formed by the radar receivers on all the ships of a task force.

Thus a high-resolution microwave aperture is intrinsically distorted. Figure 4-7 depicts a scene $s(u)$ illuminating an aperture with an electric field $e(x) = \mathscr{F}^{-1}\{s(u)\}$. The current in the aperture $i(x) = e(x)w(x)$. The complex image $\hat{s}(u) = \mathscr{F}\{i(x)\}$

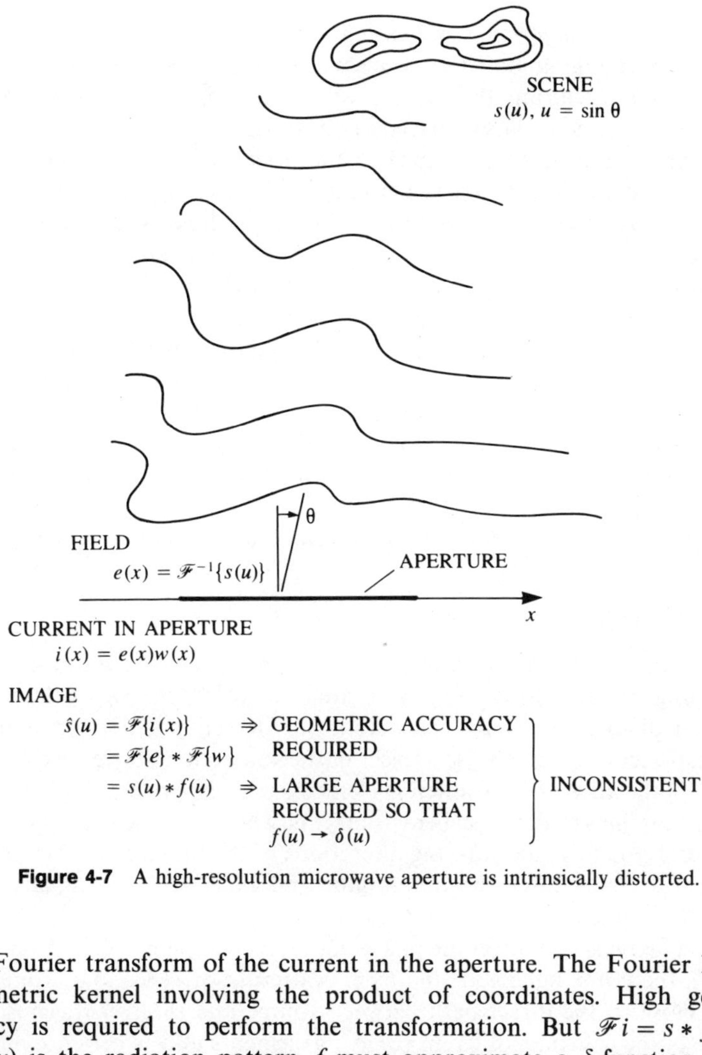

Figure 4-7 A high-resolution microwave aperture is intrinsically distorted.

is the Fourier transform of the current in the aperture. The Fourier kernel is a geometric kernel involving the product of coordinates. High geometric accuracy is required to perform the transformation. But $\mathscr{F}i = s * f$, where $f = f(u)$ is the radiation pattern. f must approximate a δ-function for $\hat{s}$ to approximate s, which in turn requires that the aperture be very large. A large aperture, in general, is unstable geometrically. Hence the two requirements are inconsistent, leading to the conclusion stated above.

4-3-2 Self-calibration

In all cases, the self-calibration design procedure must accommodate severe position uncertainty in the locations of the parts of the system. Self-calibration is the solution for the siting problem. The system must be able to cohere the signals received from the target irrespective of the lack of sufficient *a priori* knowledge of the locations of the elements that receive the signals. Self-calibration is called *self-cohering*, *self-synchronizing* and *self-focusing*. A generic term is *adaptive beamforming*, indicating the dynamic requirement for accommodation to changing position of elements in a large, nonrigid system; to multipath and scattering that varies with scan angle; and to several other phenomena endemic to large phased-array systems.

Adaptive beamforming is central to the design of a radio camera. All the high-resolution pictures shown earlier required its use. Figure 4-8 (upper) shows the diffraction-limited, high-resolution image of houses on two streets in Phoenixville. The lower figure is the image of the same data processed without the self-calibrating adaptive beamforming algorithm. It is a useless array of pixels. It is clear that echo energy is in the proper set of range of bins, but without the self-cohering feature the signal components within each range bin fail to combine to form the image. Figure 4-9 is a similar pair. The upper image is of a Boeing 727 airplane. The lower image is made from the same data but without adaptive beamforming. A plan-view sketch is in the middle. Figure 4-10 illustrates imaging with and without adaptive beamforming for a much smaller, commuter airplane.

Procedures for self-calibration are described in Chapters 7 and 8.

4-3-3 Phase Distortion

There are many other sources of phase distortion that can seriously degrade the imaging system. Some of them are illustrated in Fig. 4-11. A large distributed, distorted receiving array is shown. Element position error, indicated by the displacement $\overline{\delta r}$, was the subject of the subsection above. Because of the large size of the aperture, the antenna elements may experience different electromagnetic couplings to their local environments. This phenomenon may introduce random and possibly time-varying alterations of the driving point impedances, which in turn may alter the amplitudes and phases of the signals passing through the elements.

Indirect arrivals also affect the phase of the received signal. Commonly called *multipath*, these are signals arising from scattering or reflections from objects that are close to the direct path between source and receiver. Because of the large size of the aperture, different receiving elements may experience multipath from different reflectors. Hence the phase errors induced by the scattered energies will differ from element to element, and so no single turning or correction will suffice. An adaptive, self-calibrating procedure is required to solve this problem (see [97], [99], and Chapters 8 and 9 for the analysis of this problem).

Another phase-error-inducing problem encountered in very large systems is a nonconstant index of refraction of the propagation medium. This problem is encountered in radio and optical astronomy, long-wave underwater acoustics, ultrasound within the human body, and ionospheric propagation of HF radio waves. Large-scale variations in the refractive index will cause differential ray bending and a variation of the apparent arrival angle from element to element. Small-scale fluctuations due to random variations of electron density in the

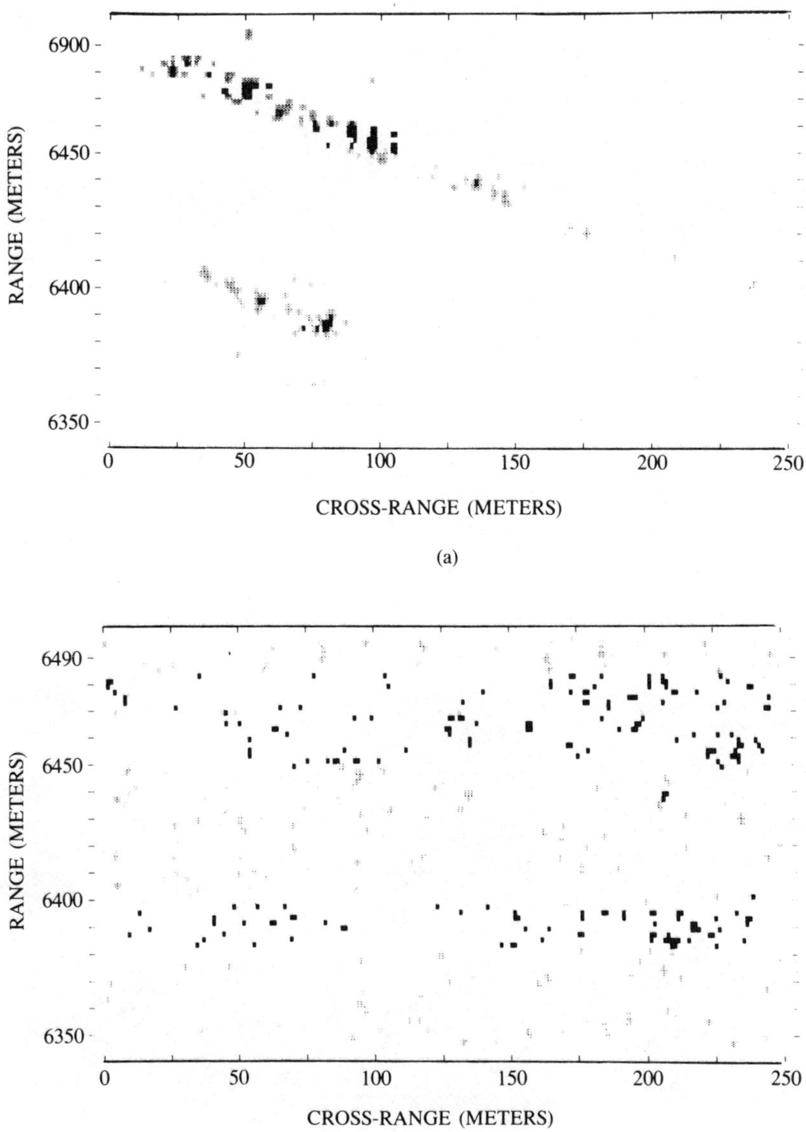

Figure 4-8 Images of High Street, Phoenixville, PA, with and without adaptive beamforming.

ionosphere will introduce random phase shifts through the medium. Both effects will randomly bias the phase of the received waves.

In addition to the external sources of phase error, there are system imperfections and mistunings that also add unwanted and unknown phase shifts to the signals passing through the system. In principle, these can be tuned out,

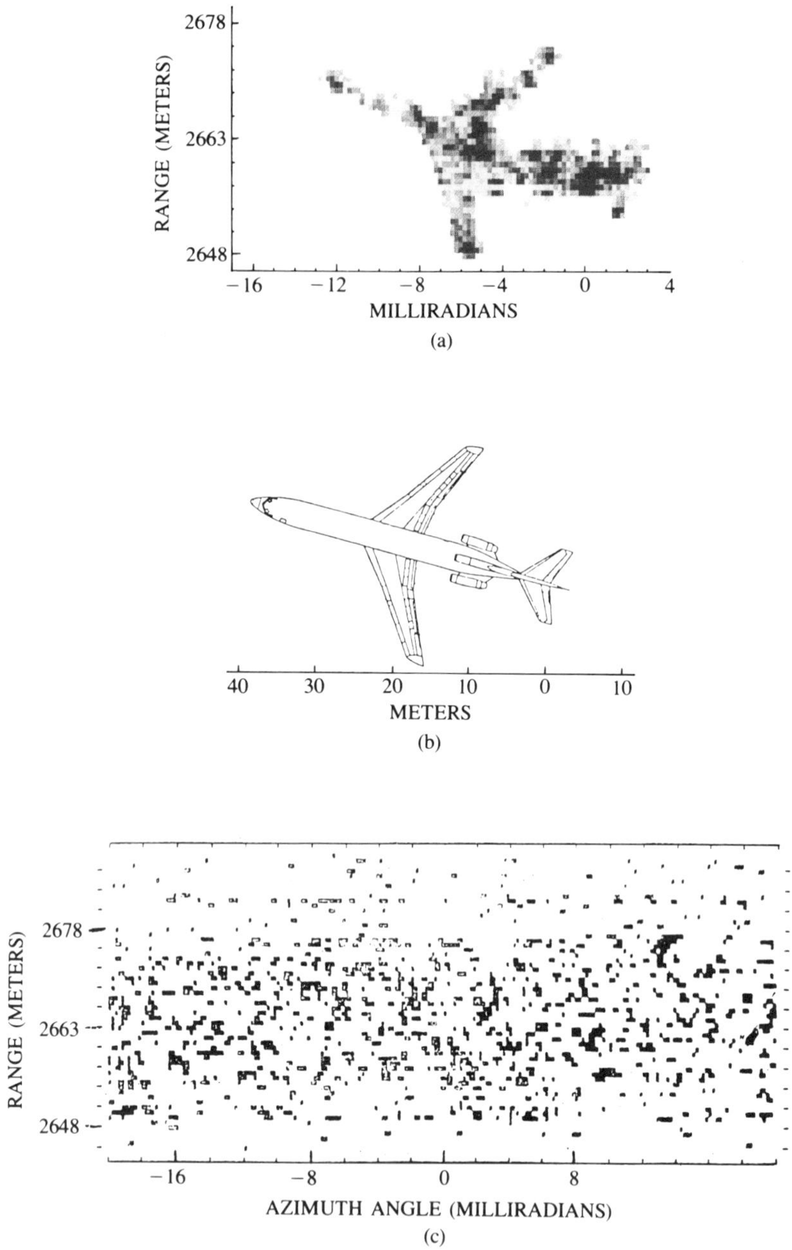

Figure 4-9 (a) Microwave image of Boeing 727 obtained at X-band with monostatic radio camera. The airplane is flying at a distance of 2.7 km from VFRC. (b) Plan view drawing of aircraft. (c) Image made without adaptive beamforming from the same data set. (From ref. [102])

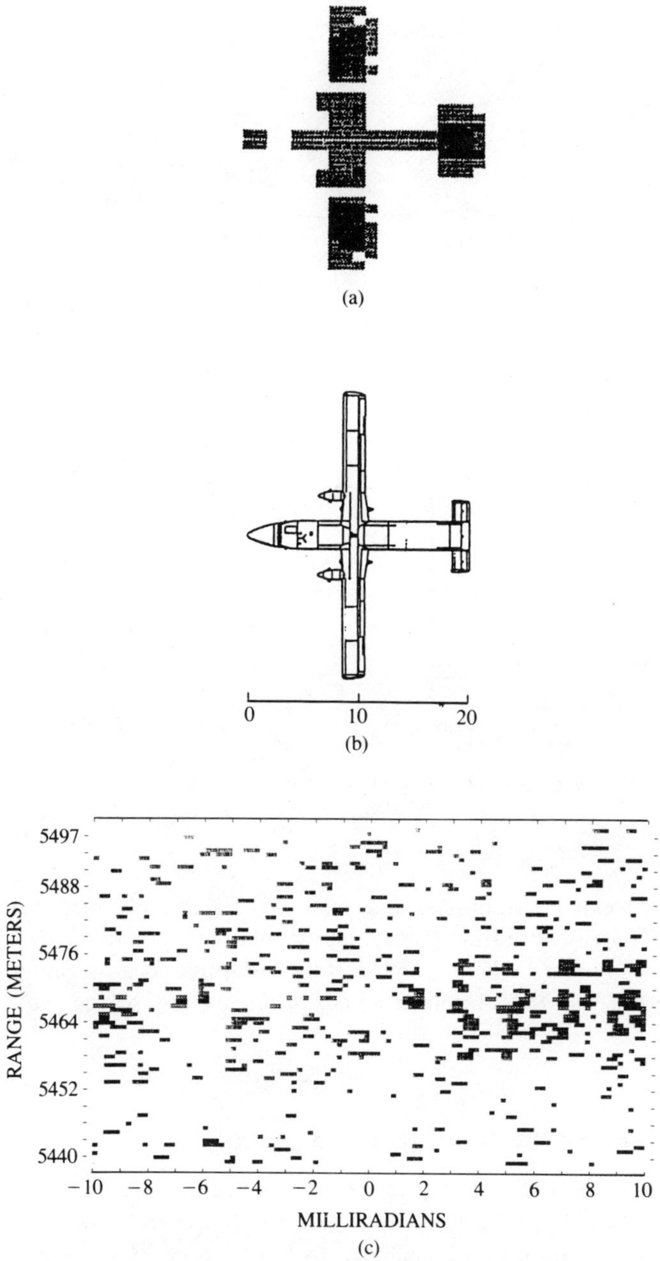

Figure 4-10 (a) Microwave image of a small commuter airplane (Shorts-330) obtained at X-band with monostatic radio camera. The airplane is flying at a distance of 5.5 km from VFRC. (b) Plan view drawing of the aircraft. (c) Image made without adaptive beamforming from the same data set.

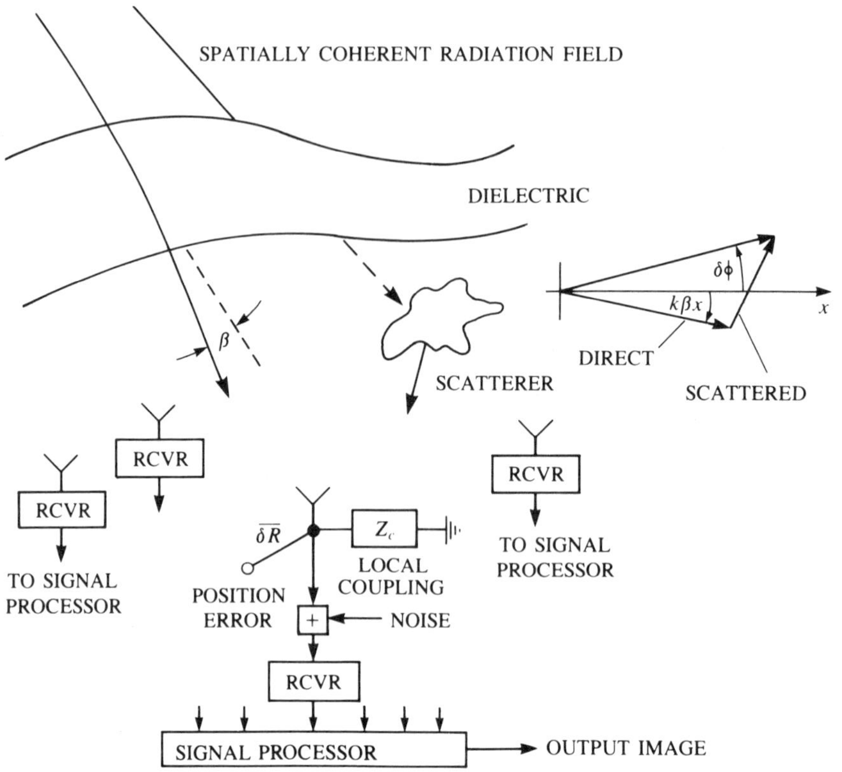

Figure 4-11 Problems in terrestrial microwave imaging.

components can be stabilized and pilot signals can be introduced into the receiver for self-calibrating components and subsystems. None of this is necessary, however, when a satisfactory adaptive beamforming system is introduced: when the aforementioned problems are solved, so too is this one.

4-3-4 Cost and the Number of Elements

The diameter of the camera lens in Fig. 4-12 is $10^4\lambda$. The phased array equivalent would be $10^4\lambda$ on each side with elements spaced every half-wavelength. The number of elements, therefore, exceeds 10^8. Associated with each element is a receiver module that includes, in addition to time delays and phase shifts, an RF amplifier, mixer, local oscillator, IF phase detectors, and so on. Because the cost of parts grows as the number of elements, the cost is one hundred times larger for a $10^5\lambda$ imaging system, which is the microwave equivalent in resolving power to a 2-inch optical telescope.

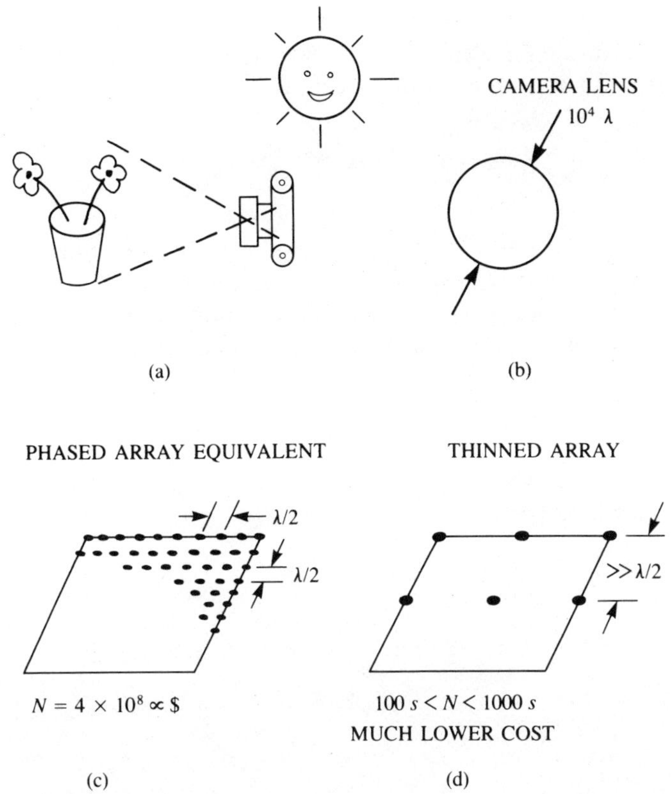

Figure 4-12 (a) A camera taking pictures in bright sunlight has (b) a lens diameter of $10^4\lambda$. (c) The phased-array equivalent has at least 4×10^8 antenna elements, phase shifters, receivers, etc. (d) Thinning the aperture is essential to constrain cost.

4-3-5 Grating Lobes

The array must be drastically thinned, for otherwise the cost would grow beyond practical limits. Element overspacing on a regular or periodic basis as in Fig. 4-12(d) can lead to grating lobes and hence multiple images. This phenomenon must be avoided.

The trouble arises when the aperture weighting function is thinned in a regular manner. The image

$$\hat{s} = s * f \tag{4-1}$$

is the convolution of the radiation pattern with the source distribution. The radiation pattern

$$f = \mathscr{F} w \tag{4-2}$$

is the transform of the weighting function. Let $w_0(x)$ be the underlying taper or aperture weighting function for which $f_0(u)$ is the design radiation pattern. The phased array equivalent is a sampled version of w_0. Let the sampling function be periodic with element spacing d. Then

$$w(x) = w_0(x) \sum_{n=-\infty}^{\infty} \delta(x - nd) \tag{4-3}$$

Because Eq. 4-3 is a product of two functions, the radiation pattern (4-2) is the convolution of the transforms of the two functions. The transform of w_0 is f_0. The transform of the periodic sampling function also is a periodic sampling function having period λ/d. Thus the radiation pattern is

$$f(u) = f_0(u) * \sum_{m=-\infty}^{\infty} \delta\left(u - \frac{m\lambda}{d}\right) \tag{4-4}$$

and the image becomes

$$\hat{s}(u) = s(u) * f_0(u) * \sum_{m=-\infty}^{\infty} \delta\left(u - \frac{m\lambda}{d}\right) \tag{4-5}$$

Identifying

$$\hat{s}_0 = s * f_0 \tag{4-6}$$

as the image in the absence of sampling, the final image

$$\hat{s}(u) = \hat{s}_0(u) * \sum_{m=-\infty}^{\infty} \delta\left(u - \frac{m\lambda}{d}\right) \tag{4-7}$$

is the image of the unsampled aperture replicated at intervals of λ/d.

Figure 4-13 illustrates the grating lobe phenomenon resulting from periodic overspacing.

Figure 4-14 illustrates how multiple images result from overspacing. Part (a)

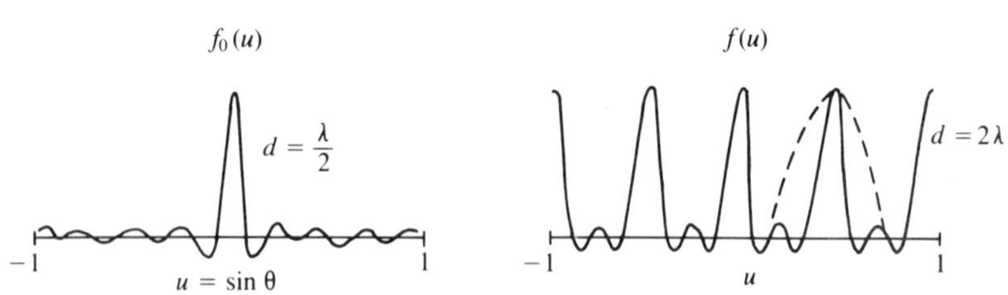

Figure 4-13 Thinning of the aperture by periodic overspacing of the elements leads to grating lobes. (a) Radiation pattern with no thinning. Element spacing $d = \lambda/2$. (b) $d = 2\lambda$: radiation pattern shows multiple lobes. Transmitting pattern (showed dashed) illuminates only a single lobe.

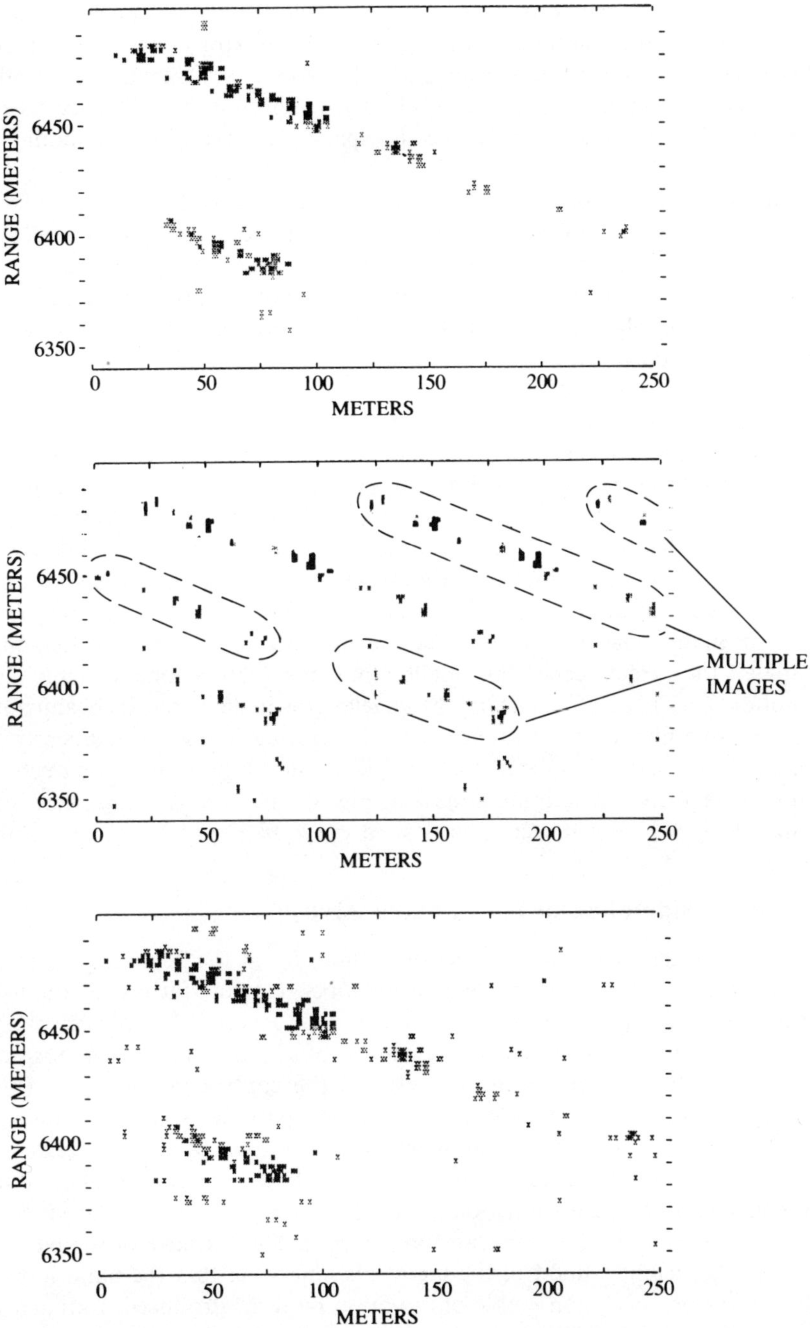

Figure 4-14 Effect of overspacing elements. (a) Original (200 elements); (b) periodic overspacing (20 elements) gives rise to multiple images, shown enclosed by dashed ovals; (c) random overspacing (90 elements) gives rise to sidelobe artifacts. Run 110, High Street, Phoenixville, Pa.

is the original of the pair of streets in Phoenixville (Fig. 4-8a). Part (b) shows the effect of periodic thinning of the aperture. Two extra images are evident. In this experiment, the element spacing $d = 2$ m and $\lambda = 0.0325$ m. The angular spacing between images should be $\lambda/d = 0.016$. The nominal distance to the houses in Fig. 4-14 is 6400 m. Hence the spacing between images should be 100 m, which it is.

Aside from the unwanted or extraneous images in Fig. 4-14b, its quality is poor. This is because the sampled energy of the radiation field is lower in proportion to the reduced number of samples, thereby lowering the output signal-to-noise ratio and requiring a higher threshold relative to peak image intensity in the display. The sample points for the original image of Fig. 4-14a were spaced by 0.2 m and 200 element positions were used. All but 20 were eliminated for the image of Fig. 4-14b.

This experiment also illustrates a powerful means for suppression of extraneous images caused by grating lobes. The element spacing for Fig. 4-14a is 6.4 wavelengths. Grating lobes should appear every 0.16 radians or 9 degrees. The fact that the field of the images is smaller than the grating lobe spacing (about 250 m/6400 m = 0.05 radians) does not account for the elimination of the higher-order images. Foldover or aliasing of the images of targets that are located outside the image field should occur. The reason that this picture is devoid of image aliasing is that the beamwidth of the transmitter that illuminated the target scene was smaller than the grating lobe spacing. The transmitter, a 1.2-m dish, was nearly 40 wavelengths in diameter. Its beamwidth was about one-fifth of the distance between grating lobes. Consequently, it illuminated only a single lobe of the receiver radiation pattern. Target echoes did not arise from the grating lobes in Fig. 4-14a because they were not illuminated by the transmitter. The dashed curve in Fig. 4-13 illustrates the situation.

4-3-6 High Sidelobes of the Random Array

The procedure described above—periodic thinning of the receiving aperture and illumination of only a single grating lobe—cannot always be applied. Very large arrays deny the designer control and accurate knowledge of the element positions. As little as a few tenths of a wavelength rms position uncertainty in the element locations destroys the controlled behavior of the sidelobes and reduces the quality of the radiation pattern to that of a random array. A random array is one in which the locations of the elements are determined by some random process. In short, the x_n are chosen at random from a probability density function $p(x)$. The expected value of the far-field radiation pattern is the Fourier transform of $p(x)$. The variance of the pattern is small in the neighborhood of the main lobe, meaning that the main lobe is stable with respect to an ensemble of radiation patterns produced from arrays defined by the same $p(x)$. The variance away from the main lobe is approximately constant. Its value is N^{-1} where N is the number of elements. This is the expected value of a power pattern in the sidelobe region. Thus the average sidelobe level is identified with N^{-1}. The peak sidelobe is of the order of 10 dB higher [94].

The virtue of the random array is that there are no periodicities in its random weighting function. Consequently, the radiation pattern of a random array exhibits no grating lobes. An array can be dramatically thinned without fear of grating lobes by randomizing the location of its parts. The penalty, of course, is the high sidelobe level. Randomizing the element positions in a thinned array destroys the grating lobes but not the energies in these lobes. The energy in the grating lobes of the periodically thinned array must equal the energy in the sidelobes of the randomly thinned array having the same average thinning factor.

Figure 4-14c illustrates the effect of high sidelobes. The same scene was imaged with a 90-element random array. False targets are evident everywhere.

Most large imaging systems will suffer from the sidelobe properties of the random array. The system design must be sufficiently robust to accommodate high, randomly located sidelobes. The diversity techniques described in Chapters 5 and 6 address this problem, as do the deconvolution procedures described in Chapter 7.

4-3-7 Data Handling Requirements

There is a marked contrast between the data handling requirements in optics and in microwaves. In optical systems a three-dimensional scene is projected onto the two-dimensional image plane (Fig. 4-15, also Fig. 1-2). The range

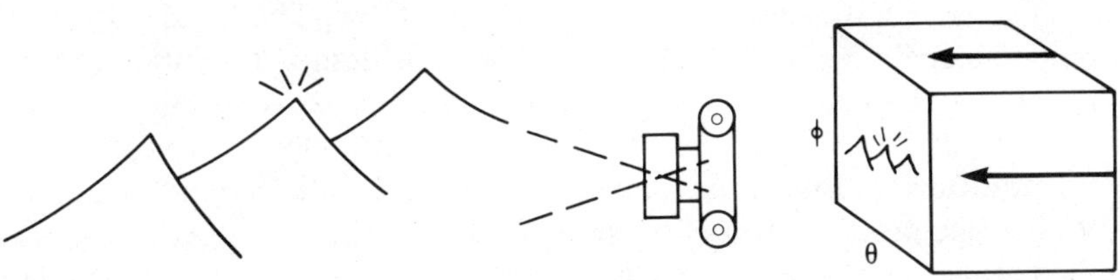

Figure 4-15 In optical imaging with a two-dimensional lens, the range or distance is collapsed into the θ–ϕ plane. A high-quality image of $1000 \times 1000 = 10^6$ pixels, each quantized to 5 or 6 bits, is defined by fewer than 10^7 bits.

dimension is collapsed and all energy arriving from direction θ, ϕ contributes to the intensity detected at that point and assigned to those coordinates. The number of bits required to define and transmit a scene is the product of the number of pixels in the image and the number of bits of quantization of the pixel intensities. An image consisting of 1000×1000 pixels, each quantized to 5 or 6 bits, has very high quality, yet it would be represented by fewer than 10^7 bits, which is a modest number. Not so with a very large microwave imaging

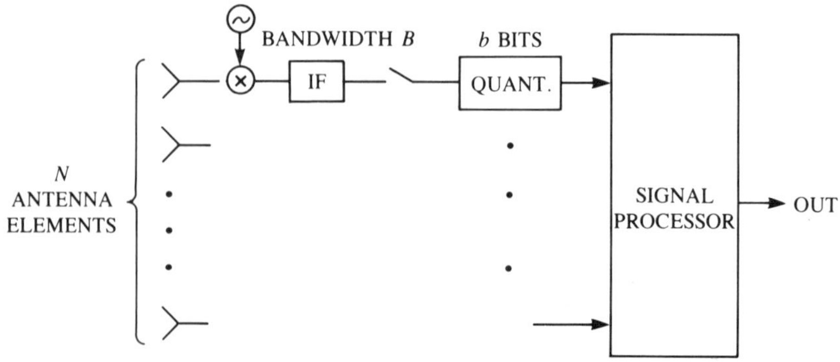

Figure 4-16 The magnitude of the data handling problem in microwave imaging is not determined by the number of pixels. It is determined instead by the number N of antenna elements, the bandwidth B of the radar signal, and the number b of quantization bits per sample. The data rate into the signal processor is NBb.

system. It has to acquire, transport, and process a huge amount of data. The dimensions in the image are range and angle. Range resolution is determined by the bandwidth of the radiated signal. High resolution demands large bandwidth, which raises the data requirement several orders of magnitude, as the following example illustrates. Let an imaging system be designed to have a resolution cell 1×1 m in range and cross-range at a distance of 10 km. The required aperture size is $10^4\lambda$. In the absence of thinning, the number of required elements for a one-dimensional aperture is about 10^4. The bandwidth of the transmitted signal must be 150 MHz to achieve the desired resolution in range. Each receiver channel must be sufficiently broadband to sustain the signal resolution. Quadrature coherent demodulation and sampling typically follow. Each quadrature component is sampled at the bandwidth rate and digitized to 6 or 8 bits. The product NBb of the number of elements, the signal bandwidth, and the number of bits per sample is the bit rate that has to be accommodated by the system (see Fig. 4-16). This value for the example given exceeds 10^{13} bits/s, which exceeds the capability of current technology. Thus bit reduction is essential.

A general formula for the required bit rate for a filled periodic array (element spacing $= \lambda/2$) is easily developed. Let ΔR be the size of the resolution cell in range and cross-range. Let R be the target distance. Then

$$\frac{R}{\Delta R} = \frac{L}{\lambda} = \frac{N}{2} \tag{4-8}$$

and

$$\Delta R = \frac{c}{2B} \tag{4-9}$$

Combining these expressions, the bit rate is

$$NBb = \frac{2R}{\Delta R} \cdot \frac{cb}{2\Delta R} = \frac{Rcb}{(\Delta R)^2} \tag{4-10}$$

Aperture thinning described above reduces not only the cost of the parts of the system but also the amount of data to be handled. It reduces the number of channels in the intra-array communication subsystem that transports the received signals to the signal processor. It also reduces the size and complexity of the signal processor. Other procedures reduce the spectral occupancy of the signal without reducing its effective bandwidth B, and bit compression techniques reduce the number of quantization bits b. These procedures are described in Chapter 10.

5

DIVERSITY

5-1 THE DIVERSITY IMAGE

The purpose of diversity-combining of signals is to enhance their desired components relative to noise, interference, and other artifacts. An early form of diversity called *sweep integration* was invented for radar during World War II [8]. Its purpose was to increase the signal-to-noise ratio of a stationary target embedded in random noise. The improvement is accomplished by summing successive radar traces. Because the target echo arrives at exactly the same time following each pulse transmission, while the receiver noise is independent from pulse to pulse, simple addition of a group of echo traces provides the desired result. Given N successive transmissions, fixed target strength from pulse to pulse, and zero-mean noise independent from pulse to pulse, the signal power increases as N^2 while the noise power grows as N. The output signal-to-noise ratio, therefore, grows linearly with the number of pulses integrated.

There are many forms of diversity-combining. In principle, any domain of the signal in which the desired component is steady and the interference is random and independent is suitable. Diversity has been used on communication circuits in several ways. A signal passing through the ionosphere on a long-distance HF (3 to 30 MHz) circuit will be refracted back to earth from different ionospheric layers and arrive at different times. Such a signal is said to experience multipath or is called a *multipath signal*. The impulse response of the channel is a sequence of bursts that vary with time because of the physical dynamics of the ionosphere. Unlike the radar case, temporal integration is not useful because smoothing over time reduces the channel capacity or data rate in proportion to the temporal averaging time. Two other forms of diversity are useful, however. The impulse response of the channel is highly sensitive both to carrier frequency and to the exact route of the signal through the ionosphere. The former fact is exploited by *frequency diversity*: the signal is transmitted at two different carrier frequencies and the received signals have independent distortions or fading properties [86]. The probability that both signals fade simultaneously is smaller than the probability of a single fade. Therein lies the potential benefit of diversity combining: any of several ways of combining the signal waveforms or their modulations from the two receivers will reduce the error rate. The simplest decision rule is to accept the stronger of the two signals. A more sophisticated procedure is to weight them in accordance with their signal-to-noise ratios [14].

The second form of diversity used in HF communication is called *space diversity*. A single transmitter and two receivers are employed. The receivers are spaced sufficiently that the signal-fading on the space diversity channels is independent [7]. Combining the signals reduces the effective fading rate.

Another form of diversity called *angle* or *beam diversity* is employed on tropospheric scatter circuits. The time-varying impulse responses of a scatter circuit varies with the elevation angle of the receiving beam. Thus multiple receiving beams provide output signals that can be combined to reduce the fading properties of the circuits [73]. On certain communication circuits the fading properties are approximately independent on orthogonal polarizations [86]. Polarization diversity can be used in such cases to improve the quality of the output signal.

Diversity combining also is valuable in microwave imaging. There are three ways in which diversity combining can be helpful. It was seen in the imagery shown earlier that manmade microwave targets appear as collections of relatively isolated dots. Any diversity technique that builds up the mass within the boundary of a target will tend to provide a better image. Second, coherent constructive interference builds the high lobes in the side radiation pattern of an array. These can result in target artifacts or false targets. Diversity techniques can decorrelate the side radiation patterns of antennas. A set of images formed from different radiation patterns in which the sidelobe characteristics are decorrelated will exhibit false targets in different locations. Averaging these images will reduce the effective level of the highest lobes and thereby reduce the artifact content in the image. Third, destructive coherent interference (speckle, Section 4-2) breaks edge and boundary continuity in manmade targets. Any diversity-combining procedure that decorrelates the speckle results in superior imagery.

The theory of diversity-combining in communications is given in [14]. The theory of sidelobe decorrelation by diversity combining is given in [97], [98]. The diversity image I_D is the weighted sum of images I_m, $m = 1, 2, \ldots, M$, obtained from a set of snapshots of the radiation field. These images are called *trial images* or *constituent images*. The trial images differ either because the diversity procedure alters the radiation pattern between images or because the procedure alters the phase relations among scatterers in the source function, or both. Thus the diversity image can be written

$$I_D = \sum_{m=1}^{M} a_m I_m = \sum_m a_m |f_m * s|^2 \tag{5-1a}$$

or

$$I_D = \sum_m a_m |f * s_m|^2 \tag{5-1b}$$

or

$$I_D = \sum_m a_m |f_m * s_m|^2 \tag{5-1c}$$

where, as before, the asterisk means convolution. Sometimes the sum of the moduli is taken as the diversity image and sometimes the square root of (5-1). The weighting coefficients a_m are sometimes all unity, sometimes 1 or 0 according to some measure of goodness or quality (such as signal-to-noise ratio) and sometimes proportional to the measure of goodness.

Although each trial image is formed by a convolution process, Eq. 5-1 is not a simple convolution; it is the sum of squares of convolutions and therefore linear filter theory does not apply. One form of diversity does produce a diversity image that is approximately a convolution of the intensity distribution of the source $|s(u)|^2$ and the power transfer function of the aperture $|f(u)|^2$. This procedure is called *transmitter location diversity.* It is a new form of diversity, introduced in Section 5-8 and described in Chapter 6.

5-2 SIDELOBE REDUCTION

Figure 5-1 plots an estimator of the peak sidelobe of a random array as a function of the diversity order M under the assumption of complete decorrelation of the side radiation pattern. The plotted values are at a confidence level of 0.9. Because of the assumption of complete decorrelation, this situation corresponds to Eq. 5-1a where

$$\begin{aligned} f_i = f_j, &\qquad |u| < \frac{\lambda}{L} \\ Ef_i f_j = 0, &\qquad |u| > \frac{\lambda}{L} \end{aligned} \tag{5-2}$$

for all i and j. As the number of images averaged increases, the improvement approaches unity (or 0 dB) asymptotically. This is because at high diversity order the peaks of the composite side radiation pattern smooth down to approximately the average value, beyond which no further improvement can

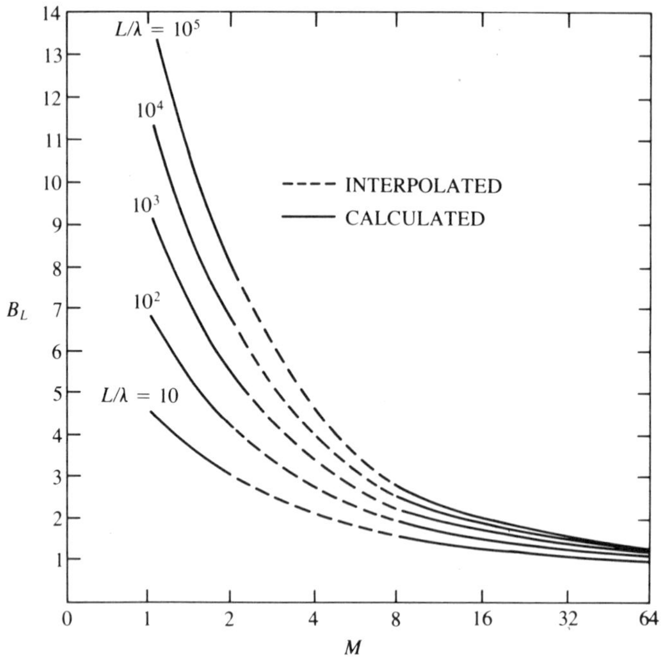

Figure 5-1 B_L is the power ratio of a downward-biased estimator of the peak sidelobe to the average value. $\beta = 0.9$ is the confidence level. M is the order of diversity. (From refs. [97] and [98])

take place. At small diversity orders the improvement is highly aperture-size dependent. This is because the highest lobes are larger for large arrays than for small arrays (about $\ln(L/\lambda)$ times larger than the average level, [94]) and therefore offer a larger available improvement. $M = 1$ corresponds to a single channel or no diversity-combining. The slope for large apertures and low diversity orders is close to 3 dB per doubling of diversity order. This implies very high efficiency in diversity reduction of the peak sidelobe because 3 dB per doubling of diversity order is the maximum available theoretical improvement.

5-3 ELEMENT-POSITION DIVERSITY

There are some situations in which it is possible to rearrange the locations of the antenna elements between snapshots. If the element locations are randomly altered over an aperture of nominally fixed size, the side radiation pattern decorrelates between observations while the main lobe hardly changes. This is the ideal condition for diversity-combining. An antenna array formed by hydrophones on freely-floating buoys is an example. Another is a radar array in space in which the receivers are deployed on different orbiting platforms. More generally, it applies to any array the parts of which are on independently moving vehicles or a connected body in which the parts undergo a sufficient amount of random motion.

Element position diversity is a useful tool whenever it is possible to form several arrays from the same set of elements. A trial image of the scene is formed from each array and the sum or average of the moduli or the squares of the moduli of the trial images is the diversity image. Let x_{jn} be the location of the nth element of the jth trial array. The jth weight function is

$$w_j(x) = \frac{1}{N} \sum_{n=1}^{N} \delta(x - x_{jn}) \tag{5-3}$$

and its radiation pattern is

$$f_j(u) = \mathscr{F}\{w_j(x)\} \tag{5-4}$$

Assuming that $s(u)$ is the scene to be imaged, the jth complex trial image is

$$\hat{s}_j(u) = s(u) * f_j(u) \tag{5-5}$$

The phase of Eq. 5-5 is removed in forming the diversity image. Summing or averaging the moduli as in Eq. 5-6,

$$I_D(u) = \frac{1}{J} \sum_{j=1}^{J} |\hat{s}_j(u)| \tag{5-6}$$

or the square of the moduli as in Eq. 5-7,

$$I_D(u) = \frac{1}{J} \sum_{j=1}^{J} |\hat{s}_j(u)|^2 \tag{5-7}$$

or taking the square root of Eq. 5-7 results in diversity images.

The key to diversity combining is decorrelating the sidelobes but not the main lobes. Following Eq. 5-2, if $f_1(u)$ and $f_2(u)$ are two radiation patterns from the set (5-4), the desired properties for diversity-combining are

$$f_1\left(|u| < \frac{\lambda}{L}\right) = f_2\left(|u| < \frac{\lambda}{L}\right) \tag{5-8}$$

and

$$E\left[f_1\left(|u| \geqslant \frac{\lambda}{L}\right) f_2\left(|u| \geqslant \frac{\lambda}{L}\right)\right] = 0 \tag{5-9}$$

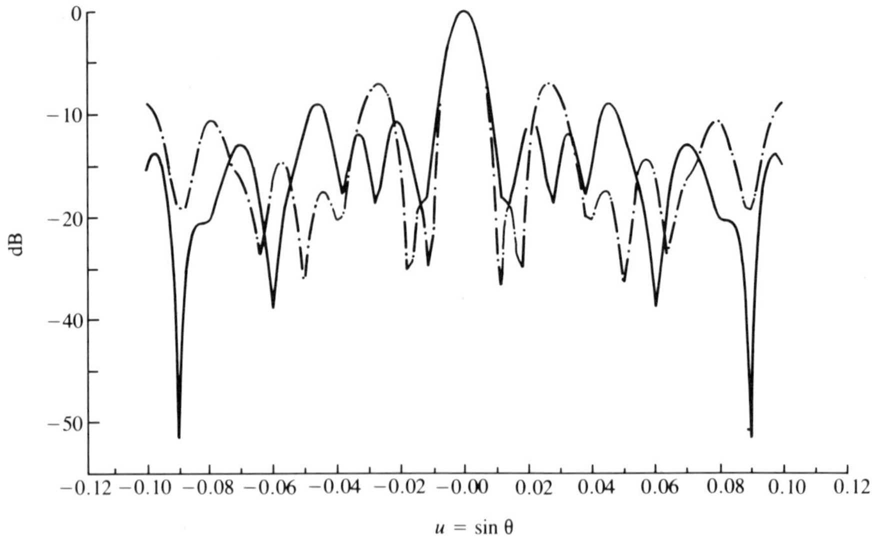

Figure 5-2 Element-position diversity simulation experiment. Element-position diversity decorrelates the sidelobes of the radiation pattern of the array. 32 elements were spread randomly over 180 wavelengths to obtain the two radiation patterns.

Changing element locations changes the locations of the sidelobe peaks without changing the statistics of the side radiation pattern. Hence this process decorrelates the side radiation pattern. The variance of the pattern of a random array in the main lobe region is very small, however [94]. Therefore the main lobe is relatively unaffected by the rearrangement of element locations provided only that the pdf of element location is unchanged. Consequently both conditions 5-6 and 5-9 are satisfied.

An example is given in Fig. 5-2. Two 32-element random arrays each of length 100λ were simulated and their radiation patterns were calculated. The main lobes are nearly identical while the correlation between the side radiation patterns is very low.

Figure 5-3, from [97], shows a one-dimensional element-position diversity experiment in which the target was a cement block. A set of 20-element, one-dimensional images were formed with the elements randomly rearranged for each experiment. Diversity orders 1, 2, 4, and 10 are shown in parts (a)–(d). The diversity-combined images show the expected reduction in the side radiation pattern of the composite. The peak value reduces approximately as the theory demands. For comparison, part (e) shows the image in which the entire set of 200 complex data samples measured by the array were combined coherently to form a single image. Experiments (d) and (e) both involve 200 data samples. They were coherently combined 20 at a time in (d) and the 10 resulting images were (noncoherently) added. In (e) the 200 samples were coherently combined to form a single image. It is evident that the sidelobe

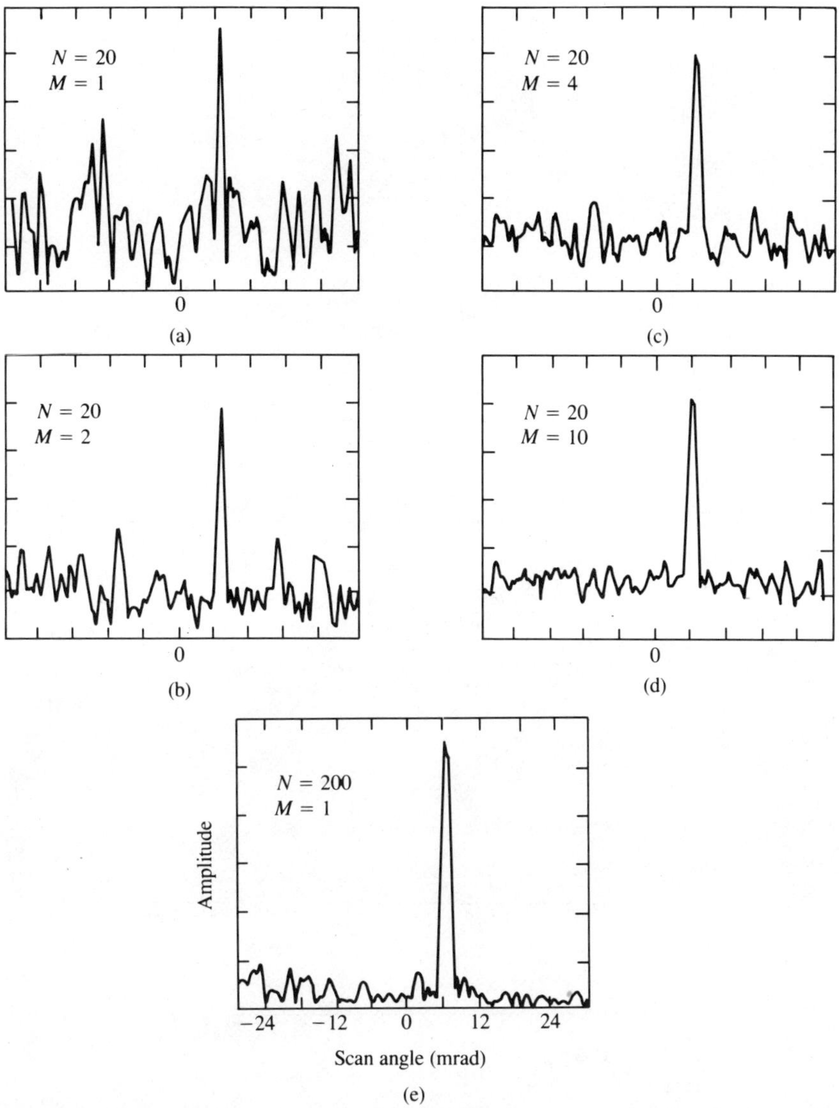

Figure 5-3 Element-position diversity experiment: 28-m array, 3-cm wavelength. Cement block target at a distance of 146 m. (a) to (d) 20 elements each, diversity orders 1, 2, 4, and 10. (e) 200 elements, diversity order 1. (From ref. [97])

excursions from the background levels are similar in both experiments, indicating efficiency in the diversity-combining process. The average background levels, of course, are different. The ASL is about 10 dB lower in (e) because ASL is inversely proportional to the number of antenna elements forming the array (200 as compared to 20 in (d)).

Figure 5-4 shows a two-dimensional diversity experiment based upon the

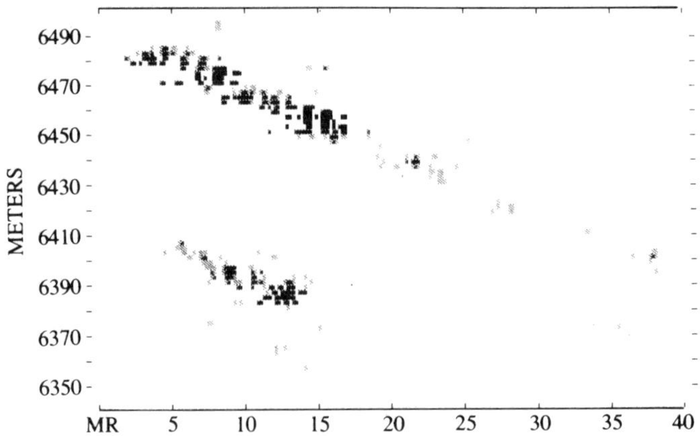
METERS
6490
6470
6450
6430
6410
6390
6370
6350
MR
5
10
15
20
25
30
35
40

(a)

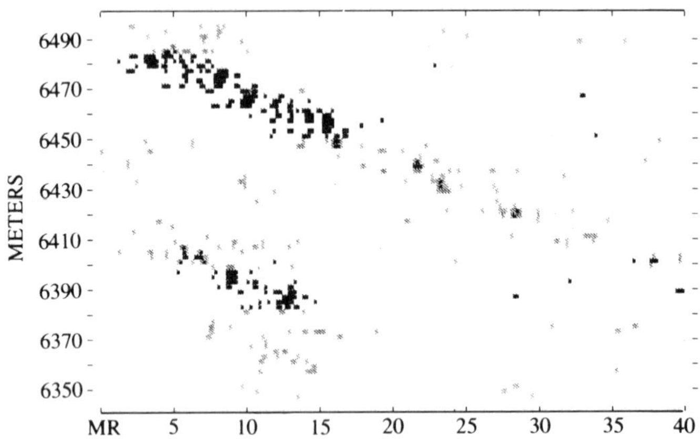
METERS
6490
6470
6450
6430
6410
6390
6370
6350
MR
5
10
15
20
25
30
35
40

(b)

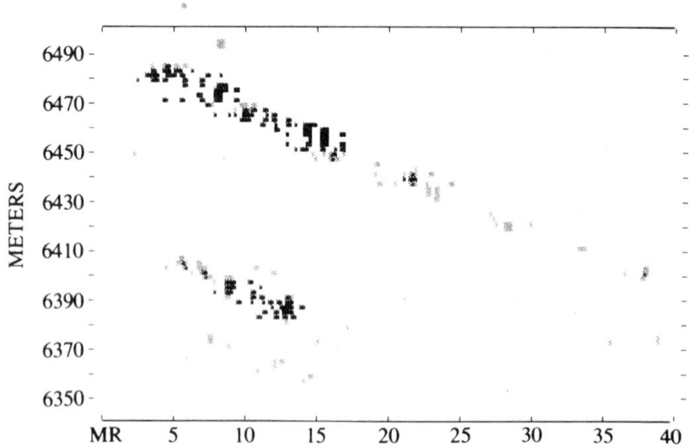
METERS
6490
6470
6450
6430
6410
6390
6370
6350
MR
5
10
15
20
25
30
35
40

(c)

same row-house scene of Fig. 4-8. The distance from the laboratory is 6.5 km. The wavelength is 3 cm. The aperture size has been increased to 83 m and the data consist of 330 samples. The upper figure is the radio camera image made from this data set. Observe the empty lot between the groups of houses on the upper street and the relative freedom from artifacts in the image.

The data were then randomly thinned by a factor of 4, resulting in the image of the middle figure (b). Artifacts abound in this image; note that the lot no longer appears empty.

Three additional randomly thinned arrays were formed, differing only in the seed fed to the random number generator. Their images were similar in nature to the middle figure. The four images were added to form a fourth-order diversity image, which is shown in the bottom of the figure. The empty lot is again empty and the image is relatively free of artifacts.

5-4 FREQUENCY DIVERSITY

Section 3-8 discussed the effect of wide signal bandwidth upon the side radiation pattern. Figure 3-16 illustrated how the peaks reduced and their valleys filled in toward the average sidelobe level. Table 3-5 listed the expected reduction in the peak sidelobe level resulting from nonmonochromatic transmission.

The discussion and analysis in that section is pertinent to frequency diversity. Frequency diversity can be achieved by transmitting a wideband waveform, as was assumed in Section 3-8, or through a sequence of narrowband signals that occupy a wide spectral interval. [98] and [97] show that both procedures lead to the same reduction in sidelobe level when the array is a random array.

Frequency diversity consists of the transmission and reception of radar signals at widely spaced frequencies. The transmissions can be simultaneous or sequential. The frequencies have to be far enough apart that the image artifacts and the speckle patterns of the images are uncorrelated.

Figure 5-5 shows one form of dual frequency diversity. A dual frequency transmitter radiates at frequencies ω_1 and ω_2. Echoes from the target scene are received by the same phased array. The radiation patterns of the receiving array at the two frequencies have the same shape but different scales, their scales being in proportion to the frequency ratio. An image is obtained at each frequency and their intensities are added. Because of the scale differences between the abscissas of the radiation pattern, the effective sidelobe peaks of the composite system smooth down toward the average background level beyond some distance from the main lobe. This phenomenon is illustrated in

Figure 5-4 Element-position diversity experiment based on diffraction-limited images of rows of houses at a distance of 6.5 km from the radio camera. $\lambda = 3$ cm. Aperture size $= 83$ m. Data set consists of 330 samples. (a) Image obtained from full data set. (b) Image obtained from random selection of one-quarter of the data set, showing large number of artifacts. (c) Fourth-order diversity image made from (b) and three others, each image formed from a different random selection of one-quarter of the data set. Observe that the image quality of (c) is comparable to that of (a).

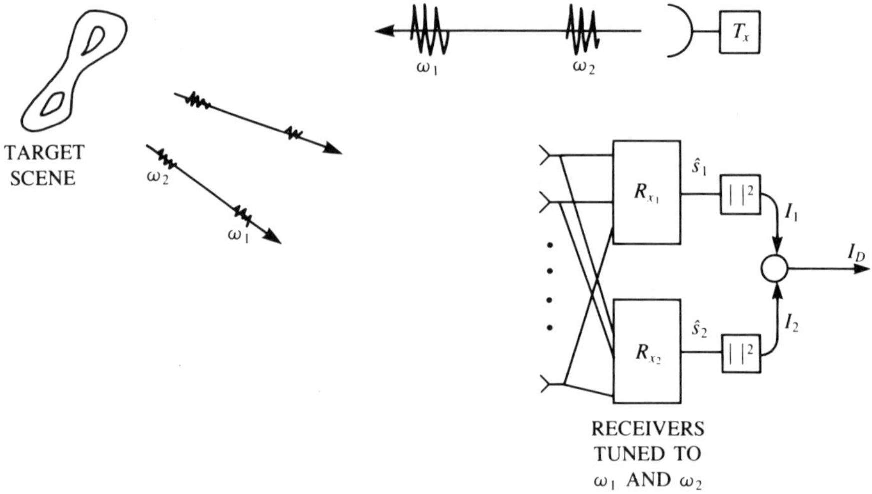

Figure 5-5 A dual-diversity transmitter alternates pulses at frequencies ω_1 *and* ω_2. Two receivers separate the echoes and form complex images $\hat{s}_1$ and $\hat{s}_2$. The diversity image $I_D = |\hat{s}_1|^2 + |\hat{s}_2|^2$.

Fig. 3-16b. Based on the analysis in Section 3-8, the side radiation patterns at the two frequencies decorrelate beyond the pth sidelobe, where p is given by

$$p = \frac{\omega_0}{\Delta\omega} \tag{5-10}$$

where ω_0 is the average of the two frequencies and $\Delta\omega$ is the frequency separation. p may be identified with Q in Section 3-8 and Table 3-5. The table may be reinterpreted as showing the extent to which frequency diversity reduces the sidelobe level of a random receiving array. A very large ($10^5\lambda$) receiving array offers an improvement of 5 dB for a frequency separation 10 percent of the mean frequency. Two X-band (3-cm) transmitters spaced by 1 GHz would produce this effect. Reducing the sidelobe level by 5 dB permits the designer to drop the input threshold level to the display, thereby passing smaller components of the image. The results of lowering the threshold is to deliver a larger target mass to the display (Chapter 12); image quality improves correspondingly.

A second factor further improves the quality of the imagery. When the separation in wavelength is sufficiently large, different target scattering characteristics may result. Objects on the target that are near resonance at one frequency will not be so at the other. Consequently the main portions of the images will exhibit different details and their addition will further increase target mass. The payment for these advantages, of course, is added system complexity, which increases by nearly the diversity order.

5-5 POLARIZATION DIVERSITY

There is a sharp distinction between the polarization properties of cultural and natural objects. The former are dominated by scattering from horizontal and vertical edges and surfaces—walls, roofs, sides of buildings, poles, fences, vehicles, etc. Natural objects exhibit less polarization sensitivity. An exception is the trunk of a tree, which acts as a vertical column of water. However, its limbs and leafy crown tend to have random orientations. Similarly for vegetation, meadowland, and extents of sand. Thus, man-made targets tend to exhibit a higher reflectivity ratio of, say, horizontal polarization to vertical polarization, or vice versa, than natural targets. Hence a strong image of a man-made target may be weak on the orthogonal polarization. Combining images obtained from both polarizations precludes loss of image due to polarization sensitivity of cultural objects and tends to increase their target masses. Its practice requires that dual orthogonal polarizations be transmitted and received.

Polarization diversity has become a significant tool in the research community in recent years. A rich and growing literature already exists, and its review is beyond the scope of this book. The interested reader is referred to [10, 25].

5-6 AXIAL-SYMMETRY DIVERSITY

Sometimes certain known geometric properties of a target permit manipulation of its image to improve it. An example is the axial symmetry of aircraft. By combining an aircraft image with its mirror image (with respect to its axis of symmetry) parts seemingly missing from one side or the other of the target reappear [28].

An example is shown in Fig. 5-6. A radio camera image of a Boeing 727 airplane flying at a distance of 3.2 km from the Valley Forge laboratory is shown in the upper section. Wings are visible on both sides of the fuselage but the tail section only appears on the far side. Combining this picture with its mirror image results in the image shown in the lower portion. The middle section is a plan-view drawing of the airplane.

5-7 ASPECT-ANGLE DIVERSITY

Imagine that the array can be moved such that successive images can be obtained from different array locations. Given sufficient displacement of the array, the distances from all elements to all scatterers on the target change sufficiently to alter the speckle pattern. The implication of the decorrelation of the speckle patterns of a pair of images obtained from different array locations is that the coordinates of the elements of target mass in one image are different from those of the second image. Consequently, adding the moduli

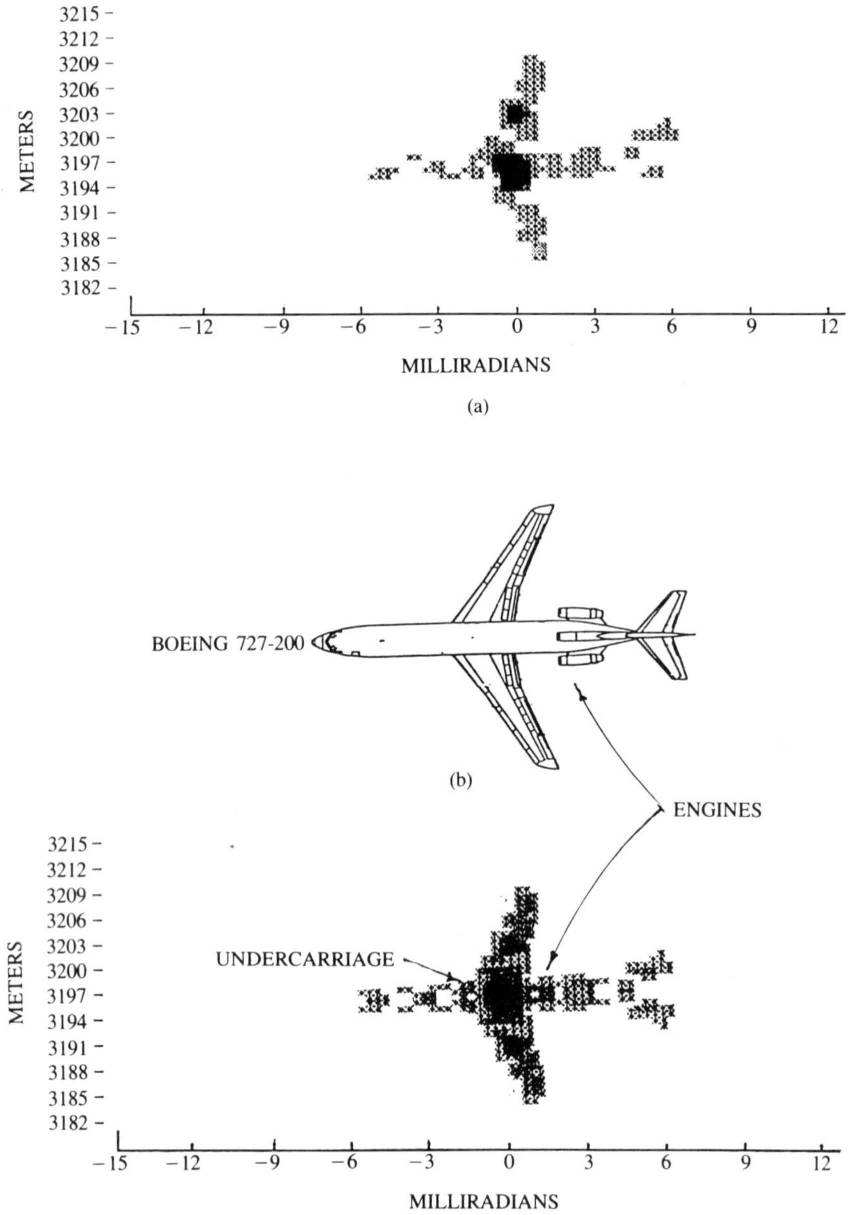

Figure 5-6 (a) Nondiversity radio camera image of a Boeing 727 airplane flying at a distance of 3.2 km from the VFRC laboratory. $\lambda = 3$ cm. Array consists of 32 X-band receivers spaced by 1 m. (b) Plan-view drawing of Boeing 727. (c) With axial symmetry applied.

or the intensities of the images tends to fill in the empty voids between the spots. This process increases the density of spots in the image and improves edge and boundary continuity.

The same improvement is obtained when a moving target is observed from a fixed antenna location. Successive snapshots of the radiation field at the array differ because the target motion alters all distance relationships between scatterers and array elements.

The geometry pertaining to aspect-angle diversity is shown in Fig. 5-7. An object of length L is viewed by an antenna element of aspect angle α. The distances from the left and right edges of the object to the element are R_1 and R_2, both being very much larger than L. R_1 can be expressed in terms of L, R_2, and α by the law of cosines:

$$R_1^2 = R_2^2 + L^2 - 2R_2 L \cos\alpha \tag{5-11}$$

and approximated by

$$R_1 = R_2 + \frac{L^2}{2R_2} - L\cos\alpha \tag{5-12}$$

The differential distance from edge scatterers to the antenna element determines the phase differences between the echoes received at the element. The phase differences are

$$\begin{aligned}\Delta\phi &= k(R_2 - R_1)\\ &\approx k\left(L\cos\alpha - \frac{L^2}{2R_2}\right)\\ &\approx kL\cos\alpha\end{aligned} \tag{5-13}$$

given the reasonable assumption that $R_2 \gg L$.

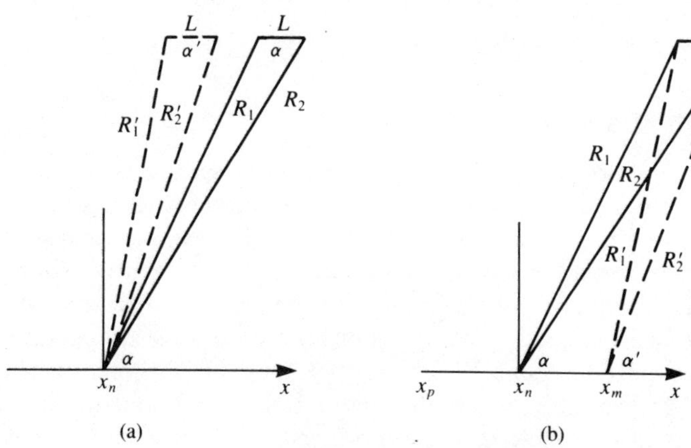

Figure 5-7 (a) Target motion or (b) displacement of the array changes aspect angle from α to α'.

Displacement of the object (as shown in Fig. 5-7a) or movement of the antenna element (Fig. 5-7b) changes the aspect angle to $\alpha' = \alpha + \delta\alpha$. The distances R_1 and R_2 change to R_1' and R_2' and the phase difference changes to

$$\Delta\phi' = \Delta\phi + \delta\phi = kL\cos\alpha' \tag{5-14}$$

The change in phase difference is

$$\begin{aligned}\delta\phi &= kL[\cos(\alpha + \delta\alpha) - \cos\alpha] \\ &\approx -kL\,\delta\alpha\sin\alpha\end{aligned} \tag{5-15}$$

for small displacements. When the magnitude of Eq. 5-15 is as large as π, the decorrelation of the echoes from a large portion of the scatterers due to the displacement is sufficient to achieve the desired speckle decorrelation. Equating the magnitude of Eq. 5-15 to π and approximating $\sin\alpha = 1$ to represent a target near broadside, the required aspect angle change is

$$\delta\alpha = \frac{\lambda}{2L} \tag{5-16}$$

Equation 5-16 is not a surprising result. It is one-half the beamwidth of an aperture of length L. In Fig. 5-7b it is easy to identify L as such an aperture and points x_n and x_m as field points at scan angles $\frac{1}{2}\pi - \alpha$ and $\frac{1}{2}\pi - \alpha'$. If the array were scanned, say, to x_n, x_m would represent a location one-half beamwidth away. Similarly, x_p, an equal distance to the left of x_n, would also be one-half beamwidth away from the scan direction. Thus the angle at the aperture L subtended by the line segment from x_p to x_m is one beamwidth. All points on the x-axis within this angles are substantially cophased across the aperture; outside they are not cophased. This phenomenon is the same as that which forms the conventional radiation pattern of an antenna, one that consists of a central main lobe surrounded by minor lobes.

5-7-1 Radio Camera Imaging

There are several imaging models in which aspect-angle diversity can be applied. The most straightforward is imaging a moving target with a large, stationary radio camera. The images obtained from a succession of snapshots of the complex radiation field of, say, an airplane correspond to a progression of aspect angles. As a consequence, image quality is enhanced in a diversity combination of the individual images. Figure 5-8 sketches a dual-diversity operating mode. Radar pulses spaced by time T each produce images of the aircraft. The aircraft speed is V. The aircraft moves through VT during T and the aspect angle changes from a to a'. Each image $|\hat{s}|$ is delayed by T and added to its successor to form the dual diversity image. Higher-order diversity is obtained by adding

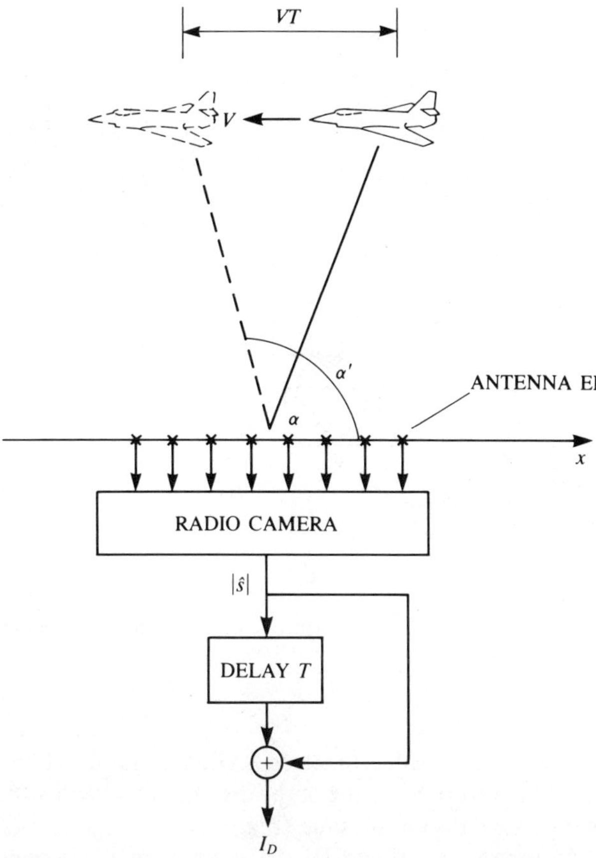

Figure 5-8 Images of an airplane obtained at different aspect angles are combined to form the diversity image.

additional images. Figure 5-13, shown later, demonstrates the value of second- and fifth-order diversity.

Figure 5-8 illustrates this combination of spatial (imaging with a radio camera) and temporal (diversity by delay and add) processing to achieve the image improvement. All other combinations of spatial and temporal imaging and combining are possible too. Figure 5-9 illustrates a spatial–spatial process involving antenna/receiver displacement. Two imaging systems observe the target scene from different aspect angles and their images are combined to form a dual-diversity image.

5-7-2 SAR and ISAR

An interesting application of temporal–temporal processing is synthetic aperture radar (SAR) and its inverse form (ISAR). The former involves receiver

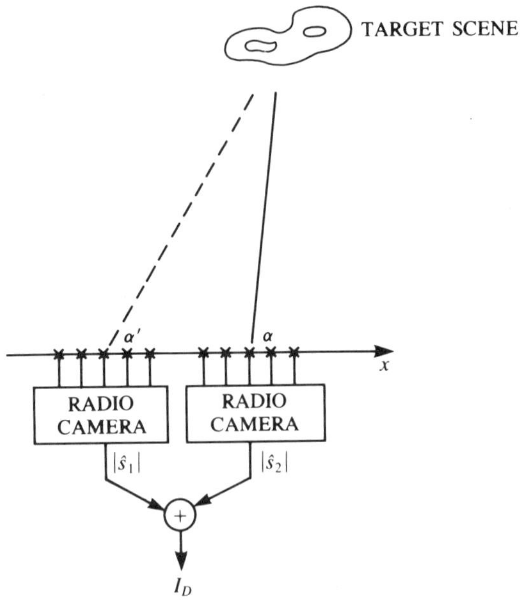

Figure 5-9 Images of target scene from different aspect angles are added to form the diversity image.

displacement, the latter target displacement. In SAR, the motion of an airborne radar forms a synthetic aperture by virtue of the distance traveled by the aircraft during the observation or data-taking time (Fig. 5-10).

Let an aircraft fly with speed V and let the radar receive a sequence of N pulses spaced by t seconds. During time $T = Nt$, the aircraft flies a distance $L = VT$. An antenna of such size would have a beamwidth λ/L. The synthetic aperture beamwidth is one-half this dimension because of an interesting subtlety. The transmitter and receiver are collocated and move together. Hence the phase delay from a scatterer on the target is $k = 2\pi/\lambda$ times the round-trip distance $2R$ from radar to target, or $2kR$. This phase delay is twice the normal phase shift of a target echo at a receiver. The effect is equivalent to doubling the wave number. The result is a nominal beamwidth of $\lambda/2L$ or $\lambda/2VT$.

The factor of 2 difference in beamwidths is irrelevant to the diversity argument. Because the aircraft is in motion, so is the synthetic aperture. Consequently, views of the target can be obtained from different aspect angles to the target. Provided that the displacement is sufficient to decorrelate the speckle pattern, the benefits of diversity combining described above will be obtained.

In ISAR the radar is on the ground and the target is the aircraft. Figure 5-11 illustrates the relation of SAR to ISAR. In both cases the coordinate frame is centered on the target. In the left part (SAR) the target is on the ground and the radar is in an airplane moving to the left. Successive pulses arrive at the

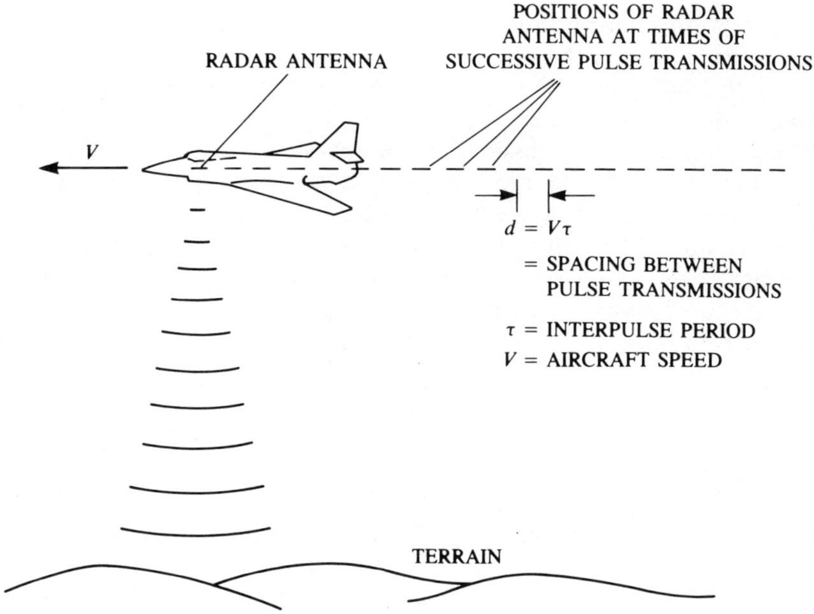

Figure 5-10 Synthetic aperture radar. (From ref. [97])

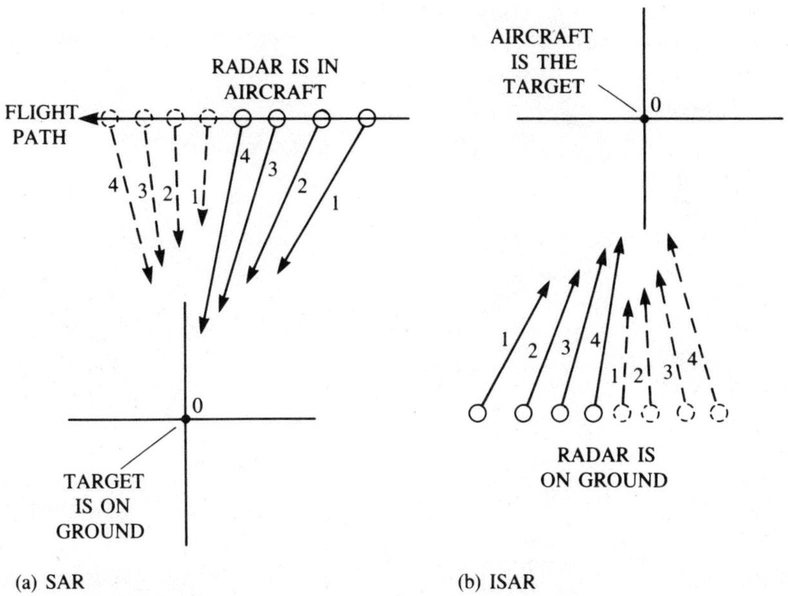

Figure 5-11 Comparisons between (a) synthetic aperture radar (SAR) and (b) inverse synthetic aperture radar (ISAR). In both cases, the first diversity image is formed by synthetic aperture represented by solid lines; the second image by dotted lines.

target from directions denoted by the rays labeled 1, 2, 3, ... The successive element locations are the open circles on the flight path at the heads of the rays. In the right part (ISAR) the radar is on the ground and the target is the airplane. By placing the frame of reference upon the target, it is easily seen that the successive radar transmissions and receptions appear to arise in the reverse order from locations on the ground that are spaced by Vt. A set of N transmissions results in a synthetic aperture of length $L = VT$ exactly as in SAR. Similarly, the transverse beamwidth is $\lambda/2L$.

In both cases, successive groups of N pulses can each provide an image. Diversity combining is effective because the apertures formed from the separate pulse sequences are spaced by L or more. The dotted circles and dashed rays in Fig. 5-11 indicate the positions of the second synthetic aperture in the two cases. The trial or constituent images are combined as in Fig. 5-7.

5-7-3 Dual-Receiver ISAR

The final processing combination is temporal (imaging) and spatial diversity. Moving target echoes from a single transmitter are received by two or more antenna elements. Each processes an ISAR image. The set of images is combined for diversity. Figure 5-12 illustrates the geometry.

Temporal–spatial combining is potentially a very powerful tool for large, ground-based radio cameras designed to image aircraft or other moving targets. The aperture can be very large, such as an airport or a city, and very sparse. Assume that the thinning factor is so great that the side radiation properties are unacceptable. Temporal imaging (ISAR) forms a synthetic aperture of each

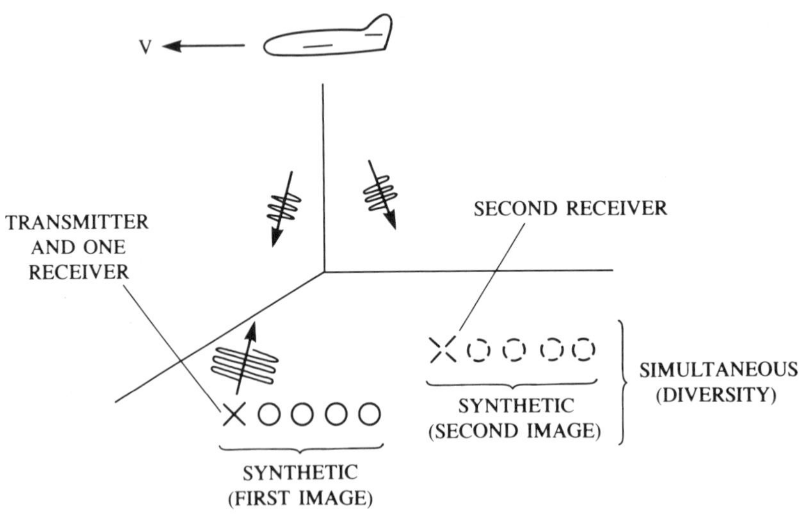

Figure 5-12 Two or more receivers form synthetic aperture images simultaneously. These images are combined for diversity enhancement. (From ref. [102])

element and, in effect, reduces the thinning factor by the amount by which a single element is effectively extended into a phased subarray. In essence, temporal processing can build the mass of the aperture to the level at which the sidelobe characteristics are acceptable.

The four combinations of spatial and temporal imaging and diversity combining are summarized in Table 5-1.

TABLE 5-1 Aspect-angle diversity combining based on combinations of spatial and temporal processing for imaging and combining

Examples of mode of operation	*Imaging*	*Diversity combining*
1 radio camera	Spatial	Temporal
2 or more radio cameras	Spatial	Spatial
SAR or ISAR with 1 receiver	Temporal	Temporal
ISAR with 2 or more receivers	Temporal	Spatial

5-7-4 Experiments

Experiments conducted at the Valley Forge Research Center illustrate each of these aspect-angle diversity modes. Figure 5-13 shows the results of a diversity experiment using a 32-m radio camera operating at X-band ($\lambda = 3$ cm). It corresponds to the spatial–temporal diversity-combining procedure illustrated in Fig. 5-8. The aperture was highly thinned (32 receivers spaced by 1 m, each with a 20-cm horn antenna) and distorted. The geometric distortion was corrected by a self-calibration procedure called *adaptive beamforming*, which is described in Chapter 8. A radar transmitter radiated 1-m pulses (7-ns duration). The target was a Boeing 727 airplane flying into Philadelphia International Airport at a distance of 3.2 km.

Each of the 32 echo traces from a group of transmitted pulses was recorded and stored as illustrated in Fig. 3-10. Each such snapshot of complex field measurements yields one image. A set of five such images was obtained. The images were spaced in time by 0.1 s, and the total movement of the airplane was about 25 m. The target range was 3.2 km.

The five images, labeled shots 5 through 9, are shown in the upper part of Fig. 5-13. Shot 7 is the best. The images of shots 5 and 7 are added (diversity order $M = 2$) and shown in the lower left. The result is significantly better. The body mass (fuselage, wings, etc.) increases and the fuselage extends to include the nose. Shot 5 contributed scattering from this region of the fuselage but little else forward of the wings. Shot 7 had no nose scattering. The forward position of the fuselage begins to take shape when these shots are combined.

The lower left-of-middle image combines all five trial images ($M = 5$). Now the fuselage and wings are well-defined and their lengths easily measured. The

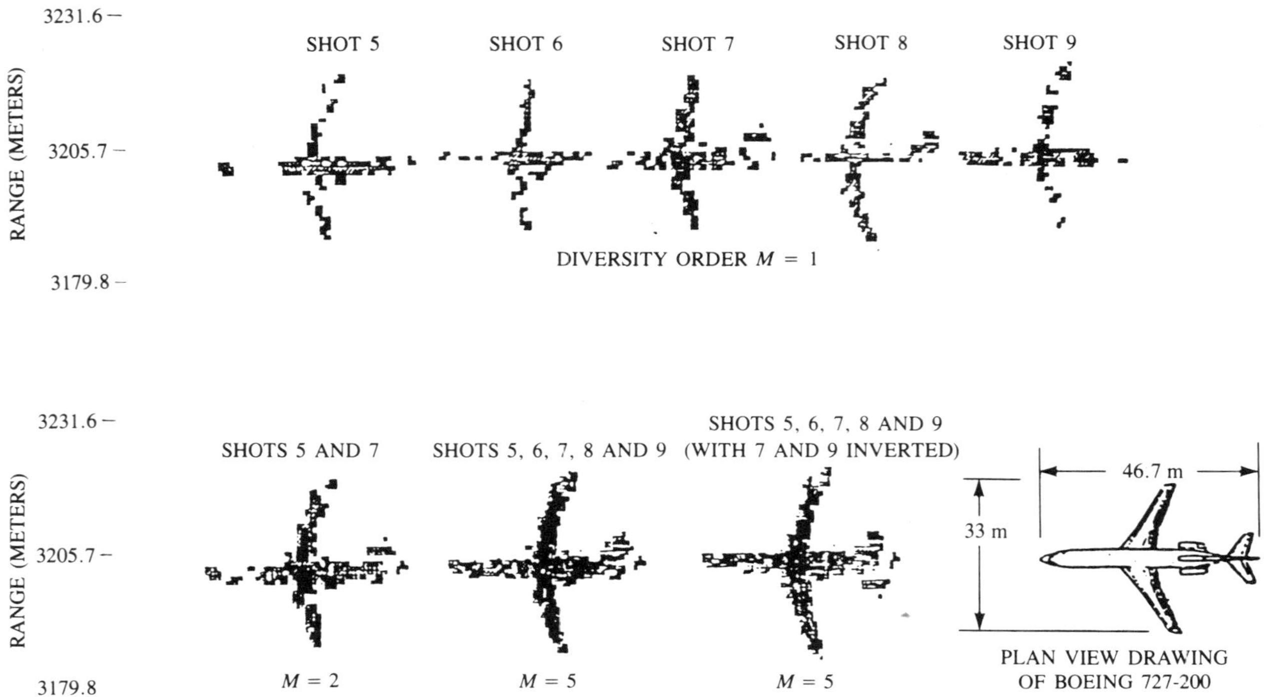

Figure 5-13 Diversity experiment showing buildup of radar image of Fig. 2-13 from five "snapshots" taken at 0.1-sec intervals. $M = 2$ is a dual-diversity image of shots 5 and 7. The first $M = 5$ is of quintuple diversity using all five shots. Shots 7 and 9 are inverted in the second $M = 5$ image in order to capitalize on the known axial symmetry of airplanes. (From ref. [102])

place on the fuselage where the wings enter the body is clear, as is the sweep of the wings. Left somewhat to the imagination is the shape of the tail structure. This is corrected to some extent by capitalizing upon the known axial symmetry in the target (see Section 5-6) and inverting the images of shots 7 and 9 prior to combining them with the other three. The result is the image in the lower right, which is the best representation of the target. A plan-view drawing of the aircraft is shown for comparison.

Figure 5-14 shows the results of an experiment that corresponds to the displaced-antenna (spatial–spatial) procedure illustrated in Fig. 5-9. An 83-m X-band radio camera array was split into two 40-m sections. Each imaged the housing development shown in Fig. 2-12. Shown in Fig. 5-14 are one of the two trial images (a) and the dual-diversity ($M = 2$) image (b). For comparison the single image ($M = 1$) from the entire array is shown in (c) and the diversity sum of (b) and (c) is shown in (d).

The nondiversity images appear to be very similar; yet they are sufficiently dissimilar in fine detail for the dual-diversity image to be superior. The single image from the entire array has finer cross-range resolution but is more highly speckled than the dual-diversity image. The reader can judge which is superior.

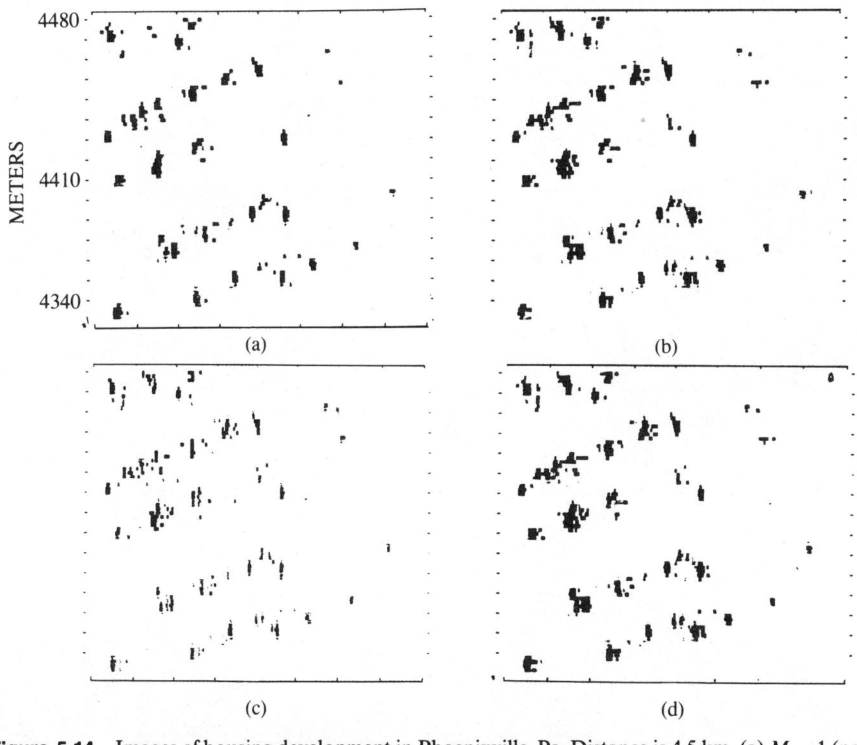

Figure 5-14 Images of housing development in Phoenixville, Pa. Distance is 4.5 km. (a) $M = 1$ (no diversity). Array length 40 m. $\lambda = 3$ cm. (b) $M = 2$. Image (a) combined with a similar image from adjacent 40-m aperture. (c) $M = 1$. Array length = 83 m. $\lambda = 3$ cm. (d) Images (b) and (c) combined.

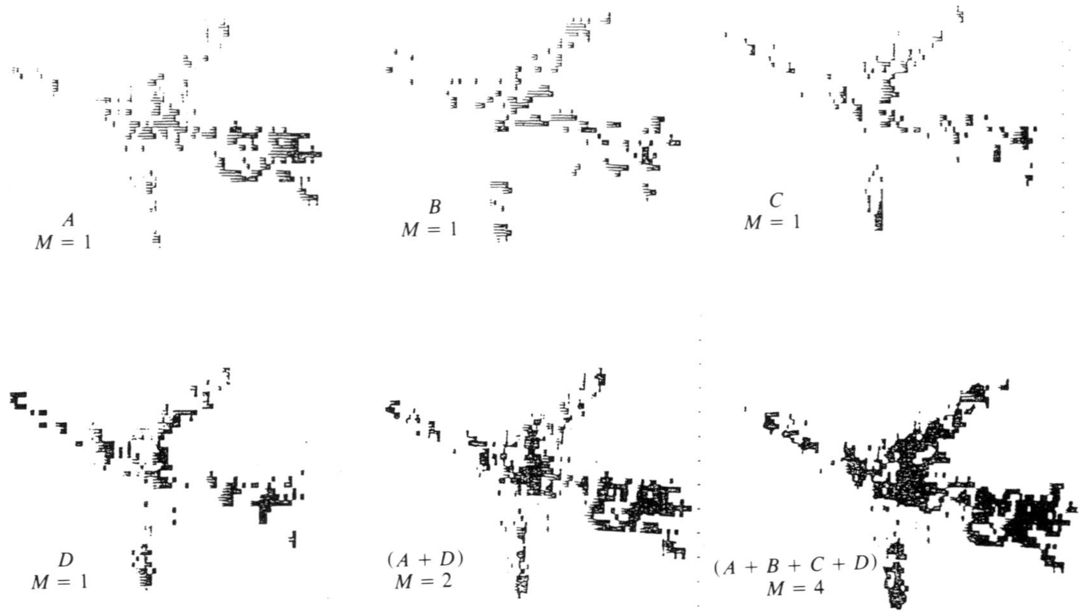

Figure 5-15 A, B, C, and D are ISAR images of a Boeing 727, formed from four independent data sets. The aircraft was flying at a distance of 2.7 km from VFRC. The diversity order is one. $A + D$ is a dual-diversity image formed by adding imaged A and D. $A + B + C + D$ is a fourth-order diversity image. The added target mass makes this image far more recognizable than any of the constituent images.

Figure 5-15 shows the results of an ISAR diversity experiment (temporal–temporal) that corresponds to the right-hand portion of Fig. 5-11. Four $M = 1$ images, one $M = 2$ image and an $M = 4$ image are shown. The plan-view drawing of the airplane is in the lower right of Fig. 5-13.

Figure 5-16 shows the result of a temporal–spatial experiment. It corresponds to the diversity procedure illustrated in Fig. 5-12. An X-band transmitter with a 1-m pulse duration and a receiver shared a common 1.2-m dish antenna. A second receiver with a 20-cm horn antenna was 25 m away. The direction of airplane flight was parallel to the base line between the receivers (rather than offset as pictured in Fig. 5-12). The airplane was an L-1011 commercial airliner flying at a distance of 2.5 km from the laboratory.

Three nondiversity options were available. Images could be formed from each set of 85 echoes received simultaneously by the two receivers. Each set constituted the equivalent of measurements made by an 85-element phased array. One such image is shown in part (a) in the upper portion of the figure. The two images have the same general characteristics. They differ in detail and therefore are good candidates for diversity combining (b). An image also could be formed from the 170 echoes, treating them as the measurements from a 170-element phased array. This image is shown in part (c). Image (c) has a finer resolving power in cross-range due to the doubling of the aperture size. Part (d) shows the three images combined with the second inverted. Part (e) shows the sum of the fourth image and its inverted version. Also shown is the plan-view drawing of the aircraft. The composite image is easily identified as an L-1011.

5-8 TRANSMITTER-LOCATION DIVERSITY

This form of diversity is particularly interesting. The diversity image formed by any of the procedures above can be written in any of the forms given in Eq. 5-1 according to whether the diversity method altered the radiation pattern or the source function or both upon successive snapshots. As noted in Section 5-1, this equation is not a simple convolution; it is the sum of squares of convolutions. The image resulting from transmitter location diversity, on the other hand, is approximately a convolution:

$$I_D(u) \approx |s(u)|^2 * H(u) \tag{5-17}$$

$|s(u)|^2$ is the intensity distribution of the source and H is the power transfer function of the aperture, that is, the squared magnitude of the radiation pattern with main lobe approximately that associated with the weighting function of the antenna and with a considerably smoothed side radiation pattern. It is a real function. $H(u)$ describes a linear system function that, when convolved with the source intensity distribution, provides the output image.

It can be shown that Eq. 5-17 describes the imaging process when the

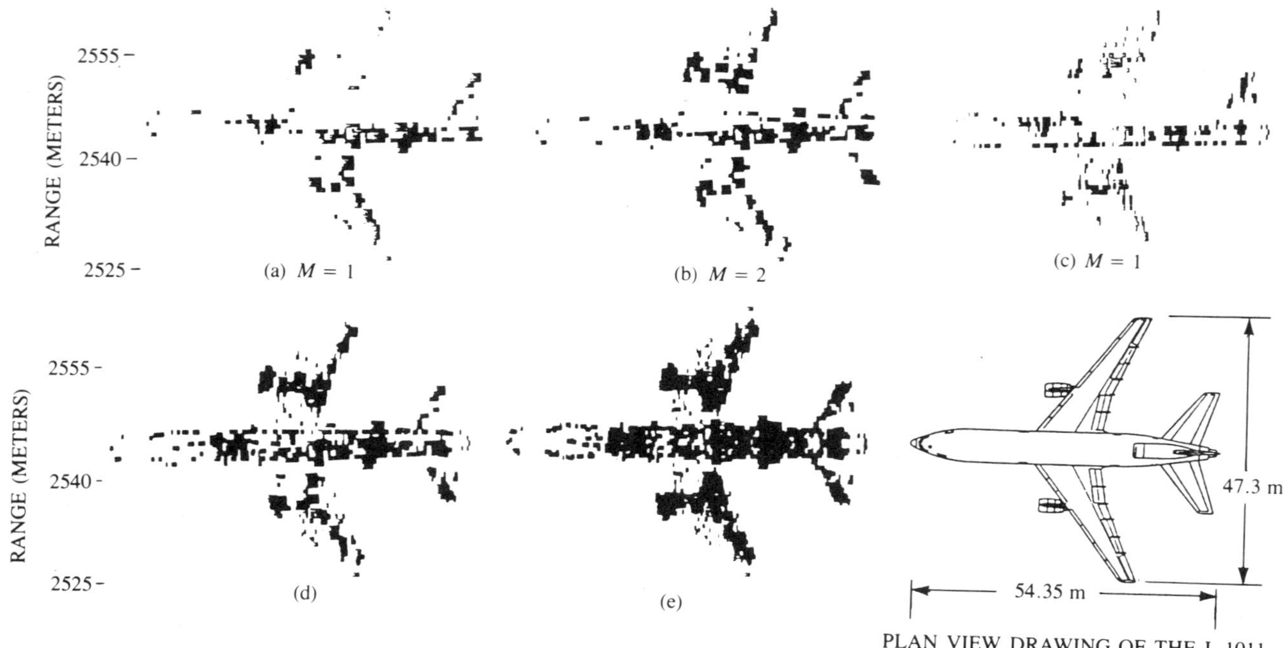

Figure 5.16 Spatial and temporal diversity-combining, showing buildup of the radar image of Fig. 3-11a from 85 echoes received on two receivers separated by 25 m. (a) Image from single receiver. (b) Combined image from the two receivers; one image inverted. (c) Image from the two receivers; data combined coherently; 2:1 resolution improvement. (d) Sum of images (a), (b), and (c). (e) Image (d) plus (d) inverted. (From ref. [102])

radiation field is incoherent [12]. This equation is the Van Cittert–Zernicke theorem. The distinction between the coherent radiation field experienced in microwave imaging and the noncoherent or incoherent field of radio astronomy or passive underwater acoustics is described in Section 1-1 and Fig. 1-2.

The advantage of this form of diversity combining is that the main lobe of the effective pattern is substantially unchanged from that derived from the weight vector of the array, while the effective side radiation pattern is highly smoothed and as a consequence the sidelobe level is greatly reduced. Target artifacts are minimized in number and intensity. The process is not achieved by decorrelating the side radiation patterns of the receiving antenna. It is accomplished instead by destroying the coherent relations between the phases of the echoes radiated from the scattering centers on the target. One importance of this procedure is that the image processing tools developed in radio astronomy now become applicable to terrestrial microwave imaging. An important example is given in Chapter 7.

The decohering procedure is illustrated in Fig. 5-17. Two targets are located at $\bar{r}_1$ and $\bar{r}_2$ relative to some arbitrary origin. A receiving element in the array is located at $\bar{r}_3$. Two transmitters are at $\bar{r}_4$ and $\bar{r}_5$. The targets are designated T_1 and T_2, R is the receiver and X_1 and X_2 are the transmitters.

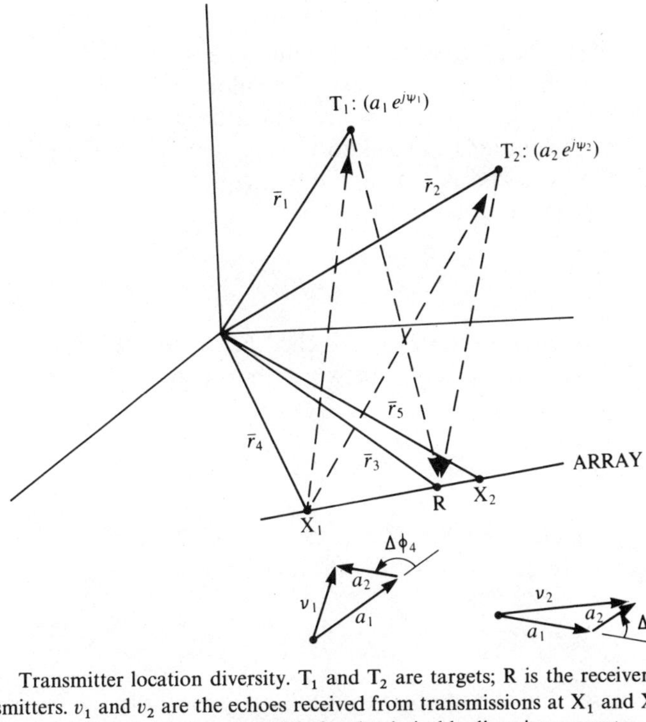

Figure 5-17 Transmitter location diversity. T_1 and T_2 are targets; R is the receiver, and X_1 and X_2 are transmitters. v_1 and v_2 are the echoes received from transmissions at X_1 and X_2. The phase difference between $\Delta\phi_4$ and $\Delta\phi_5$ is responsible for the desirable diversity property.

When X_1 transmits, the echo received at R due to the reflection from T_1 is

$$v_{1,1} = a_1 \exp[-jk(|\bar{r}_4 - \bar{r}_1| + \psi_1 + |\bar{r}_1 - \bar{r}_3|)] \tag{5-18}$$

where the phase is measured relative to the transmitted wave. a_1 is the amplitude of the reflection coefficient of target T_1 and ψ_1 is its phase. The dashed lines from X_1 to T_1 to R illustrate the path. Similarly,

$$v_{1,2} = a_2 \exp[-jk(|\bar{r}_4 - \bar{r}_2| + \psi_2 + |\bar{r}_2 - \bar{r}_3|)] \tag{5-19}$$

is the echo received from T_2. The sum of Eqs. 5-18 and 5-19,

$$v_1 = v_{1,1} + v_{1,2} \tag{5-20}$$

is the total signal at the receiver due to the transmission from X_1.

When X_2 transmits, the received wave has complex amplitude

$$\begin{aligned} v_2 &= v_{2,1} + v_{2,2} \\ &= a_1 \exp[-jk(|\bar{r}_5 - \bar{r}_1| + \psi_1 + |\bar{r}_1 - \bar{r}_3|)] \\ &\qquad + a_2 \exp[-jk(|\bar{r}_5 - \bar{r}_2| + \psi_2 + |\bar{r}_2 - \bar{r}_3|)] \end{aligned} \tag{5-21}$$

The phase of v_2 relative to v_1 has two components. One is due to the distance from the target to the receiver. Its effect is common to all points in the receiving array and therefore is of no consequence. The other is due to the change in the distances from transmitter locations to targets. This alters the phase relations between target echoes and is of material consequence.

The phase difference between the echoes following the first transmission is

$$\Delta\phi_4 = k(|\bar{r}_4 - \bar{r}_1| + \psi_1 + |\bar{r}_1 - \bar{r}_3| - |\bar{r}_4 - \bar{r}_2| - \psi_2 - |\bar{r}_2 - \bar{r}_3|) \tag{5-22}$$

The phasor representing echo v_1 is sketched below the first transmitter X_1 in Fig. 5-17. The phase difference between the echoes following the second transmission is

$$\Delta\phi_5 = k(|\bar{r}_5 - \bar{r}_1| + \psi_1 + |\bar{r}_1 - \bar{r}_3| - |\bar{r}_5 - \bar{r}_2| - \psi_2 - |\bar{r}_2 - \bar{r}_3|) \tag{5-23}$$

The echo v_2 is sketched below X_2 in the figure. The difference between the phase differences (5-22) and (5-23) is the significant quantity:

$$\delta\phi_{4,5} = k(|\bar{r}_4 - \bar{r}_1| - |\bar{r}_5 - \bar{r}_1| - |\bar{r}_4 - \bar{r}_2| + |\bar{r}_5 - \bar{r}_2|) \tag{5-24}$$

The target phases and the receiver location are gone; only the transmitter and target locations remain. The target locations are arbitrary, unknown *a priori* and assumed to be fixed. In the absence of transmitter displacement, target echoes from successive transmissions bear identical phase relations to each other. Equation 5-24 indicates that moving the transmitter between pulses alters the phase relations. This is exactly what happens to the phase relation between signals arriving from sources that radiate independent waveforms, the common case in radio astronomy and in passive underwater acoustics. Thus the set of snapshots of the radiation field sampled by a phased array in response to transmission from different locations has the characteristics of a set of snapshots from the same array that receives signals from noncoherent radiation sources. This alteration of the phases among the target echoes is a necessary condition for achieving a set of snapshots that have the statistical properties of a noncoherent radiative process. It proves to be a sufficient condition as the number of transmitter locations and the size of the transmitter array grow without limit.

Of great practical interest are the statistical properties of a set of snapshots obtained from a finite set of transmitters confined to the receiver array. That is, given an array of N locations of which $K < N$ are transmitter/receiver modules and the remaining $N - K$ are receivers only, to what extent does the set of measurements in the array represent an incoherent radiation field? More specifically, how does the diversity image formed from the set of snapshots compare to an image formed from independent sources?

The theory necessary to answer these questions has been developed in [108], [112] and is presented in Chapter 6.

6

TRANSMITTER-LOCATION DIVERSITY

6-1 INTRODUCTION

Diversity combining of signals is useful for enhancing their desired components at the expense of noise and other artifacts. Element-position, frequency, polarization, axial-symmetry, and aspect-angle diversity were described in Chapter 5. All these forms of diversity were used in their incoherent form; that is, the individual snapshots were squared in magnitude and averaged to form the diversity image. Squaring is a nonlinear process; hence the diversity images are not simple linear convolutions. As shown in Chapter 5, they are the sums of squares of convolutions. This precludes the use of standard image processing techniques for further enhancing the diversity image because many such techniques require a linear convolutional relationship between the object and the image. It is clear from the discussions in Chapter 5 that incoherent diversity techniques by themselves cannot sufficiently enhance image quality without using extremely large bandwidths (in frequency diversity) or large diversity orders. Hence it is logical to attempt to develop diversity techniques that lead to an image that is approximately the convolution of the source distribution with a known transfer function so that the diversity image can be further enhanced using image processing.

Incoherent transmitter location diversity, described later in this chapter, leads to an image that is approximately the convolution of the target scenario's intensity with the power pattern of the array. Thus it satisfies the requirement of many image processing techniques—viz. that the image and object be related via a linear convolution. The convolutional relationship arises because incoherent transmitter diversity alters the phase relationships between the various illuminated targets from snapshot to snapshot. Recall that the phase difference between any two targets stays constant from snapshot to snapshot if the targets

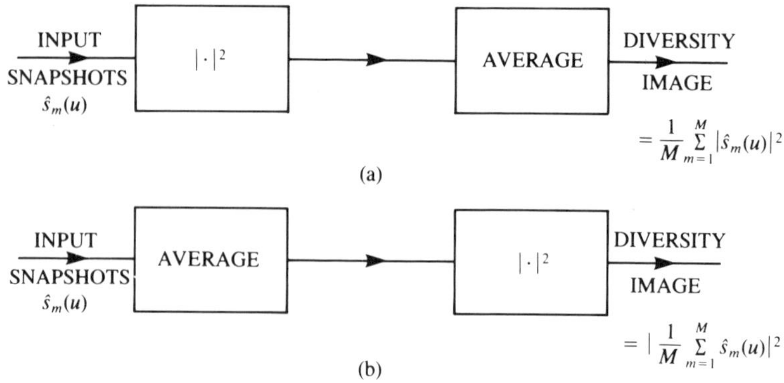

Figure 6-1 Illustrations of the difference between incoherent and coherent diversity. (a) Incoherent diversity-combining. (b) Coherent diversity-combining.

are coherent, and is a uniformly distributed random variable between $-\pi$ and π when the targets are noncoherent [9].⋆ Thus incoherent transmitter location diversity makes the coherent microwave targets appear noncoherent to the receiving array. Noncoherent targets appear commonly in optics, radio astronomy, and passive sonar. Hence many of the image processing techniques developed in those fields can be used to further enhance the incoherent diversity image. The application of the CLEAN algorithm, an image processing technique developed in radio astronomy, to further enhance the diversity image is described in Chapter 7. Incoherent transmitter location diversity in Chapter 9 is used for developing a self-cohering algorithm that is *independent* of the target scenario.

All the diversity techniques discussed previously were incoherent because they averaged the squared magnitudes of the individual snapshots. A second form of diversity, called coherent diversity, adds the complex individual snapshots before squaring in order to obtain the coherent diversity image. The difference between coherent and incoherent diversity is illustrated in Fig. 6-1. The next section develops preliminary notation.

6-2 NOTATION AND ASSUMPTIONS

Consider a linear phased array antenna with N receivers located at $x = x_n$, $1 \leq n \leq N$. Let T targets having amplitudes A_t and phases ψ_t be located in the far field of the receiving array at $u = u_t = \sin \theta_t$, $1 \leq t \leq T$. Let the field of view be illuminated sequentially by M transmitters located at $x = X_m$, $1 \leq m \leq M$. The transmitters and receivers could be collocated; however, it is convenient to

⋆ There is a slight difference between the adjectives *incoherent* and *noncoherent*, as used here. The adjective incoherent is used to qualify the type of diversity and the adjective noncoherent is used to qualify the property of the target scenario (e.g., incoherent diversity and noncoherent targets).

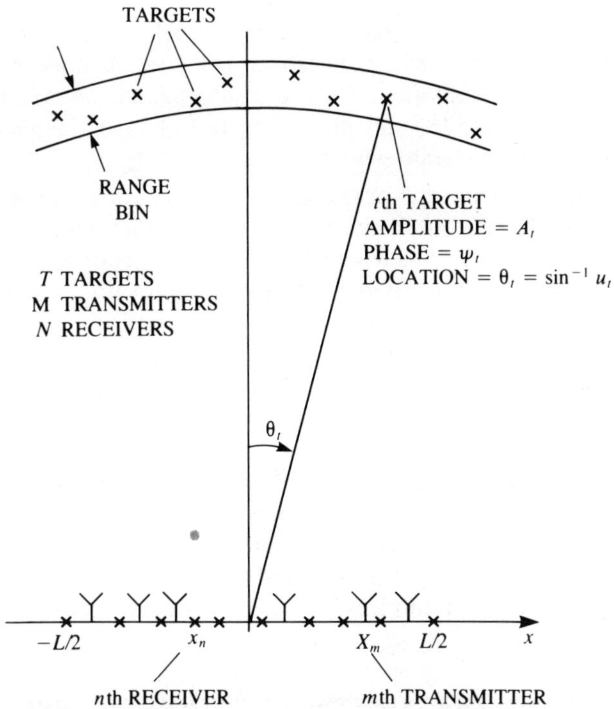

Figure 6-2 Array and target geometry.

differentiate between their locations for the purpose of this discussion. Further, the far-field assumption has no bearing on the results of this chapter; it is made for mathematical simplicity. The relevant geometry is depicted in Fig. 6-2.

The signal at the nth receiver due to a transmission from the mth antenna, denoted $e_{m,n}$, is the superposition of the echoes received from all the targets in the field of view. The phase of the echo received from the tth target consists of the following three terms: (1) a phase corresponding to the distance from the transmitter to the target, (2) a phase corresponding to the distance from the target to the receiver and (3) the phase of the reflection coefficient of the target ψ_t. Under the far-field assumption, the first two terms are ku_tX_m and ku_tx_n, respectively. Hence the echo from the tth target is $A_t \exp\{j[ku_t(x_n + X_m) + \psi_t]\}$ and

$$e_{m,n} = \sum_{t=1}^{T} A_t e^{j[ku_t(X_m + x_n) + \psi_t]} \tag{6-1}$$

Implicit in Eq. 6-1 are three assumptions:

1. The gain patterns of the transmitters and receivers are unity over the field of view.

2. The amplitudes and phases of the target echoes do not change with transmitter location. Many microwave scatterers appear as specular targets over the small angle the array subtends at the target. Thus their backscatter amplitudes and phases do not change significantly due to variation of the transmitter location.
3. The phases of the signals transmitted from each location are known or can be determined from the measured data. In other words, the equivalent of coherent transmissions from the M different locations can be obtained through either sophisticated hardware or innovative signal processing techniques. A novel method for achieving coherent transmissions from different locations in real time is described in [1]. Adaptive beamforming techniques, described in Chapters 8 and 9, can be used to adaptively achieve the equivalent of coherent transmissions.

The assumptions are stronger than are required by the theory, which imposes only the following conditions. Neither the amplitude nor the phase of the transmitted pulses need be known to implement incoherent transmitter location diversity. Only the phase of each transmitted pulse should be known (or inferred from the measured data) to implement coherent transmitter location diversity.

6-3 INCOHERENT TRANSMITTER-LOCATION DIVERSITY

6-3-1 Introduction

This form of diversity averages the squared magnitudes of the images obtained from each of the M transmissions. The estimated image due to the mth transmission, denoted $\hat{s}_m(u)$, is obtained by multiplying the received data $e_{m,n}$ with the Fourier kernel $\exp(-jkux_n)$ and summing:

$$\hat{s}_m(u) = \frac{1}{N} \sum_{n=1}^{N} e_{m,n} e^{-jkux_n} \qquad (6\text{-}2)$$

At this point, it is convenient to separate the phase of the echo received from each of the T targets into two parts. The first part is the sum of the target phase ψ_t and the phase corresponding to the distance of the target from the transmitter ku_tX_m. Call this "overall" phase of the tth target during the mth transmission $\phi_{m,t}$. The second part is the phase corresponding to the distance between the target and the receiver ku_tx_n. Equation 6-1 becomes

$$e_{m,n} = \sum_{t=1}^{T} A_t e^{j[\phi_{m,t} + ku_tx_n]} \qquad (6\text{-}3)$$

where

$$\phi_{m,t} = \psi_t + ku_tX_m \qquad (6\text{-}4)$$

Substitution of Eq. 6-3 into 6-2 leads to

$$\hat{s}_m(u) = \sum_{t=1}^{T} A_t e^{j\phi_{m,t}} H(u - u_t) \tag{6-5}$$

where

$$H(u - u_t) = \frac{1}{N} \sum_{n=1}^{N} e^{-jk(u-u_t)x_n} \tag{6-6}$$

is the radiation pattern of the receiving array steered to $u = u_t$.

The image $\hat{s}_m(u)$ can be recognized to be the convolution of the target scenario as viewed from the mth transmitter, $s_m(u) = \sum_t A_t \exp(j\phi_{m,t})\delta(u - u_t)$ with the radiation pattern of the array $H(u)$. The estimated image due to the mth transmission is the squared magnitude of $\hat{s}_m(u)$:

$$\hat{I}_m(u) = |\hat{s}_m(u)|^2 = \sum_{r=1}^{T} \sum_{t=1}^{T} A_r A_t \operatorname{Re}\{H(u - u_r)H^*(u - u_t)e^{j(\phi_{m,r} - \phi_{m,t})}\} \tag{6-7}$$

where Re$\{\cdot\}$ denotes the real part of a complex number.

The complex image $\hat{s}_m(u)$ is the convolution of $s_m(u)$ with $H(u)$; however, the intensity image $\hat{I}_m(u)$ is not related to the target intensity $I(u) = \sum_t A_t^2 \delta(u - u_t)$ via a linear convolution. The following section demonstrates that averaging M intensity estimates leads to a diversity image that is approximately the linear convolution of the $I(u)$ with the power pattern of the array $|H(u)|^2$. It will be shown that as few as five transmissions are sufficient for achieving this approximation. In order to do so, first note that $\hat{I}_m(u) = \sum_t A_t^2 |H(u - u_t)|^2$ if all the terms in the double summation in Eq. 6-7 are zero for $r \neq t$. Second, note that the $r \neq t$ terms will average out to a small quantity if (1) the phase difference between any two targets $(\phi_{m,r} - \phi_{m,t})$ at the mth transmission is independent of their phase difference at any other transmission, and (2) enough diversity orders are used. The following section evaluates the probability density function (pdf) of the phase difference between any two targets as a function of transmitter location.

6-3-2 Randomization of Target Phase

This section develops a technique that randomizes the phase relationships between the coherent targets by transmitting sequentially from several locations. The transmitter locations X_m are considered to be random variables uniformly distributed between $-L/2$ and $L/2$: such a pdf will be denoted $U(-L/2, L/2)$. This assumption will help derive the statistical properties of the diversity image. It is judged that the results obtained under this assumption will approximate the results for any set of transmitter locations $|X_m| \leqslant L/2$. This is because each set will be a realization of the random process defined by the $U(-L/2, L/2)$ pdf.

The phase of the tth target during the mth transmission $\phi_{m,t}$, defined by Eq. 6-4, is $ku_t X_m + \psi_t$. The phase difference between the rth and tth targets $(\phi_{m,r} - \phi_{m,t})$ is $k(u_r - u_t)X_m + \psi_r - \psi_t$. Because the targets are coherent, their phase difference will remain constant from snapshot to snapshot as long as the transmitter location does not change. However, changing the transmitter location does change the phase difference between the two targets. This section demonstrates that the pdf of the phase difference approaches a $U(-\pi, \pi)$ pdf as the separation between the targets $|u_r - u_t|$ increases. The phase difference between noncoherent targets also has the same pdf; hence transmitter location diversity tends to make the coherent targets appear noncoherent to the receiving array.

The term $\psi_r - \psi_t$ in the phase difference does not depend on transmitter location and hence can be subtracted from $\phi_{m,r} - \phi_{m,t}$ without altering its statistics. Define the random variable

$$\chi_{r,t} = [\phi_{m,r} - \phi_{m,t} - (\psi_r - \psi_t)]_{\text{mod } 2\pi} = [kX_m(u_r - u_t)]_{\text{mod } 2\pi} \qquad \text{(6-8)}$$

which is the phase difference, modulo 2π, between the rth and tth targets. Because X_m has a $U(-L/2, L/2)$ pdf, $kX_m(u_r - u_t)$ has a $U(-kL(u_r - u_t)/2, kL(u_r - u_t)/2)$ pdf. The modulo 2π nature of $\chi_{r,t}$ leads to

$$\Pr\{\chi_{r,t} = \theta\} = \sum_{l=-\infty}^{\infty} \Pr\{k(u_r - u_t)X_m = 2l\pi + \theta\}, \qquad -\pi < \theta \leqslant \pi \quad \text{(6-9)}$$

where $\Pr\{X = a\}$ is the probability that the random variable X equals a. Thus the pdf of $\chi_{r,t}$ can be obtained by translating the pdf of $k(u_r - u_t)X_m$ over intervals separated by 2π to the interval $(-\pi, \pi]$ and superposing all these functions. Let

$$|k(u_r - u_t)L/2| = m\pi + \alpha; \qquad m = 0, 1, 2, \ldots \qquad 0 \leqslant \alpha < \pi \qquad \text{(6-10)}$$

which defines the target separation as $(\alpha/\pi + m)$ beamwidths of the transmitting array. Equations 6-8 to 6-10 can be used to show that the pdf of $\chi_{r,t}$, denoted $P(\chi_{r,t})$, is

$$\begin{aligned} P(\chi_{r,t}) &= \frac{m+1}{2(m\pi + \alpha)}, \qquad 0 \leqslant |\chi_{r,t}| \leqslant \alpha \\ &= \frac{m}{2(m\pi + \alpha)}, \qquad \alpha \leqslant |\chi_{r,t}| < \pi \end{aligned} \qquad \text{(6-11)}$$

This pdf is depicted in Fig. 6-3. Note that the pdf tends to a uniform distribution for large values of m, that is, when the targets are separated by a large number of beamwidths. Hence coherent targets separated by more than

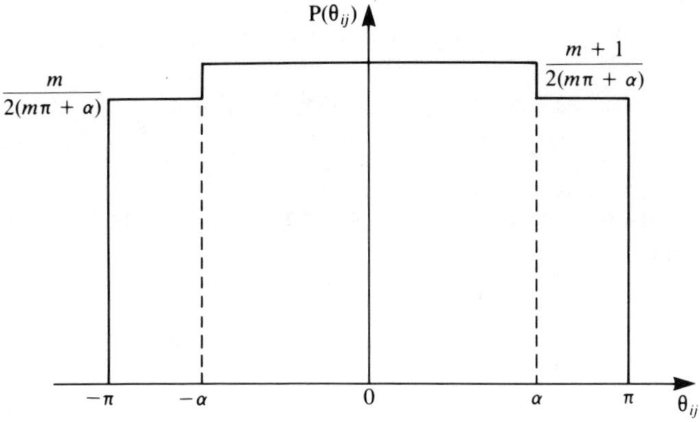

Figure 6-3 The pdf of the phase difference between two coherent targets after application of incoherent transmitter location diversity. The two targets are separated by $(m\pi + \alpha)$ beamwidths, that is, $|(u_r - u_t)| = (m\pi + \alpha)\lambda/L$. (From ref. [108])

a few beamwidths can be made to appear noncoherent to the receiving array. The number of transmissions required to achieve this objective needs to be computed. The next section shows that the diversity image is the sum of the convolution of the target intensity $I(u) = \sum_t A_t^2 \delta(u - u_t)$ with the power pattern of the array $|H(u)|^2$ plus an error term $E(u)$. The statistics of the error term are examined in order to derive a lower bound on the number of transmitters required for achieving incoherence.

6-3-3 The Incoherent Diversity Image

An expression for the incoherent diversity image, which is the average of the intensity estimates due to each of the M transmissions, is derived in this section. The measured data, contaminated by additive noise, are

$$\hat{e}_{m,n} = e_{m,n} + w_{m,n} \tag{6-12}$$

where the $e_{m,n}$ are defined by Eq. 6-1 and the real and imaginary parts of the noise samples $w_{m,n}$ are zero-mean and independent Gaussian random variables each having a variance $\sigma^2/2$, that is,

$$E\{w_{m,n}\} = 0; \qquad E\{w_{m,n} w_{o,p}^*\} = \sigma_n^2 \delta_{m,o} \delta_{n,p} \tag{6-13}$$

where $E\{\cdot\}$ denotes expectation and $\delta_{m,n} = 1$ for $m = n$ and is zero otherwise.*

* The symbol w was used to represent the aperture weighting function in the previous chapters and is used for representing the noise samples in this chapter. This should not lead to any ambiguity because unit aperture weights are assumed throughout this chapter.

The incoherent diversity image, denoted $\hat{I}_D(u)$, is

$$\hat{I}_D(u) = \frac{1}{M} \sum_{m=1}^{M} |\hat{s}_m(u)|^2 = \frac{1}{M} \sum_{m=1}^{M} \left| \frac{1}{N} \sum_{n=1}^{N} \hat{e}_{m,n} e^{-jkux_n} \right|^2 \tag{6-14}$$

Substituting Eqs. 6-3 and 6-12 into 6-14 yields the following expression for the diversity image:

$$\hat{I}_D(u) = \sum_{m=1}^{M} \hat{I}_m(u) = I_{IC}(u) + E(u) \tag{6-15}$$

where

$$I_{IC}(u) = \sum_{t=1}^{T} A_t^2 |H(u - u_t)|^2 = I(u) * |H(u)|^2 \tag{6-16a}$$

and

$$E(u) = E_{s,s}(u) + E_{s,n}(u) + E_{n,n}(u) \tag{6-16b}$$

In other words, the diversity image differs from the convolution of the target intensity $I(u) = \sum_t A_t^2 \delta(u - u_t)$ with the power pattern of the array $|H(u)|^2$ by an error term $E(u)$. The following section shows that the magnitude of the error term becomes small compared to the mean value of $I_{IC}(u)$ when as few as five transmissions are used for decorrelating the target phases. It is useful to develop an intuitive feeling for the two components of Eq. 6-15 before proceeding to the next section. The first term $I_{IC}(u)$ is the image that would have been obtained had all the targets been noncoherent. The error term $E(u)$ consists of three components. The first one, $E_{s,s}(u)$, arises because the pdf of the differences between the target phases after application of transmitter-location diversity differs from a $U(-\pi, \pi)$ pdf by a small amount (see Section 6-3-2 and Fig. 6-3). The second term, $E_{s,n}(u)$, arises due to mixing of the noise-free data $e_{m,n}$ with the noise samples $w_{m,n}$, and the third term, $E_{n,n}(u)$, arises due to the noise samples alone. The three components of $E(u)$ are enumerated below:

$$E_{s,s}(u) = \sum_{r=1}^{T} \sum_{t=1_{t \neq r}}^{T} A_r A_t H(u - u_r) H^*(u - u_t) \left(\frac{1}{M} \sum_{m=1}^{M} e^{j(\phi_{m,r} - \phi_{m,t})} \right) \tag{6-17a}$$

$$E_{s,n}(u) = \frac{1}{N^2 M} \sum_{m=1}^{M} \sum_{n=1}^{N} \sum_{l=1}^{N} e_{m,n} w_{m,l}^* e^{-jku(x_n - x_l)} \tag{6-17b}$$

$$E_{n,n}(u) = \frac{1}{N^2 M} \sum_{m=1}^{M} \left| \sum_{n=1}^{N} w_{m,n} e^{-jkux_n} \right|^2 \tag{6-17c}$$

The last two terms arise due to the additive noise in the measured data, and would have been present even if the targets were noncoherent. The first term $E_{s,s}(u)$ does not depend on the noise samples. Had the targets been noncoherent, the mean value of each of the terms in the summation over m in Eq. 6-17a, that is, $\exp(j(\phi_{m,r} - \phi_{m,t}))$ would have been zero because $\phi_{m,r} - \phi_{m,t}$ would have a $U(-\pi, \pi)$ pdf. Hence the mean value of $E_{s,s}(u)$ would have been zero for noncoherent targets. However, incoherent transmitter location diversity leads to a pdf that is close to but not exactly equal to $U(-\pi, \pi)$; hence the mean value of $E_{s,s}(u)$ will not be zero when transmitter diversity is used for decorrelating coherent targets. The following section uses results from [108] in order to show that both the mean and standard deviation of $E_{s,s}(u)$ become small compared to the mean of $I(u) * |H(u)|^2$ when as few as five transmissions are used for decorrelating the coherent targets.

6-3-4 Comparison of the Incoherent Diversity Image with the Image of Noncoherent Targets

The object of using incoherent transmitter location diversity is to make the diversity image approximate the convolution (6.16a) of the true target intensity with the power pattern of the array. Clearly, the difference $\hat{I}_D(u) - I_{IC}(u) = E(u)$ will depend on the noise power and the number of transmitters. The closeness of the diversity image to the noncoherent image $I_{IC}(u)$ can be measured in several ways; this section uses a statistical measure to compute the magnitude of $E(u)$. An expression for the number of transmitters required for making both the mean and standard deviation of $E(u)$ smaller than the mean value of $I_{IC}(u)$ for high confidence levels is derived in this section. Results from [108] will be used to accomplish this task.

The following discussion assumes the receiving arrays to be thinned and random. First the image is divided into two regions. The first region, called the *target region*, consists of those values of u that are close to the target locations u_t. The second region, called the *artifact region*, consists of *all image points that are at least three beamwidths away from all targets*. The reason for choosing the artifact region in such a fashion is that the first few lobes in the radiation pattern of thinned random arrays are large and follow a predictable pattern that depends mainly on array size and aperture taper [94]. Therefore the image close to the target locations will closely follow $|H(u)|^2$ scaled by the target intensity, and will hardly be affected by the presence of other targets located several beamwidths away. Hence both the diversity and incoherent images will closely follow the shape of $A_t^2|H(u - u_t)|^2$ close to $u = u_t$. However, the image points that are not close to any target location will contain energy due to leakage of all targets through the sidelobes of the radiation pattern; hence the artifact region is defined as stated previously.

It can be shown that the mean of both $\hat{I}_D(u)$ and $I_{IC}(u)$ for u near the target locations u_t is equal to the target intensities A_t^2 [108]. Further, the standard deviation of both images at $u = u_t$ is approximately $[A_t^2(\sum_{r \neq t} A_r^2)/MN]^{0.5}$, which is much smaller than the mean value A_t^2. Hence both the diversity and incoherent images can be considered to be nonrandom at the target locations having the same mean value A_t^2. In other words, the magnitude of $E(u)$ will be much smaller than the value of $I_{IC}(u)$ close to the target locations irrespective of the number of transmitters used. This result is derived in [108].

The following paragraphs show that the magnitude of $E(u)$ will be much smaller than the mean of $I_{IC}(u)$ in the artifact region when at least five transmitters are used. Note that $I_{IC}(u)$ is the convolution of the target intensity with the power pattern of the array; hence its mean value in the artifact region is the product of the total target power and the average sidelobe level of the array $1/N$. That is, the mean of $I_{IC}(u)$ in the artifact region, denoted μ_{IC}, is

$$\mu_{IC} = \frac{1}{N} \sum_{t=1}^{T} A_t^2 \tag{6-18}$$

The mean and variance of the error term $E(u)$ in the artifact region, denoted μ_E and σ_E^2, are [108]

$$\mu_E = \frac{\sigma_n^2}{N} \tag{6-19a}$$

$$\sigma_E^2 = \frac{1}{N^2} \sum_{r=1}^{T} \sum_{t=1_{t \neq r}}^{T} A_r^2 A_t^2 \left(\frac{1}{M} + \frac{M-1}{M} \operatorname{sinc}^2\left(\frac{\pi L(u_r - u_t)}{\lambda} \right) \right) + \frac{\sigma_n^4}{N^2 M} \tag{6-19b}$$

where $\operatorname{sinc}(x) = \sin x/x$.

Note that the mean value of $E(u)$ depends only on the noise power σ_n^2 and not on the number of transmitters. Consider the two terms in the double summation in Eq. 6-19b, viz. $1/M$ and $(M-1)\operatorname{sinc}^2(\pi L(u_r - u_t)/\lambda)/M$. For small M, $1/M \gg \operatorname{sinc}^2(\cdot)$; hence σ_E^2 decreases with increasing M when M is small. For large M, $\operatorname{sinc}^2(\cdot)$ will dominate $1/M$; hence, the variance of $E(u)$ will not decrease appreciably if the number of transmitters is increased beyond a certain value. This implies that even an infinite number of transmissions cannot completely decorrelate the coherent targets. This can be inferred from the deviation of the pdf of the phase difference between the targets (see Fig. 6-3) from $U(-\pi, \pi)$. For small values of M, σ_E^2 can be approximated as

$$\sigma_E^2 \simeq \frac{(\sum_{t=1}^{T} A_t^2)^2 + \sigma_n^4}{(N^2 M)} \tag{6-20}$$

The value of σ_E^2 will be small when $|E(u)|$ is small. The value of M for which σ_E^2 is small can be obtained by constraining $|E(u)|$ to be smaller than the mean μ_{IC} of $I_{IC}(u)$ with very high probability. This can be accomplished if the pdf of $E(u)$ is known. Results from [108] indicate that the pdf of $E(u)$ is approximately Gaussian with mean and variance given by Eqs. 6-19a and 6-20, respectively. For such a pdf, the probability of $|E(u)|$ exceeding the mean of $I_{IC}(u)$ $(= \mu_{IC})$, denoted $\Pr\{|E(u)| \geqslant \mu_{IC}\}$, is

$$\Pr\left\{|E(u)| \geqslant \frac{\sum A_t^2}{N}\right\} = 2\,\text{erfc}_* \frac{\sum A_t^2/N - \mu_E}{\sigma_E} \tag{6-21}$$

where

$$\text{erfc}_*(a) = \frac{1}{\sqrt{2\pi}} \int_a^\infty e^{-x^2/2}\, dx$$

Constraining $|E(u)|$ to be smaller than μ_{IC} with a high confidence level will make $\hat{I}_D(u)$ approximate $I_{IC}(u)$ for most values of u in the artifact region, especially those u for which $I_{IC}(u) \gg \mu_{IC}$, that is, especially at the peak artifacts. This constraint operation can be accomplished for a confidence level of 0.95 by setting $2\,\text{erfc}_*((\mu_{IC} - \mu_E)/\sigma_E) \leqslant 0.05$. From the tables of the Gaussian distribution, $2\,\text{erfc}_*(2) \simeq 0.05$; hence, constraining $\Pr\{|E(u)| \geqslant \mu_{IC}\}$ entails

$$\frac{\mu_{IC} - \mu_E}{\sigma_E} \geqslant 2 \tag{6-22}$$

Assuming that the signal power $\sum A_t^2$ is much larger than the noise power σ_n^2 and substituting Eqs. 6-18, 6-19a, and 6-20 into Eq. 6-22 yields

$$M \geqslant 4\left(\frac{2\sigma_n^2}{\sum A_t^2} + 1\right) \tag{6-23}$$

Equation 6-23 is an important result. It states that as few as five transmitters are sufficient for making the magnitude of the difference between the diversity and incoherent images smaller than the mean value of the incoherent image with a confidence level of 0.95. In other words, it is adequate to use five transmissions and expect the diversity image to closely resemble the convolution of the true target intensity with the power pattern of the array. The following section uses computer simulations to verify this result.

6-3-5 Computer Simulations

In the first simulation, six targets of equal strengths were placed at $u = -0.70$, -0.44, -0.10, 0.2, 0.46, and 0.79, respectively. Twenty-five receivers and seven transmitters, located randomly between -12.5λ and 12.5λ, were used for imaging this target scenario. Figure 6.4a is the image of the target scenario obtained using one transmission (solid line). This first-order diversity (no diversity) image is overlaid by the image of the target scenario obtained by assuming that the targets were noncoherent (dashed line). In other words the dashed line represents $I_{IC}(u)$ defined by Eq. 6-16a. Note that the difference between the two images is substantial. Figures 6-4b, 6-4c, and 6-4d show the third-, fifth-, and seventh-order diversity images overlaying $I_{IC}(u)$ (which is independent of the number of transmissions). The diversity image resembles $I_{IC}(u)$ more closely as the number of transmitters increases, although there is not much difference between the fifth- and seventh-order diversity images. This result was predicted by Eq. 6-23.

The next computer simulation quantifies the difference between $I_D(u)$ and $I_{IC}(u)$ as a function of the number of transmissions. A correlation metric, proposed in [101], will be used for comparison purposes. This metric has been successfully used in the microwave imaging context in order to determine the closeness between different images of the same target scenario (Chapter 10). It is equal to the correlation between the two sample images normalized by the product of the energy in the two images, that is,

$$\rho = \frac{\int \hat{I}_D(u) I_{IC}(u)\, du}{[(\int \hat{I}_D^2(u)\, du)(\int I_{IC}(u)\, du)]^{0.5}} \tag{6-24}$$

Clearly $0 \leqslant \rho \leqslant 1$. Empirical results from [101] indicate that the two images being compared will appear to be almost identical when $\rho \geqslant 0.9$.

In the following computer simulation, 50 coherent targets were uniformly distributed in the region $|u| \leqslant 0.5$. Their amplitudes were independent and identically distributed (iid) random variables having a Rayleigh pdf, and their phases were iid $U(-\pi, \pi)$ random variables. This target scenario was in the far field of a 200-wavelength, 50-element thinned random receiving array. Twenty transmitters were also distributed randomly over 200 wavelengths in order to compute the diversity image for different diversity orders. Figure 6-5 shows the correlation between the diversity and incoherent images, evaluated according to Eq. 6-24, for different diversity orders.

Observe that the correlation between the diversity and incoherent images increases rapidly for small diversity orders and stays almost constant thereafter. This behavior was predicted in Section 6-3-4.

Figure 6-4a Comparison of the first-order incoherent transmitter-location diversity image (no diversity) with the incoherent image. Six targets were located at $u = -0.70, -0.44, -0.10, 0.2, 0.46$, and 0.79, respectively. A 15-element receiving array distributed over 25 wavelengths was used for imaging purposes. The solid line represents the first-order diversity image and the dashed line the incoherent image. Note that the two images are substantially different.

Figure 6-4b Comparison of the third-order incoherent transmitter-location diversity image (solid line) with the incoherent image (dashed line). Target and receiver geometry are the same as in Fig. 6-4a. Three transmitters, distributed randomly over 25λ, were used for computing the diversity image. Note that the third-order diversity image resembles the incoherent image more closely than the first-order diversity image.

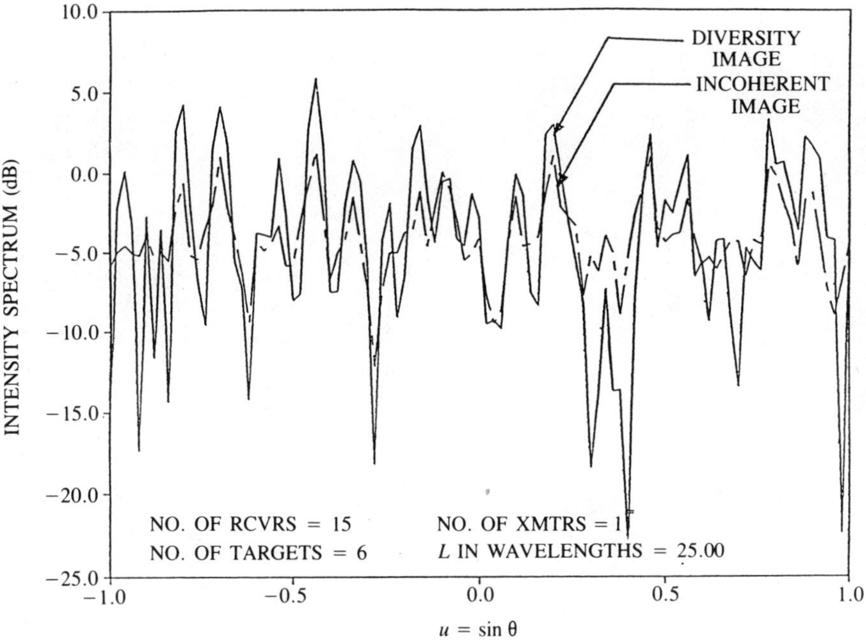

DIVERSITY IMAGE
INCOHERENT IMAGE
INTENSITY SPECTRUM (dB)
10.0
5.0
0.0
−5.0
−10.0
−15.0
−20.0
−25.0
NO. OF RCVRS = 15
NO. OF TARGETS = 6
NO. OF XMTRS = 1
L IN WAVELENGTHS = 25.00
−1.0
−0.5
0.0
0.5
1.0
u = sin θ

(a)

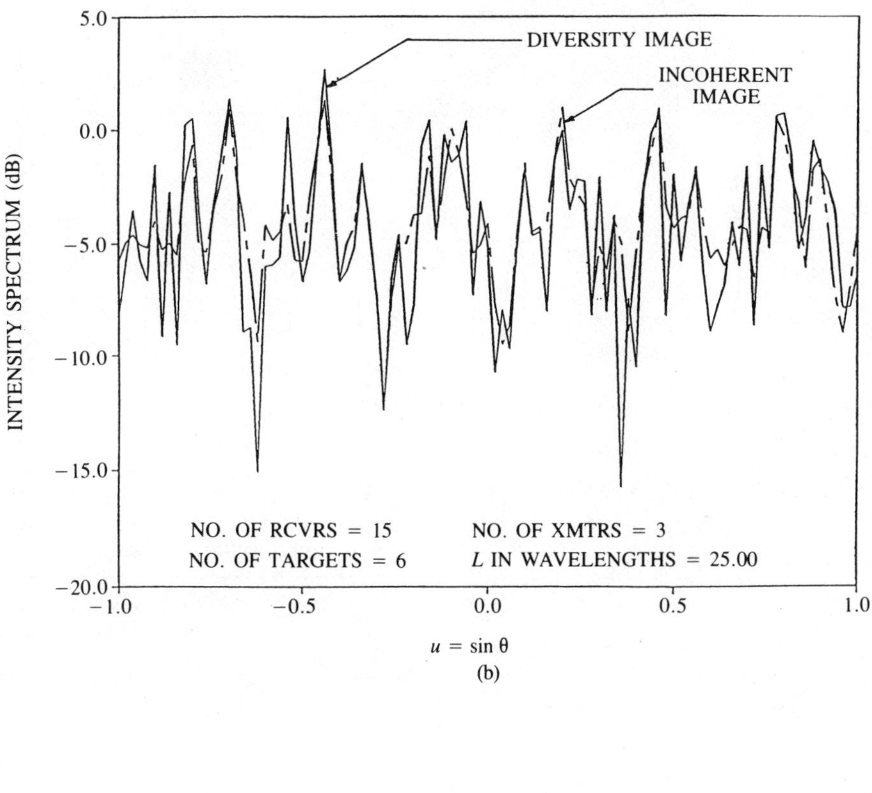

DIVERSITY IMAGE
INCOHERENT IMAGE
INTENSITY SPECTRUM (dB)
5.0
0.0
−5.0
−10.0
−15.0
−20.0
NO. OF RCVRS = 15
NO. OF TARGETS = 6
NO. OF XMTRS = 3
L IN WAVELENGTHS = 25.00
−1.0
−0.5
0.0
0.5
1.0
u = sin θ

(b)

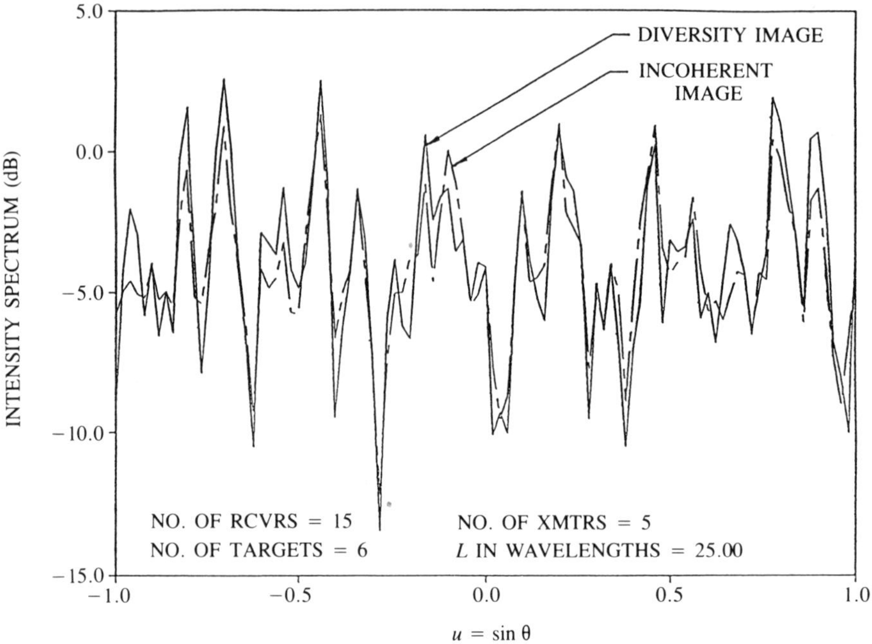
DIVERSITY IMAGE
INCOHERENT IMAGE
INTENSITY SPECTRUM (dB)
5.0
0.0
−5.0
−10.0
−15.0
NO. OF RCVRS = 15
NO. OF TARGETS = 6
NO. OF XMTRS = 5
L IN WAVELENGTHS = 25.00
−1.0
−0.5
0.0
0.5
1.0
u = sin θ

(c)

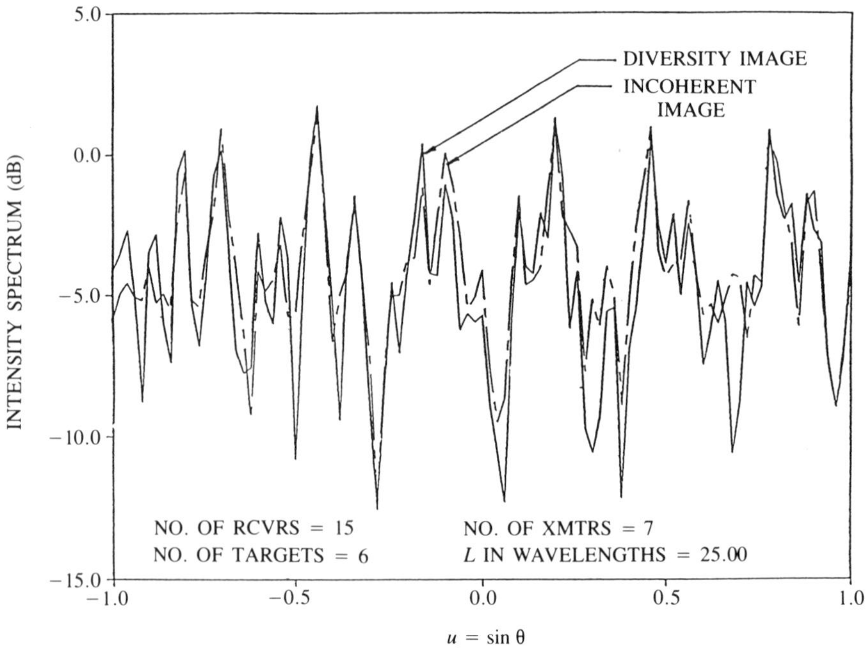
DIVERSITY IMAGE
INCOHERENT IMAGE
INTENSITY SPECTRUM (dB)
5.0
0.0
−5.0
−10.0
−15.0
NO. OF RCVRS = 15
NO. OF TARGETS = 6
NO. OF XMTRS = 7
L IN WAVELENGTHS = 25.00
−1.0
−0.5
0.0
0.5
1.0
u = sin θ

(d)

Figure 6-4c Comparison of the fifth-order incoherent transmitter-location diversity image (solid line) with the incoherent image (dashed line). Target and receiver geometry are the same as in Fig. 6-4a. Five transmitters, distributed randomly over 25λ, were used for computing the diversity image. Note that the fifth-order diversity image resembles the incoherent image more closely than the third-order diversity image.

Figure 6-4d Comparison of the seventh-order incoherent transmitter-location diversity image (solid line) with the incoherent image (dashed line). Target and receiver geometry are the same as in Fig. 6-4a. Seven transmitters, distributed randomly over 25λ, were used for computing the diversity image. Note that the fifth-order and seventh-order diversity images are similar.

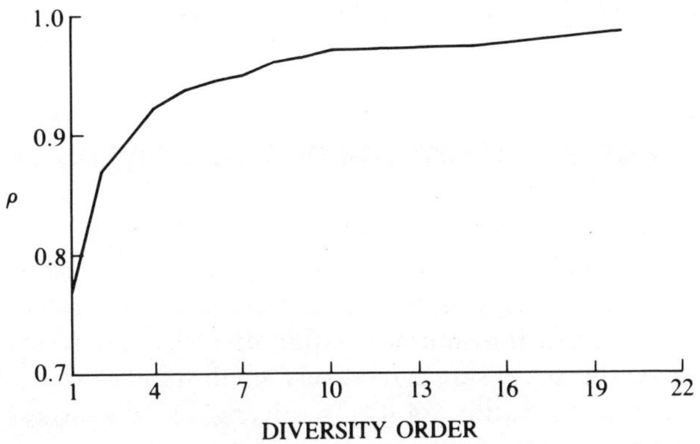

Figure 6-5 Correlation between the diversity and incoherent images as a function of diversity order. Fifty targets having random amplitudes and phases were placed in the region $|u| \leq 0.5$. Twenty transmitters and 50 receivers were used for computing the diversity images. Note that the correlation between the diversity and incoherent images increases rapidly for small diversity orders and stays almost constant thereafter.

6-3-6 Discussion of Assumptions

Section 6-2 described three assumptions made while deriving the relationship between the measured data with the target scenario and the transmitter and receiver locations. This subsection demonstrates that none were necessary for decorrelating coherent targets using incoherent transmitter location diversity; they were made only to reduce mathematical complexity.

The assumptions were: (1) the gain patterns of the transmitters and receivers are unity over the field of view, (2) the amplitudes and phases of the target echoes are independent of transmitter location, and (3) the phase of the signal transmitted from each location is known.

Distortions in the amplitude patterns of the antennas do not pose a problem because they do not prevent coherent addition of target echoes. However, phase distortion of the elemental radiation patterns does prevent coherent addition of target echoes; hence this discussion will be entirely focused on phase distortion and will ignore amplitude distortion. Note that incoherent transmitter location diversity was used to change the phase relationships between the targets from one snapshot to another in order to decorrelate the coherent targets. Therefore, any change in the phase relationships between the targets that occurs due to reasons other than a change in the transmitter location will only help in the process of decorrelation, not weaken it. Thus incoherent transmitter location diversity does not require (1) transmitters with identical radiation patterns, (2) a small transmitting array (so that the targets have a reflection coefficient that is independent of transmitter location), and (3) coherent transmissions from each location. In other words, the assumptions made in Section 6-2 were necessary only to simplify the development of the theory of incoherent transmitter location diversity, they are not necessary for its application.

6-4 COHERENT TRANSMITTER-LOCATION DIVERSITY

This section describes a method for coherently combining the images obtained by sequential transmissions from different locations. In particular, it is shown that a MN-element array can be synthesized using M transmitters and N receivers using coherent transmitter location diversity. Thus very low sidelobe arrays can be synthesized using a relatively small number of transmitters and receivers. Further, the locations of the "receivers" in the synthesized array will not overlap when the transmitter and receiver locations are random; hence, filled aperiodic arrays can be synthesized from thinned arrays using coherent transmitter-location diversity. The technique has been described in [105], [114]. Its application to multi-dimensional arrays is described in [41].

When the three assumptions of Section 6-2 hold, the measured data $e_{m,n}$ are related to the source and array parameters according to Eq. 6-1, that is,

$$e_{m,n} = \sum_{t=1}^{T} A_t e^{j[ku_t(X_m + x_n) + \psi_t]} \tag{6-25}$$

Inspection of Eq. 6-25 reveals that the target locations u_t are related to the transmitter and receiver locations X_m and x_n by the kernel $\exp(jku_t(X_m + x_n))$. Hence an inverse transform that uses the kernel $\exp(-jku(X_m + x_n))$ will yield an estimate of the target scenario that combines all the MN data samples coherently, that is,

$$\hat{s}(u) = \frac{1}{MN} \sum_{m=1}^{M} \sum_{n=1}^{N} e_{m,n} e^{-jku(X_m + x_n)} \tag{6-26}$$

Substitution of Eq. 6-25 into 6-26 yields

$$\hat{s}(u) = \sum_{t = A_t e^{j\psi_t}}^{T} \left\{ \frac{1}{MN} \sum_{m=1}^{M} \sum_{n=1}^{N} e^{-jk(u - u_t)(X_m + x_n)} \right\} = s(u) * H(u) \tag{6-27}$$

where $*$ denotes convolution, and $s(u)$ and $H(u)$ represent the target scenario and the radiation pattern of the synthesized array respectively, that is,

$$s(u) = \sum_{t=1}^{T} A_t e^{j\psi_t} \delta(u - u_t) \tag{6-28}$$

and

$$\begin{aligned} H(u) &= \frac{1}{MN} \sum_{m=1}^{M} \sum_{n=1}^{N} e^{-jku(X_m + x_n)} \\ &= H_R(u) H_T(u) = \left(\frac{1}{N} \sum_{n=1}^{N} e^{-jkux_n} \right) \left(\frac{1}{M} \sum_{m=1}^{M} e^{-jkuX_m} \right) \end{aligned} \tag{6-29}$$

In other words, the radiation pattern of the synthesized array is the product of the radiation patterns of the transmit and receive arrays. Further, note that the radiation pattern $H(u)$ is the same as the radiation pattern of a MN-element array whose receivers are located at $(X_m + x_n)$, $1 \leqslant m \leqslant M$, $1 \leqslant n \leqslant N$. In other words, coherent transmitter-location diversity synthesizes a MN-element array using M transmitters and N receivers.

If the transmitting and receiving arrays are periodic and collocated with $X_n = x_n = nd$, $1 \leqslant n \leqslant N$, the length of the synthesized array is $2Nd$. Further, the location of many synthesized elements overlap because both the transmitters and receivers occur on a periodic grid. A little thought will show that there are N synthesized elements for $X_m + x_n = (N + 1)d$, and that there is only one synthesized element at $X_m + x_n = 2d$ and $X_m + x_n = 2Nd$. Hence, for periodic and collocated transmit and receive arrays, coherent transmitter-location diversity synthesizes a $(2N - 1)$-element array with a triangular aperture weighting function. Further, the radiation pattern of the synthesized array is the square of the radiation pattern of the periodic array. Thus, for periodic arrays, coherent transmitter diversity improves resolution (by a factor of $\sqrt{2}$) and decreases sidelobe levels by squaring the radiation pattern of the array. This improvement is illustrated by the following computer simulation.

The solid line in Fig. 6-6 shows the radiation pattern of a 16-element periodic array with an interelement spacing of $\lambda/2$ obtained using a single transmitter and uniform weighting. Note that the first sidelobe is at -13 dB as expected. The dashed line shows the pattern of the array synthesized by sequentially transmitting from each element and combining the received data according to Eq. 6-27. Note that the peak sidelobe has decreased by 13 dB and the average sidelobe level by about 30 dB. Further improvement in the sidelobe levels can be obtained using nonuniform weights on transmit and receive.

Coherent transmitter diversity also improves resolution and sidelobe levels in the images obtained using thinned random transmit and receive arrays. The improvement in resolution occurs because the length of the synthesized array is $L_t + L_r$ for a receive array of length L_r and a transmit array of length L_t. In order to see the improvement in the sidelobe levels, assume that the thinning

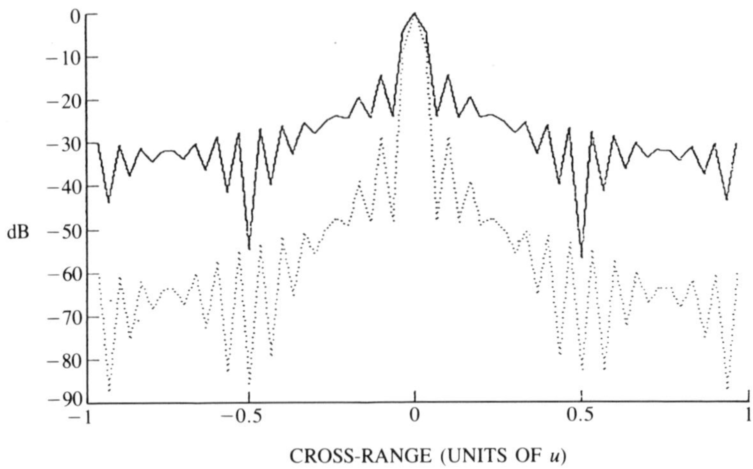

Figure 6-6 Illustration of improvement in sidelobe structure obtained using coherent transmitter-location diversity with periodic arrays. The solid line represents the pattern of a 16-element filled periodic array. The dotted line is the pattern of the array synthesized using coherent transmitter-location diversity.

ratio of the receive array $2L_r/(N\lambda)$ is much larger than unity (a filled periodic array has a thinning ratio of unity). The locations of the element in the synthesized array are $(X_m + x_n)$, $1 \leqslant m \leqslant M$, $1 \leqslant n \leqslant N$. Clearly, two of the synthesized elements are unlikely to be at the same location if both X_m and x_n are randomly distributed over the transmit and receive apertures. Thus the pdf of the MN synthesized elements is approximately trapezoidal in the region $0 \leqslant x \leqslant L_t + L_r$ with a peak at $(L_t + L_r)/2$. The thinning ratio of the synthesized array is therefore $2(L_t + L_r)/(MN\lambda)$, which is $(1 + L_t/L_r)/M$ times smaller than the thinning ratio of the receiving array. For $L_t = L_r$, the density of the elements due to coherent transmitter diversity is increased by a factor of $M/2$, where M is the number of transmitters. Hence, coherent transmitter-location diversity can be used to fill thinned random arrays, thereby causing a tremendous improvement in the sidelobe structure of the imaging system. This improvement is illustrated by the following computer simulation.

The solid line in Fig. 6-7 shows the radiation pattern of a 15-element thinned random array of length 14λ. As expected, this radiation pattern has a large peak sidelobe (approximately −4 dB). The dashed line shows the pattern of the array synthesized by sequentially transmitting from 4 of the 15 elements and combining the received data according to Eq. 6-27. Note the much improved sidelobe structure and the increase in angular resolution.

In conclusion, both coherent and incoherent transmitter-location diversity can be used for suppressing artifacts in microwave images.

Incoherent diversity leads to a small improvement in the artifact level; however, its primary benefits are (a) the incoherent transmitter-location

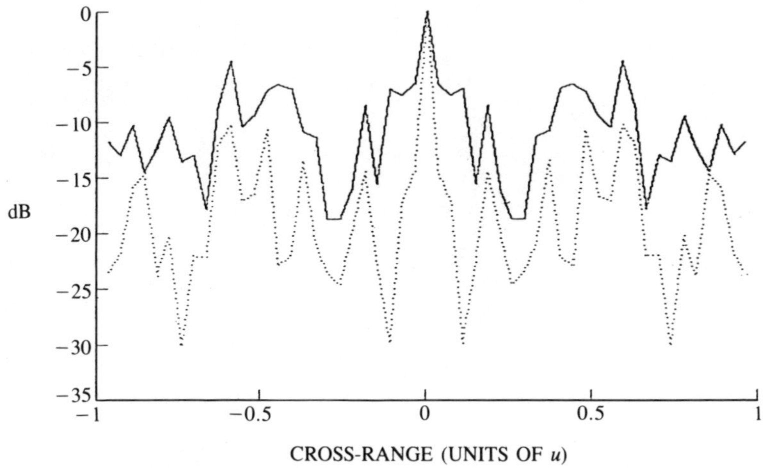

Figure 6-7 Illustration of improvement in sidelobe structure obtained using coherent transmitter-location diversity with thinned random arrays. The solid line represents the pattern of a 15-element thinned random array of length 14λ. The dotted line is the pattern of the array synthesized using fourth-order coherent transmitter-location diversity.

diversity image is the linear convolution of the actual intensity spectrum with the power pattern of the array, and (b) incoherent transmitter-location diversity makes coherent targets appear noncoherent to the receiving array. The first property can be exploited to suppress sidelobe artifacts using image processing techniques, as described in the following chapter. The second property of incoherent transmitter-location diversity is used in Chapter 9 for developing a scene-independent adaptive self-cohering algorithm.

Coherent transmitter diversity synthesizes an array whose radiation pattern is the product of the radiation patterns of the transmitting and receiving arrays. This property of coherent transmitter diversity can be used to obtain dramatic improvement in the dynamic range of microwave images.

7

DECONVOLUTION

7-1 INVERSE FILTERING

Diversity techniques described in Chapter 5 improve image quality by increasing target mass, reducing speckle and reducing image artifacts due to the high sidelobes of the radiation pattern. Diversity is an additive process and therefore is quite stable. Cancellation or neutralization processes can also be used to improve image quality. They offer greater potential reward but usually require a delicate balance between signal or image components being canceled or balanced against each other.

The earliest deconvolution procedure was inverse filtering. When a signal $x(t)$ is passed through a filter having response $h(t)$, the output $y(t)$ is the convolution $y = x * h$. Convolution is a low-pass or smearing process; consequently the features in the input waveform $x(t)$ can easily be lost. In principle, inverse filtering can restore the waveform characteristics of a wave that has been distorted by a linear process. Its measure is a fidelity criterion; this is in contrast, for example, to the maximization of signal-to-noise ratio, which is the objective in matched filter design practice [115].

The Laplace transform of $h(t)$ is the transfer function of the linear filter or system or, more simply, the system function. Its conventional designation is $H = H(s) = H(\sigma + j\omega)$. The inverse filter is H^{-1}. It is placed in cascade with the distorting filter H. The overall system function is their product, which is unity. Thus the impulse response of the combination is an impulse, and signals pass through distortion-free.

Inverse filtering suffers from two serious drawbacks, however: instability and noise. Both severely limit the realm of application of this simple procedure. The system function is the ratio of two polynomials in s, that is, $H(s) = N(s)/D(s)$.

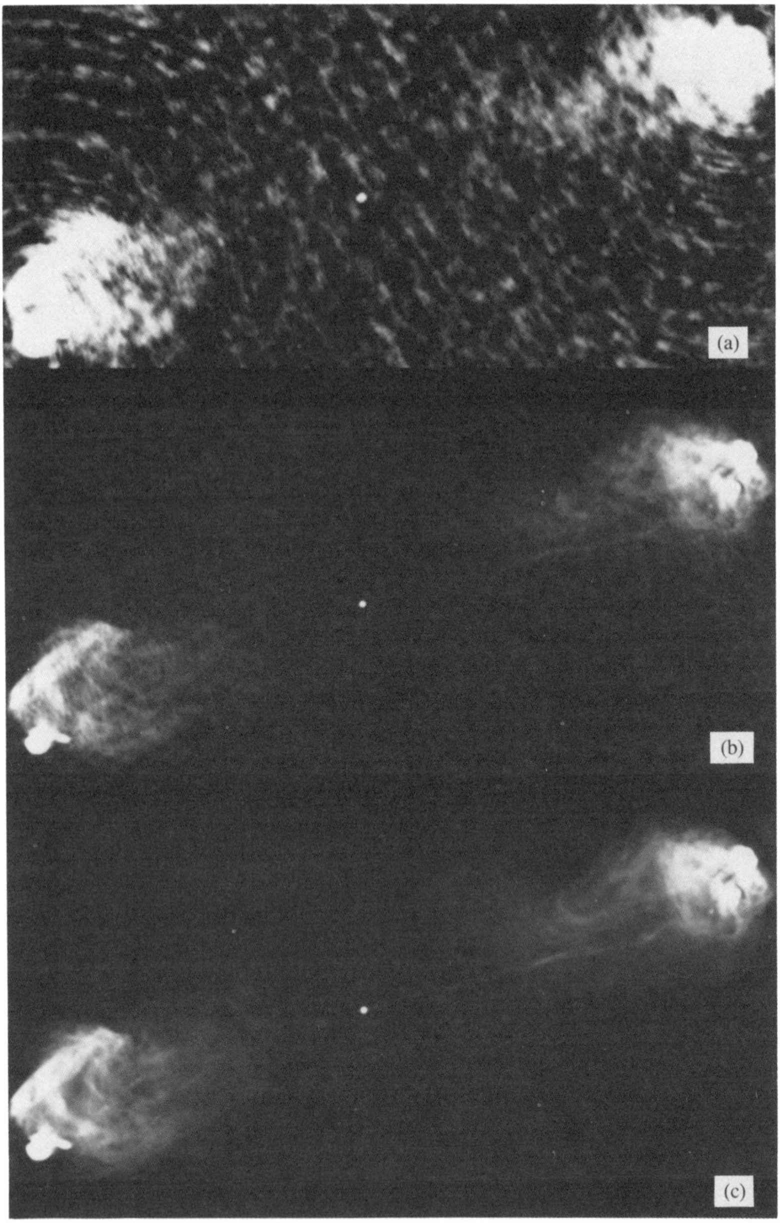
(a)
(b)
(c)

Reciprocating a transfer function interchanges the roots of the numerator and denominator polynomials, thereby turning poles into zeros and vice versa. For a network to be stable, all poles must lie in the left half s-plane ($\sigma < 0$). Thus all the zeros of $D(s)$ are in the left half plane. There is no such restriction for the zeros of the numerator polynomial, however. Consequently, some of the zeros of $N(s)$ can be in the right half s-plane: these zeros become poles in H^{-1} and make the inverse filter unstable.

A partially successful procedure for which experiments have been conducted on HF ionospheric circuits decomposes H into the product $H_1 H_2$. The first factor contains no zeros on the imaginary axis and on the right half s-plane. Its inverse H_1^{-1} is stable. The cascade of H and H_1^{-1} is $HH_1^{-1} = H_2 H_1 H_1^{-1} = H_2$. Provided that the transfer function H_2 induces less distortion than the original system function H, the fidelity of the output wave is superior.

The second problem with inverse filtering is noise. Imagine that H has nulls in the neighborhoods of frequencies ω_i. When the signal arrives at the receiver, its frequency components in these neighborhoods are weak. It is necessary to restore them to full strength to recreate a facsimile of the original signal. This is exactly what the inverse filter does. By cascading H with H^{-1}, it exposes the signal components that have undergone severe fading to high gain, thereby compensating for the nonuniform passband of the transmission filter. The noise background in which the signal is embedded, however, is white or slightly colored. Raising the gain in the neighborhood of ω_i not only restores the signal components to their proper relative strengths but provides high amplification to the noise at those frequencies as well.

7-2 CLEAN

Deconvolution has been used in seismological exploration for oil [88] and in radio astronomy [20], [42], [85]. The latter field offers highly relevant experience for microwave imaging. The most important deconvolution procedure used in radio astronomy is called CLEAN [42]. It was invented in the middle 1970s and has been in use in radio astronomy observatories ever since. Figure 7-1 is a CLEANed image of the radio galaxy Cygnus A obtained by the Very Large Array (VLA) of the National Radio Astronomy Observatory at Socorro, New Mexico [116].

Figure 7-1 Three images of the radio galaxy Cygnus A. Data taken by the 27-antenna Very Large Array (VLA) of the National Radio Astronomy Observatory (NRAO) at Socorro, New Mexico. Wavelength is 6 cm. Resolution is 1.75 μradian (0.35 arc-sec). Each picture size is 2 × 1 arc-min. Courtesy NRAO/AUI. Observers: R. A. Perley and J. W. Dreher. (a) Original aperture-synthesis image. Sparseness of the array results in high and irregular sidelobe distribution, which is evidenced by the sidelobe artifacts that flood this image. (b) Cleaned or deconvolved image using a maximum entropy algorithm that provided the smoothest, all-positive image consistent with the original data. (c) Final improved image following self-calibration, showing the jets of Cygnus A and the filamentary structure throughout both lobes. (See Chapters 8 and 9 for theory of self-calibration).

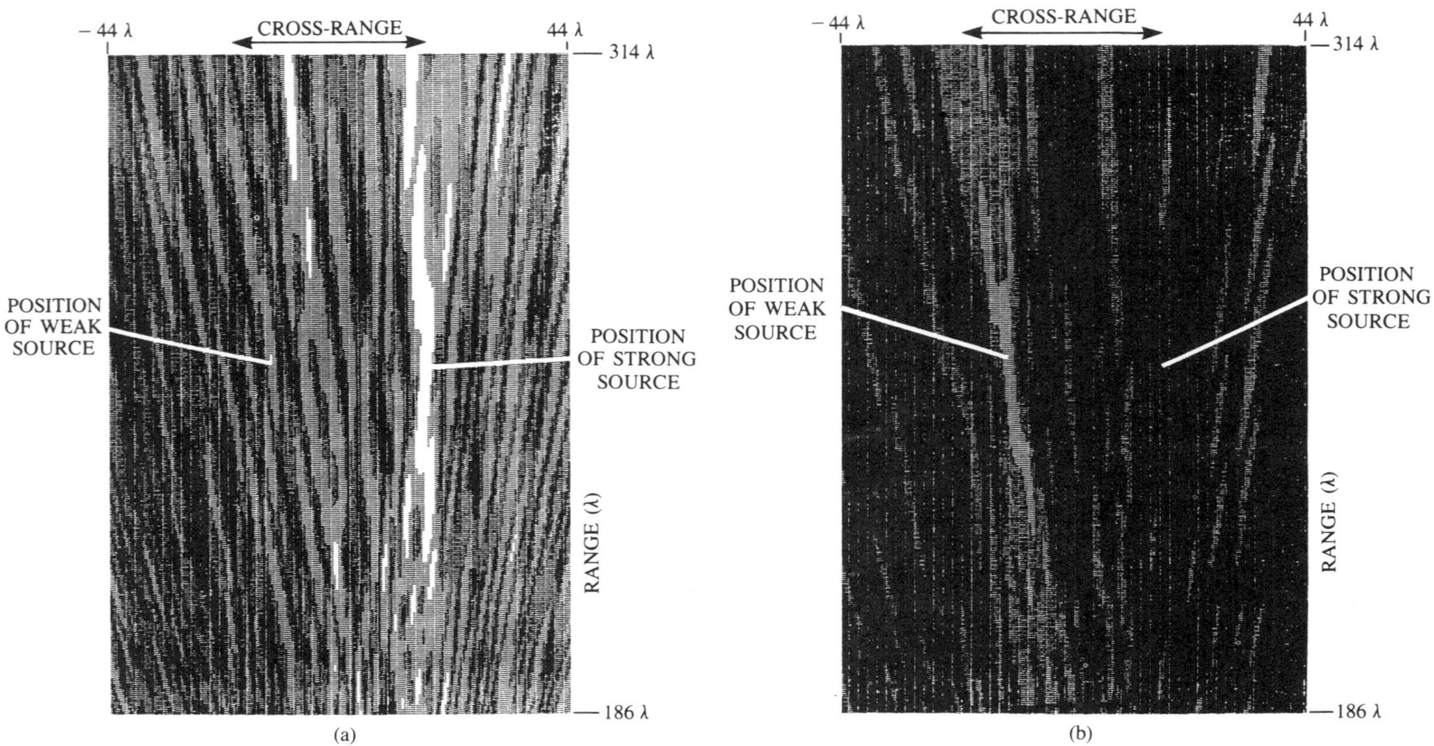

Figure 7-2 Illustration of the improvement in image quality using CLEAN. The target scenario consists of a weak source at $(-15\lambda, 250\lambda)$ and an interfering source at $(20\lambda, 250\lambda)$. Both sources are active. A 32-element, two-dimensional passive random array whose elements are randomly placed in the region $|x| \leqslant 50\lambda$, $|y| \leqslant 50\lambda$ is used to image the target scenario. (a) Dirty image. Note that the weak target is obscured by the sidelobe artifacts of the interfering target. (b) Image after one iteration of CLEAN, that is, after subtracting out the interfering target. Now the weak target is clearly visible.

Producing the image is a two-step process. First, an image is obtained by a method called *aperture synthesis* [113]. In this method a beam is formed from a highly thinned array with high sidelobes. The image is called a "dirty" image. It is then "cleaned" by the deconvolution procedure. Both steps are now described.

Aperture synthesis is based upon the incoherence of radio astronomical sources and the Van-Cittert–Zernicke theorem. Because stars are independent energy sources, their radiation fields bear no fixed phase relations to each other. Consequently, successive snapshots of the radiation field at the array differ in detail. They do not differ in statistics, however. The Van-Cittert–Zernicke theorem, which is the spatial analog of the Weiner–Khintchine theorem [115], equates the angular power density function to the Fourier transform of the two-dimensional correlation function of the radiation field. Thus by measuring the correlogram in the aperture plane, a single, two-dimensional Fourier transformation produces an image.

Figure 1-4 showed how the rotation of the earth is used to build up the sampling density in the correlogram. Consider two elements separated by a vector distance $\vec{d}$. Their respective signal voltages are $v_i(t)$ and $v_j(t)$. The correlation coefficient of lag distance $\vec{d}$ is

$$R(\vec{d}) = E\{v_i v_j^*\} \tag{7-1}$$

where E means expectation as usual. It is estimated by

$$\hat{R}(\vec{d}) = \frac{1}{T_i} \int_0^{T_i} v_i(t) v_j^*(t) \, dt \tag{7-2}$$

The integration time T_i is small compared to the time required for the earth's rotation to significantly alter the direction of the baseline $\vec{d}$. All N elements in the array are interconnected to form the cross-products required by Eq. 7-2. In this way $(N^2 - N)/2$ samples of the correlation function are measured. Over a period of hours, each multiplier traces out a locus of the correlogram of the radiation field. Let the set of values resulting from these measurements be the set $\{\hat{R}\}$. Then the image of the source distribution is

$$\hat{I} = \mathscr{F}\{\hat{R}\} \tag{7-3}$$

where $\mathscr{F}$ is the two-dimensional Fourier transform operator.

Figure 7-2a shows a "dirty" map formed in this manner. Its companion in Fig. 7-2b is a "cleaned" map of the same data. The CLEAN procedure begins by identifying the brightest point in the "dirty" map and assumes that it represents a star rather than an artifact due to the sum of sidelobe responses of the array to stars in other locations. This is not a restrictive assumption because the procedure will self-correct if wrong. Let the strongest intensity in the image be $i_1 = i_1(u_1, v_1)$. Let the two-dimensional radiation pattern be $f = f(u, v)$ and its squared modulus $|f|^2$ be the point spread function of the aperture. The first stage in the CLEAN process removes a fraction γ of the energy of the brightest source by forming

$$I_1 = \hat{I} - \gamma i_1 |f_1|^2 \tag{7-4}$$

where f_1 is shorthand notation for

$$f_1 = f(u - u_1, v - v_1) \tag{7-5}$$

The parameter γ, $0 < \gamma < 1$, is called the *loop gain.* It represents the amount of source energy subtracted out by each iteration of CLEAN. It is desirable to choose a value of γ smaller than unity so that an inaccurate estimate of the source intensity i_1 does not introduce a corresponding error in residual image I_1. Thus a loop gain smaller than unity improves the stability property of the CLEAN procedure. Varying the value of γ between 0.25 and 0.5 yields a huge savings in computation, as described in [42].

Carrying out Eq. 7-4 not only removes part of the energy from the brightest spot in the image ($= \gamma i_1$) but also the sidelobe energy introduced by the brightest spot. To the extent that the sidelobe responses to i_1 masked weaker sources, these smaller targets may now be observable in the image. The process is repeated by looking for the strongest source i_2 in I_1. Another image

$$I_2 = I_1 - \gamma i_2 |f_2|^2 \tag{7-6}$$

is formed. The function f_2 is the radiation pattern of the array displaced to source coordinates u_2, v_2. This image is largely devoid of the energies from the two brightest sources for γ close to unity. The source subtraction procedure should be repeated a few more times when γ is smaller than unity in order to remove the residual energy from the two strongest sources.

The procedure continues until the residual image is devoid of target source structure. This occurs when the global maximum of the residual image is not much larger than the general noise level in the image. A second criterion for stopping the source subtraction procedure uses the fact that the images I_k must be nonnegative. Therefore, regions of negative intensity in the image indicate that an artifact has been subtracted by the CLEAN procedure. Thus the CLEAN procedure is stopped when (a) the global maximum is comparable to the noise level (indicating that all sources have been subtracted out) and/or (b) the residual image consists of regions of negative intensity, implying that artifacts are being subtracted out by CLEAN.

Assume that T targets having intensities i_t, $1 \leqslant t \leqslant T$, are discovered before target subtraction is stopped. Let (u_t, v_t) denote the coordinates of these targets. The final or "cleaned" image is obtained by convolving the set of T delta functions represented by these sources, that is, $\sum_t i_t \delta(u - u_t, v - v_t)$, with a "clean" beam consisting of only the mainlobe of the radiation pattern. Designating $I_c = I_c(u, v)$ as the final or "cleaned" image and $f_c = f_c(u, v)$ as the "clean" beam used to form the output image yields

$$I_c(u, v) = C(u, v) * \sum_{t=1}^{T} \gamma i_t \delta(u - u_t, v - v_t) \tag{7-7}$$

where $C(u, v) = |f_c(u, v)|^2$.

Figure 7-2 compares "dirty" and "cleaned" images made from the same data.

7-3 INCOHERENT CLEAN IN MICROWAVE IMAGING

This extraordinary technique cannot be applied directly to terrestrial microwave imaging. The source distribution is not incoherent as in radio astronomy. Instead, the phase relations between the echoes from the coherent targets are fixed by the geometry. This distinction was illustrated in Fig. 1-2. The phase of the echo from each target measured at the array is the phase of the reflection coefficient of the target plus the phase delay due to the round trip from the transmitter to the receiver. Provided that the geometry is fixed, the phase difference between any two targets does not change from pulse to pulse. Hence, the fields of these targets at the array are also unchanged from pulse to pulse and as a consequence the vector sum of these fields does not change.

Transmitter location diversity, described in Chapter 6, can be used to make the coherent targets appear incoherent to the receiving array. This is achieved by transmitting from different locations from pulse to pulse, thereby changing the geometry and the phase relations between the coherent targets. The CLEAN procedure described in the previous section can be applied to the incoherent transmitter diversity image in order to enhance image quality.

The diversity image $\hat{I}_D(u)$ is the starting point of the procedure described in Eqs. 7-3 to 7-7. From Eqs. 6-15 and 6-16, observe that $\hat{I}_D(u)$ is the convolution of the true intensity $I(u)$ with the array power pattern $|H(u)|^2$ to within an error term $E(u)$. As a deconvolution procedure, CLEAN can be used to suppress the sidelobe artifacts generated by the convolution $I(u) * |H(u)|^2$ in $\hat{I}_D(u)$ when the error term $E(u)$ is identically zero. In practice, the presence of $E(u)$ will lead to imperfect target cancellation, and not all artifacts of $\hat{I}_D(u)$ can be suppressed. The estimated image, obtained after application of CLEAN to the diversity image, denoted $I_c(u)$, can be expressed as

$$I_c(u) = I(u) * C(u) + E(u) + E_1(u) \tag{7-8}$$

where the "clean beam" $C(u)$ is identical to $|H(u)|^2$ in the main lobe of $|H(u)|^2$ and is zero elsewhere. The error term $E(u)$ in the diversity image is not effected by CLEAN; hence it will be present in $I_c(u)$. A second error term $E_1(u)$ is introduced because of imperfect target cancellation by the CLEAN procedure.

The statistics of the error terms after application of CLEAN were developed in [108]. In addition, methods for designing thinned random arrays in order to effectively suppress sidelobe artifacts using incoherent transmitter diversity and CLEAN were also developed. The results are summarized below.

Let T represent the number of targets in the field of view, M the number of transmitters used for decohering the T coherent targets, and N the number of receivers in the array. Further, let $A_{\max}^2$ and $A_{\min}^2$ be the maximum and minimum target intensities, and let σ_n^2 be the noise power per receiving element per pulse transmission.

There are two array design criteria that need to be met for effective implementation of CLEAN. The first criterion requires that the number of transmitters should be large enough so that the coherent targets are sufficiently decorrelated by the process of transmitter location diversity. This constraint is met with a confidence level of 0.95 when Eq. 6-23 is satisfied, that is, when

$$M > 4\left(\frac{2\sigma_n^2}{\sum A_t^2} + 1\right) \tag{7-9}$$

In other words, the number of transmitters M has to exceed the lower bound of Eq. 7-9 for effective implementation of CLEAN.

The second design criterion prevents CLEAN from picking out artifacts caused by noise and sidelobe leakage as targets. One way to ensure that CLEAN does not confuse sidelobe artifacts with targets is to choose the number of receivers so that the largest sidelobe artifact is smaller than the lowest target intensity $A_{\min}^2$. Such a criterion is too stringent because it leads to a larger number of array elements than required for effective artifact suppression. This is because the intensity of the largest sidelobe artifact decreases as CLEAN successively cancels targets. However, this constraint will be used in this section in order to derive worst-case design criteria. A more realistic design criterion will be described in the following sections. Forcing the intensity of the largest sidelobe artifact to be smaller than the lowest target intensity in a probabilistic sense leads to the following lower bound on the number of receivers [108]

$$N \geqslant \left\{T + 2.3\sqrt{\frac{T(T+M-1)}{M}} + \frac{\sigma_n^2}{A_{\max}^2}\right\}\frac{A_{\max}^2}{A_{\min}^2} \tag{7-10}$$

This lower bound was derived using the statistics of the radiation pattern of thinned random arrays. The derivation is omitted due to its complexity. If N exceeds the lower bound of Eq. 7-10, the peak artifact will be smaller than the smallest target intensity $A_{\min}^2$ with a confidence level of 0.99.

Equations 7-9 and 7-10 provide a means for designing thinned random phased array systems that effectively suppress sidelobe artifacts using the combination of incoherent transmitter location diversity and CLEAN. The following computer simulations demonstrate the improvement in image quality.

The target scenario consists of four targets of strength -3.1, 0, -1.9, and -4.4 dB located at $u = -0.251$, $u = -0.118$, $u = 0.103$, and $u = 0.288$, respectively. The targets were imaged using a quasilinear random array 100 wavelengths long. The element positions were known to within one-tenth of a wavelength both along and perpendicular to the array. The SNR with respect to the largest (0 dB) target was fixed at 10 dB per element per transmission. Substitution of the target parameters and the SNR into Eq. 7-9 shows that five transmitters ($M = 5$) are required to effectively decorrelate the coherent targets. Using $M = 5$, $T = 4$, and $10 \log(A_{\text{max}}/A_{\text{min}})^2 = 4.4$ dB in Eq. 7-10 shows that at least 27 receivers are required to effectively suppress the sidelobe artifacts in the diversity image using CLEAN. Note that 24 receivers were used in the simulation.

Figures 7-3 and 7-4 are plots of the diversity and cleaned images for one and five orders of transmitter location diversity, respectively. The ordinate scale is the estimated intensity in dB and is thresholded at the average target artifact level $\sum_t A_t^2/N$. The target locations and intensities are shown by the symbol "$\ast$".

The number of artifacts or extra lobes in Fig. 7-3 larger than -10 dB is reduced from 33 to 11 by CLEAN and the largest artifact is reduced by 2 dB. Thus CLEAN reduced the sidelobes in the first-order diversity (no diversity) image. However, it could not fully suppress the artifacts because transmitter location diversity was not used to decorrelate the coherent targets. In contrast, the fifth order "cleaned" image (Fig. 7-4) does not have large sidelobe artifacts. The number of artifacts larger than about -10 dB is reduced from 37 to 2, and the largest artifact is about 4 dB smaller.

The statistics of the sidelobe patterns of thinned random arrays were used to estimate the peak artifact level of the diversity and CLEANed images in [108]. The difference between the peak artifact estimators of the diversity and CLEANed images is a measure of the image quality improvement obtained using CLEAN. The peak artifact estimators were determined for high confidence levels β, where $0.9 \leqslant \beta < 1$. In other words, the peak artifact levels of sample images would be smaller than their estimated values with probability β. Only the approximate results are shown below; additional second-order terms derived in [108] are not included. The peak artifact estimator of the diversity image with confidence level β, denoted P_d, is obtained by solving the following equation:

$$\operatorname{erfc}_{\ast}\left(\frac{P_d - \mu_d}{\sigma_d}\right) = 1 - \beta^{1/n}$$

where μ_d and σ_d^2 are the mean and variance of the diversity image intensity. They are obtained by sampling the diversity image at n points that are separated by an array beamwidth λ/L (L = array length), where n is the number of beamwidths in the field of view. It is equal to $2L/\lambda$ if the illuminated patch

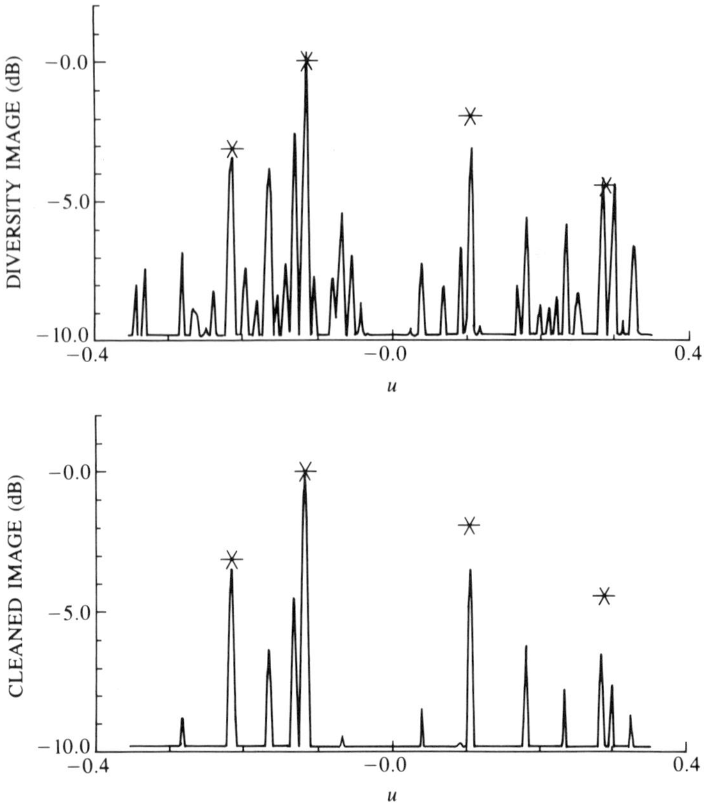

Figure 7-3 Application of incoherent CLEAN without using transmitter-location diversity. The target scenario consists of four targets whose locations and amplitudes are indicated by the symbol "$\ast$". A 24-element quasilinear array of length 100λ was used to image the target scenario. The element positions were known to within one-tenth of a wavelength both along and perpendicular to the array. Only a single transmitter was used to illuminate the target scenario. The abscissa represents u and the ordinate represents intensity in dB. The -10 dB threshold is approximtely equal to the average artifact level in the uncleaned image. (a) The first-order diversity image—no diversity. (b) Image after applying CLEAN to (a). Observe that CLEAN removed about two-thirds of the image artifacts greater than the average artifact level.

occupies the region $|u| \leqslant 1$. The function $\text{erfc}_*(\cdot)$ is defined as

$$\text{erfc}_*(A) = \frac{1}{\sqrt{2\pi}} \int_A^\infty e^{-x^2/2}\, dx$$

The value of A can be obtained using tables of the Gaussian distribution for different values of $\text{erfc}_*(A)$. Thus for a given β, P_d can be obtained after computing μ_d and σ_d from the diversity image. The expression for P_d shown above is approximate and holds only for large confidence levels $0.9 \leqslant \beta < 1$.

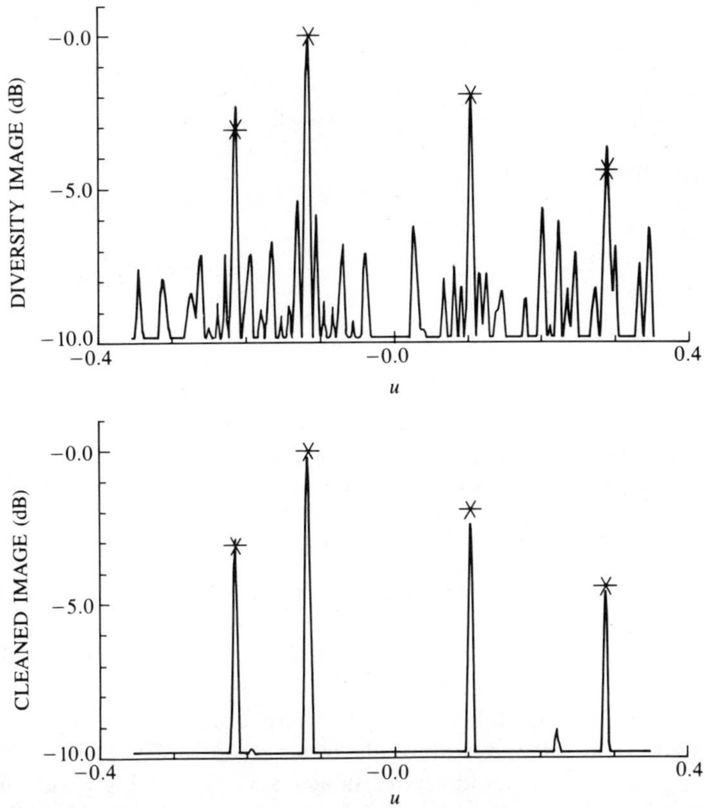

Figure 7-4 Application of incoherent CLEAN to a fifth-order transmitter-location diversity image. Target scenario and array parameters are the same as in Fig. 7-3 except that five transmitters were used. (a) The fifth-order diversity image. (b) Image after applying CLEAN to (a). Note that the CLEANed image is nearly artifact free.

An approximate value of the peak artifact estimator of the cleaned image derived in [108], denoted P_c, is obtained by solving

$$\operatorname{erfc}_*\left(\frac{P_c - \mu_c}{\sigma_c}\right) = 1 - \beta^{1/n}$$

where μ_c and σ_c^2 are the mean and variance of the residual image intensity after application of CLEAN. They are obtained by sampling the residual image at n points that are separated by an array beamwidth.

The ratio of P_d and P_c, usually expressed in dB as $10\log(P_d/P_c)$, is the improvement in the peak artifact level obtained using CLEAN. The theoretical values of the peak artifact estimators were validated using Monte Carlo methods in [108]. The results for a particular target scenario are shown in Fig. 7-5. Four targets having intensities of -1, 0, -3 and -2 dB, were located at

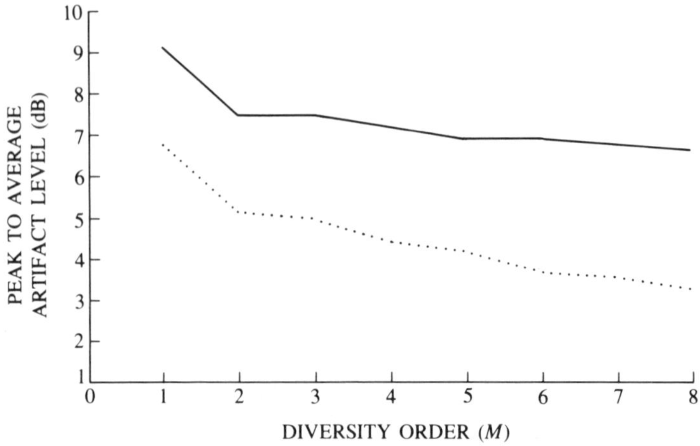

Figure 7-5 Improvement in peak artifact levels obtained using incoherent CLEAN. Four targets having intensities of -1.0, 0, -3, and -2 dB, and located at $u = -0.66$, -0.32, 0.07, and 0.55, were imaged using a 24-element random array of length 100λ for various orders of transmitter-location diversity. The abscissa represents diversity order. The order represents empirical estimates of the peak artifact level of the diversity (solid line) and cleaned (dotted line) images, obtained using 100 Monte Carlo runs. The ordinate scale is normalized to the average artifact levels of the diversity image.

$u = -0.66$, -0.32, 0.07, and 0.55 in the simulation. A 24-element random array of length 100λ was used to image the target scenario. The SNR per element was 8 dB for the first-order diversity image and $(8 - 10 \log K)$ dB for the Kth-order diversity image. In other words, the noise level was identical for all orders of diversity.

Figure 7-5 shows empirical estimates of the peak artifact level of the diversity and cleaned images as a function of diversity order. The abscissa represents diversity order. The ordinate represents the peak artifact level of the diversity and CLEANed images normalized to the average artifact level of the diversity image $(\sum_t A_t^2)/N$. The peak artifact levels were obtained by averaging the results from 100 computer simulation trials. In each trial, the transmitter and receiver locations were chosen randomly. The theoretical values derived in [108] agreed with the empirical estimates to within 1 dB.

Three observations can be made from Fig. 7-5. First, the peak artifact level of the first-order diversity image is 5 dB larger than the fifth-order cleaned image. Thus the combination of fifth-order transmitter diversity and incoherent CLEAN provides a 5 dB improvement in the peak artifact level for this particular target scenario. Secondly, when CLEAN is applied without diversity, the peak artifact drops by 2 dB. Thirdly, when fifth-order diversity is employed, the improvement increases to 5 dB.

It is interesting to compute the number of elements required to achieve the same peak artifact level without using transmitter location diversity and CLEAN. The average artifact level of the diversity image for this particular target scenario is $(\sigma_n^2 + \sum_t A_t^2)/N = -8.9$ dB (σ_n^2 = noise power per element, intensity of the largest target = 0 dB). The peak artifact of the fifth-order cleaned image is 4 dB larger than -8.9 dB, that is, -4.9 dB. Let N' represent the number of elements required to achieve the same peak artifact level without using diversity or CLEAN. The peak artifact level of an array with N' elements is approximately $B_p(\sigma_n^2 + \sum_t A_t^2)/N'$, where B_p is approximately equal to 8 [94]. Equating this quantity to the peak artifact level of the fifth-order CLEANed image results in $N' = 76$. The above solution for N' is approximate; a stricter derivation in [108] yielded $N' = 64$.

Thus, a 64-element array that does not use diversity and CLEAN is equivalent to a 24-element array that uses five transmitters and CLEAN for this particular target scenario. In other words, using transmitter location diversity and CLEAN reduces the requirement on the number of receivers by a factor of 2.7. The price that is paid is that the dwell time on the target scenario is five times longer. The procedure is useful when incoherent addition of several images is used to increase target SNR, that is, when peak power requirements of the transmitter demand that several snapshots of the target scenario are required in order to achieve reasonable probabilities of detection and false alarm.

This concludes the discussion on the application of incoherent CLEAN on the image obtained using incoherent transmitter diversity. The technique is described in detail in [108]; only the important results were presented in this section. A variation of CLEAN that offers greater reduction in the sidelobe artifact levels, called *coherent CLEAN*, is now described.

7-4 COHERENT CLEAN

This section presents the coherent CLEAN procedure in its simplest form. A detailed analysis of the procedure is given in the following sections.

The coherent CLEAN procedure was first developed for use in microwave imaging by Tsao in [118], [121]. It does not differ from incoherent CLEAN in principle; however, the criteria used for determining whether all targets have been canceled are different. The stopping criteria for incoherent CLEAN, described earlier, are simple because the image is known to be real and positive. Incoherent CLEAN cancels targets until (a) the residual images $I_1, I_2, \ldots$ (see Eqs. 7-4 and 7-6) contain large regions of negative intensity, and/or (b) the ratio of the peak to average image intensity in the residual images is small, implying that the residual images consist of noise alone. Thus the *a priori* knowledge that the incoherent image is real and positive is of great value when implementing incoherent CLEAN. This kind of *a priori* knowledge is not available for coherent imaging because coherent images are complex. Consequently, different stopping criteria need to be developed when deconvolving coherent microwave images using coherent CLEAN. These criteria were developed by Tsao [121]. They will be described in the following sections; first the basic procedure will be described.

Let T targets having amplitudes A_t, phases ψ_t, and locations u_t, $1 \leqslant t \leqslant T$, lie in the far field of a linear array. Assume that the targets are ordered so that $A_1 \geqslant A_2 \geqslant \cdots \geqslant A_T$. Let x_n represent the locations of the N elements in the random linear array. Under the far-field assumption, the field at the nth element, denoted $e_n(0)$, is

$$e_n(0) = \sum_{t=1}^{T} A_t \exp(j(\psi_t + ku_t x_n)) \tag{7-11}$$

where noise has been ignored for the present. The image of the target scenario, denoted $\hat{s}_0(u)$, is

$$\begin{aligned} \hat{s}_0(u) &= \frac{1}{N} \sum_{n=1}^{N} e_n(0) \exp(-jkux_n) \\ &= \sum_{t=1}^{T} A_t e^{j\psi_t} \frac{1}{N} \sum_{n=1}^{N} e^{-jk(u-u_t)x_n} \\ &= s(u) * H(u) \end{aligned} \tag{7-12}$$

where

$$s(u) = \sum_{t=1}^{T} A_t e^{j\psi_t} \delta(u - u_t) \tag{7-13}$$

$$H(u) = \frac{1}{N} \sum_{n=1}^{N} e^{-jkux_n} \tag{7-14}$$

Thus the estimated image $\hat{s}_0(u)$ is the convolution of the target scene $s(u)$ with the radiation pattern of the array. The subscript 0 in $\hat{s}_0(u)$ is used to differentiate it from other images formed by the CLEAN procedure.

If the array is well designed, the global maximum of $|\hat{s}_0(u)|$, at $\hat{u}_1$, will be close to the location u_1 of the largest target. Coherent CLEAN first removes a fraction γ of the largest target and its associated sidelobes from the image $\hat{s}_0(u)$. This is accomplished by subtracting out a portion of the field generated by the largest target at the aperture. This modifies the received data $e_n(0)$ in the following fashion:

$$e_n(1) = e_n(0) - \gamma \hat{s}_0(\hat{u}_1)\, e^{jk\hat{u}_1 x_n} \tag{7-15}$$

where $\hat{u}_1$ is the location of the global maximum of $|\hat{s}_0(u)|$ and $0 < \gamma \leqslant 1$ is called the *loop gain.* A value of γ smaller than unity should be used when the estimate of the target strength and location is significantly distorted by noise. Incoherent CLEAN used $0.25 \leqslant \gamma \leqslant 0.5$ because such a variation in γ provided huge savings in computation [42]. Choosing such values for γ in coherent CLEAN does not significantly affect the number of computations; however, it is usually safe to choose $\gamma = 0.5$. The quantities $e_n(1)$ represent the data after being modified during the first iteration of CLEAN. Use of Eq. 7-11 in the above yields

$$\begin{aligned} e_n(1) &= \left(\sum_{t=1}^{T} A_t \exp j(\psi_t + ku_t x_n) \right) - \gamma \hat{s}_0(\hat{u}_1) \exp(jk\hat{u}_1 x_n) \\ &= A_1 \exp j(\psi_1 + ku_1 x_n) - \gamma \hat{s}_0(\hat{u}_1) \exp(jk\hat{u}_1 x_n) \\ &\qquad + \sum_{t=2}^{T} A_t \exp j(\psi_t + ku_t x_n) \end{aligned} \tag{7-16}$$

The difference between the first two terms is the uncanceled residue of the largest target. Had the estimates of u_1 *and* $A_1\, e^{j\psi_1}$, which are $\hat{u}_1$ and $\hat{s}_0(\hat{u}_1)$ been error-free, this term would have been zero for $\gamma = 1$ and the echoes due to the largest target would be absent in the $e_n(1)$.

The image after the first CLEAN iteration, denoted $\hat{s}_1(u)$, is

$$\hat{s}_1(u) = \frac{1}{N} \sum_{n=1}^{N} e_n(1) \exp(-jkux_n) \tag{7-17}$$

The global maximum of $|\hat{s}_1(u)|$ can be used to infer the complex amplitude and location of the second-largest target. The echoes due to this target can then be subtracted out from $e_n(1)$ in a similar fashion. Ideally, the procedure can be continued until the echoes due to all the T targets are subtracted from the measured data. Thus at the end of T iterations, $e_n(T) = 0$, all n, provided that the estimates of the target locations and amplitudes were determined exactly by CLEAN and $\gamma = 1$.

Unfortunately, in practice, none of the target location and amplitude estimates will be devoid of errors. Further, the element positions x_n will not be known exactly to the radar designer. The process of adaptive beamforming (Chapters 8 and 9), which approximately corrects for element-position uncertainty, will introduce some multiplicative noise in the measured data. This is in addition to the noise present before adaptive beamforming was applied. Deviation of the array elements from a straight line introduces errors in the image formation process, thereby distorting the target location and amplitude estimates. Thus the target echoes cannot be exactly subtracted out from the measured data. Imperfect cancellation adds noise, and the amount of noise added increases as more targets are subtracted out by CLEAN. Target subtraction has to be stopped when (a) all targets have been subtracted out and/or (b) the noise power introduced by CLEAN becomes so large that further target subtraction becomes meaningless. The development of the stopping criteria is the subject of the following section. It and the following sections are based on the theory developed in [104], [118], [121].

7-5 STOPPING CRITERIA FOR COHERENT CLEAN

It is desirable that the peak artifact intensity in the images at each CLEAN iteration be smaller than the intensity of the largest target in that image with high probability. This ensures that (a) the global maximum of the image corresponds to a target location and (b) a sidelobe artifact is not subtracted out by CLEAN.

Targets create artifacts because of leakage of target energy through the sidelobes of the radiation pattern. Additive noise in the measured data also creates artifacts because the noise samples at the various elements could conceivably combine coherently at certain look directions. The following discussion assumes that the locations of the artifacts caused by target and noise energy are independent. In this case, CLEAN can distinguish between targets and artifacts if the global maximum of the image is larger than (a) the peak artifact level caused by leakage of target energy through the sidelobes and (b) the peak artifact level caused by coherent build-up of the noise samples.

These two conditions can be quantified using two thresholds denoted T_s and T_n. The first threshold, T_s, represents the nominal peak artifact level caused by leakage of target energy through the high sidelobes in the radiation pattern of the random array. The second threshold, T_n, represents the nominal peak artifact level due to additive noise.

Mathematically, the two conditions that allow CLEAN to continue target cancellation after the kth iteration are

$$\hat{s}_k(u_{\max}) > T_s \qquad \text{with large probability } P_s \tag{7-18}$$

and

$$\hat{s}_k(u_{\max}) > T_n \qquad \text{with large probability } P_n \tag{7-19}$$

where $u_{\max}$ is the value of u at the global maximum of the image after the kth iteration, and P_s and P_n are close to unity. The CLEAN iterations are stopped when either or both of the two inequalities (7-18) and (7-19) do not hold.

Methods for determining and updating T_s and T_n during the course of the CLEAN iterations are presented in the following sections. The behavior of T_s and T_n as the iterations progress is discussed in a qualitative fashion in the remaining portion of this section.

Assume that the total target energy is much larger than the noise power in the image at the start of the CLEAN iterations. This condition can be ensured by using a large number of receivers. Thus $T_s \gg T_n$ at the start of the CLEAN procedure. As the iterations progress, target energy gets subtracted out. This decreases T_s because the peak artifact level due to leakage of target energy through the sidelobes is roughly proportional to the target energy. However, the noise power increases as the iterations progress because of imperfect target cancellation, thereby increasing T_n. If (a) the scene consists of only a few targets, (b) the initial noise power is small, and (c) errors caused by element position uncertainty and adaptive beamforming corrections are negligible, then T_s will be larger than T_n after all targets have been subtracted out. Otherwise CLEAN may not be able to detect all targets.

The locations of the peak image artifacts caused by sidelobe leakage and noise were assumed to be independent random variables in order to develop the inequalities (7-18) and (7-19). In the highly unlikely event that the signal and noise peak artifacts combine coherently at the same value of u, the stopping criterion changes from $\hat{s}_k(u_{\max}) < T_s$ and/or T_n to $\hat{s}_k(u_{\max}) < (T_s + T_n)$. Thus the criteria given by Eqs. 7-18 and 7-19 for continuing target subtraction using CLEAN are approximate. However, extensive computer simulations have shown that these criteria are adequate to effectively suppress sidelobe artifacts for most target scenarios. In other words, the criterion that the global maximum of the image be larger than both T_s and T_n rather than their sum $T_s + T_n$ appears reasonable for continuing CLEAN iterations. The following analysis uses a compromise: some of the noise energy is used for computing T_s.

7-6 ANALYSIS OF COHERENT CLEAN

In what follows, the signal and noise thresholds T_s and T_n are computed before application of CLEAN and updated after each iteration of CLEAN. Clearly, computation of T_s and T_n should use only a small amount of *a priori* knowledge of the target scenario, for otherwise their usefulness vanishes. The following discussion assumes the availability of the following quantities: (a) the power of the additive noise present in the measured data; (b) the variance of the phase errors introduced after element position uncertainty is approximately compensated using adaptive beamforming procedures; (c) the variance of the element position errors along the direction of the array; and (d) the variance of the displacement of the element positions from the X-axis (the array is assumed to lie along the X-axis for simplicity; this assumption has no bearing on the results of this and the subsequent sections). A thinned random linear array is used in the analysis and results from random array theory are freely used [94].

First, data measurement and image formation in the presence of noise and systemic errors are described. Then the noise added due to imperfect target cancellation is quantified. The result is used to compute and update the signal and noise thresholds T_s and T_n.

Consider a thinned random quasilinear array of length L whose N elements are located at (x_n, y_n), $1 \leqslant n \leqslant N$. Here, the x_n are assumed to be uniformly distributed between $-L/2$ and $L/2$. The y_n are assumed to be zero-mean random variables having common variance σ_Y^2. Estimates of (x_n, y_n), denoted by the pair $(\hat{x}_n, 0)$, are available to the radar designer. $\hat{y}_n = 0$ means that the array elements are assumed to lie along the X-axis. The variances

$$E\{(\hat{x}_n - x_n)^2\} = E\{\delta_n^2\} = \sigma_X^2; \qquad E\{y_n^2\} = \sigma_Y^2 \tag{7-20}$$

are known. δ_n is shorthand for $\hat{x}_n - x_n$. In addition to the element-position errors, phase errors β_n, uniformly distributed between $-\pi$ and π, are assumed to be present at each receiver. The adaptive beamforming procedures described in Chapters 8 and 9 can be used to partially compensate for element-position uncertainty and the phase errors.

The quasilinear array is used to image a target scene consisting of T targets at $u = u'_t$, having amplitudes and phases denoted A_t and ψ'_t, respectively. The datum measured at the nth element, denoted e_n, is

$$e_n = \left\{ \sum_{t=1}^{T} A_t \exp j\left[\psi'_t + k\left(u'_t x_n + \sqrt{1 - u'^2_t}\, y_n\right)\right] \right\} e^{j\beta_n} + w'_n$$

$$= \left\{ \sum_{t=1}^{T} A_t \exp j[\psi'_t + k(u'_t x_n + v'_t y_n)] \right\} e^{j\beta_n} + w'_n \qquad (7\text{-}21)$$

where w'_n denotes additive noise whose real and imaginary parts each have variance $\sigma_n^2/2$, and $v'_t = \sqrt{1 - u'^2_t}$. The braces are used to indicate that the phase errors β_n change the phases of the measured data. The phase errors also change the phases of the noise samples w'_n; however, changing phase of w'_n does not alter its statistics. Thus the effect of the phase errors on the noise samples is ignored.

The phase errors β_n and element-position uncertainty can be partially compensated using the adaptive beamforming procedures described in the next two chapters. These procedures estimate the phase errors at each of the N elements. The phase error estimates, denoted $\hat{\beta}_n$, are

$$\hat{\beta}_n = \psi_0 + k(u_0 x_n + v_0 y_n) + \beta_n - \varepsilon_n \qquad (7\text{-}22)$$

where u_0 is the location of the beamforming source, $v_0 = \sqrt{1 - u_0^2}$, ψ_0 is a constant independent of n, and ε_n is an error introduced by the adaptive beamforming process. Adaptive beamforming compensates for the β_n (and also element-position uncertainty) by multiplying the e_n with $\exp(-j\hat{\beta}_n)$. Performing this multiplication and denoting the result by $e_n(0)$ leads to*

$$e_n(0) = \left\{ \sum_{t=1}^{T} A_t \exp j[\psi'_t - \psi_0 + k(u'_t - u_0)x_n + k(v'_t - v_0)y_n] \right\} e^{j\varepsilon_n} + w_n$$

$$= \left\{ \sum_{t=1}^{T} A_t \exp j\{\psi_t + k(u_t x_n + v_t y_n)] \right\} e^{j\varepsilon_n} + w_n \qquad (7.23)$$

where $\psi_t = \psi'_t - \psi_0$, $w_n = w'_n e^{-j\beta_n}$, and

$$u_t = u'_t - u_0; \qquad v_t = v'_t - v_0 = \sqrt{1 - u'^2_t} - \sqrt{1 - u_0^2} \qquad (7.24)$$

* The inclusion of zero within the parentheses states that the $e_n(0)$ represents the data before the start of the CLEAN iterations. Data after the first iteration are represented by $e_n(1)$, and so on.

Note that the phase distortion of the signal part of $e_n(0)$ *is* ε_n, which is much smaller than the initial phase errors β_n. The image is formed using the assumed element locations. Let $\hat{s}_0(u)$ denote the "dirty" image. Using the assumed element positions $\hat{x}_n$ in the Fourier kernel leads to*

$$\hat{s}_0(u) = \frac{1}{N} \sum_{n=1}^{N} e_n(0) \exp(-jku\hat{x}_n) = \hat{s}_{0,s}(u) + \hat{s}_{0,n}(u) \tag{7-25}$$

The image has been divided into two parts: the first, $\hat{s}_{0,s}(u)$, is due to echoes from the T targets and the second, $\hat{s}_{0,n}(u)$, is due to additive noise. Using $\hat{x}_n = x_n + \delta_n$ and Eq. 7-23 in Eq. 7-25 leads to

$$\hat{s}_{0,s}(u) = \sum_{t=1}^{T} A_t e^{j\psi_t} \left\{ \frac{1}{N} \sum_{n=1}^{N} \exp\{-j[k(u-u_t)x_n - kv_t y_n + k\delta_n u - \varepsilon_n]\} \right\} \tag{7-26}$$

$$\hat{s}_{0,n}(u) = \frac{1}{N} \sum_{n=1}^{N} w_n \exp(-jku\hat{x}_n) \tag{7-27}$$

Computation of the signal and noise thresholds requires estimates of the peak artifact level of the signal and noise images.

7-6-1 Peak Artifact Levels of Signal and Noise Images

The average artifact level of the signal image is equal to the amount of target energy divided by the number of elements N [94], that is,

$$AAL = \frac{1}{N} \sum_{t=1}^{T} A_t^2 \tag{7-28}$$

where AAL is shorthand for average artifact level.† The peak artifact level can be estimated for high confidence levels by multiplying AAL with the quantity B_p, where B_p is approximately 10 for thinned random arrays [94]. This peak artifact level is equal to the threshold T_s, that is,

$$T_s(0) = PAL = \frac{B_p}{N} \sum_{t=1}^{T} A_t^2 \tag{7-29}$$

* The subscript 0 in $\hat{s}$ states that this is the image before the start of the CLEAN iterations. The image after the first iteration is represented by $\hat{s}_1(u)$, and so on.

† The popular notation ASL represents average sidelobe level of the array radiation pattern. The notation AAL is chosen here in order to distinguish the ASL from the average intensity of an image consisting of many targets.

where PAL is shorthand for peak artifact level and $T_s(0)$ represents the peak artifact of the signal part of the image at the zeroth iteration. The global maximum of the image $|\hat{s}_0(u)|$ should be larger than $T_s(0)$ if CLEAN is to be used to subtract out the largest target and its associated sidelobe artifacts from $\hat{s}_0(u)$. An estimate of the target energy $\sum A_t^2$ can be obtained from the data using $\sum A_t^2 = (\sum_n |e_n(0)|^2)/N$. This is because the zero lag correlation is equal to the total energy (target plus noise energy) from Fourier theory. Thus $(\sum |e_n(0)|^2)/N = \sum A_t^2 + \sigma_n^2 \approx \sum A_t^2$ when the noise energy is much smaller than the target energy. The estimate of the signal threshold, obtained using the measured data, is therefore

$$T_s(0) = \frac{B_p}{N}\left(\frac{1}{N}\sum_{n=1}^{N} |e_n(0)|^2\right) \tag{7-30}$$

This estimate of the peak signal artifact will be close to its actual value when the noise power per element is much smaller than the total target energy. It will be an overestimate if the noise power is comparable to the target energy.

Note that $T_s(0)$ is the estimate of the peak signal artifact before the start of the CLEAN iterations. It will decrease as the iterations progress; the amount it diminishes by is proportional to the target energy subtracted during its iterations.

The peak artifact due to noise can be computed in a similar fashion. The average artifact level due to noise is σ_n^2/N, where $\sigma_n^2 = E\{|w_n|^2\}$ is the mean energy in the noise samples. The peak artifact level is B_p times this quantity, that is,

$$T_n(0) = \frac{B_p}{N}\sigma_n^2 \tag{7-31}$$

Note that $T_n(0)$ can be estimated because the noise energy σ_n^2 in the data samples is known or can be obtained using the data from range bins that have no appreciable radar cross-section.

Also note that $T_n(0)$ is the estimate of the peak noise artifact before the start of the CLEAN iterations. It will increase as the iterations progress; the amount it increases is proportional to the noise energy added due to imperfect target cancellation during the iterations.

7-6-2 Target Parameter Estimation

Phase errors and element position uncertainty degrade the estimates of the target amplitude, phase, and location. The variance of the errors in these estimates is now calculated using random array theory [94, Ch. 13].

Consider the image $\hat{s}_0(u)$ for u in the vicinity of the location of the largest target u_1. This portion of the image can be split into three terms as follows:

$$\hat{s}_0(u_1) = A_1\, e^{j\psi_1}\left\{\frac{1}{N}\sum_{n=1}^{N} \exp(-j[k(u_1\delta_n - v_1 y_n) - \varepsilon_n])\right\}$$
$$+\frac{1}{N}\sum_{n=1}^{N} e^{j\varepsilon_n}\left\{\sum_{t=2}^{T} A_t \exp(j[\psi_t + k(u_t x_n + v_t y_n - u_1\hat{x}_n)])\right\}$$
$$+\frac{1}{N}\sum_{n=1}^{N} w_n \exp(-jku_1\hat{x}_n) \qquad (7\text{-}32)$$

where $\hat{x}_n = x_n + \delta_n$ is an estimate of x_n.

The image $\hat{s}_0(u_1)$ has been split into three parts: the first represents the contribution of the largest target to the image. The second and third parts represent the energy due to the other targets and noise, respectively. Note that all the quantities in the first term are cophased when the system is error-free, that is, $\delta_n = y_n = \varepsilon_n = 0$, and the magnitude of that term becomes A_1. The second and third terms are sums of randomly phased phasors; their magnitudes are approximately $A_{\text{rms}}/\sqrt{N(T-1)}$ and $w_{\text{rms}}/\sqrt{N}$. Thus the first term will generally dominate the remaining two terms in the vicinity of $u \approx u_1$ and

$$\hat{s}_0(u_1) \approx A_1\, e^{j\psi_1}\left\{\frac{1}{N}\sum_{n=1}^{N} e^{-j\alpha_n}\right\} \qquad (7\text{-}33a)$$

where

$$\alpha_n = k(u_1\delta_n - v_1 y_n) - \varepsilon_n \qquad (7\text{-}33b)$$

The estimates of the amplitude, phase, and location of the largest target are obtained by searching for the peak of $|\hat{s}_0(u)|$ in the vicinity of u_1. The nonzero phase errors α_n introduce errors in these estimates. The mean value of $\hat{s}_0(u_1)$ can be estimated as the loss in gain times $A_1\, e^{j\psi_1}$ [94], or

$$E\{\hat{s}_0(u_1)\} = A_1\, e^{j\psi_1}\, e^{-\sigma_\alpha^2/2} \qquad (7\text{-}34)$$

where σ_α^2 is the variance of the phase errors α_n. Assuming independence of the δ_n, ε_n, and y_n,

$$\sigma_n^2 = E\{\alpha_n^2\} = k^2(u_1^2\sigma_X^2 + v_1^2\sigma_Y^2) + \sigma_\varepsilon^2 \qquad (7\text{-}35)$$

where σ_X^2 and σ_Y^2 are defined by Eq. 7-20 and σ_ε^2 is the variance of the residual phase errors after adaptive beamforming. These three quantities are assumed to be known in the following analysis.

The estimate of the location of the largest target is also distorted by the phase errors. From random array theory, an estimate of the rms pointing error, denoted σ_u, is [94]

$$\sigma_u^2 = E\{(\hat{u}_1 - u_1)^2\} = \sqrt{\frac{12}{N}}\left(\frac{\sigma_\alpha}{kL}\right) \tag{7-36}$$

Note that both the expected gain–loss and the rms pointing error can be computed after the amplitude and location of the largest target have been estimated. This can be done by substituting the estimates $\hat{u}_1$ and $\hat{v}_1$ into 7-35 and using the result in Eqs. 7-34 and 7-36.

7-6-3 Target Cancellation

The amount of target energy that is subtracted by one iteration of CLEAN is estimated in this section. In addition, the amount of noise energy that is added due to imperfect target cancellation is also estimated. These two quantities will be used to update the two thresholds T_s and T_n after completion of the target cancellation procedure.

Imperfect target cancellation arises because element position uncertainty and phase errors introduce errors in the estimates of the target amplitude, phase, and location. Subtraction of a fraction γ of the amplitude of the largest target at the aperture results in

$$e_n(1) = e_n(0) - \gamma \hat{s}_0(\hat{u}_1)\, e^{jk\hat{u}_1\hat{x}_n} \tag{7-37}$$

This subtraction is similar to the one in Eq. 7-15; the only difference is that the assumed element locations $\hat{x}_n$ are used to compute the target phase at each element. The amount of target energy subtracted and the amount of noise introduced can be estimated by examining the difference between the energy of the largest target before and after target cancellation, which is called the *residue* $r_n(1)$:

$$r_n(1) = A_1\, e^{j\psi_1} \exp\{j[k(u_1 x_n + v_1 y_n) + \varepsilon_n]\} - \gamma \hat{s}_0(\hat{u}_1) \exp(jk\hat{u}_1 \hat{x}_n) \tag{7-38}$$

Using $\hat{x}_n = x_n + \delta_n$ and Eqs. 7-34 in 7-38 leads to

$$r_n(1) = A_1\, e^{j(\psi_1 + ku_1 x_n)} \times \{\exp[j(kv_1 y_n + \varepsilon_n)] - \gamma e^{-\sigma_\alpha^2/2} \exp[jk(u_1 \delta_n + (\hat{u}_1 - u_1)x_n]\} \tag{7-39}$$

This is the residual noise introduced by subtracting out the largest target from the data. Its energy is $E\{|r_n(1)|^2\}$, which evaluates to

$$t_n(1) = E\{|r_n(1)|^2\} = A_1^2(1 + \gamma^2 e^{-\sigma_\alpha^2} - 2\gamma e^{-\sigma_\alpha^2(1+1/N)})$$
$$\approx A_1^2(1 + \gamma^2 e^{-\sigma_\alpha^2} - 2\gamma e^{-\sigma_\alpha^2}) \tag{7-40}$$

The target energy subtracted is A_1^2 minus the above quantity, which is

$$t_s(1) = A_1^2\, e^{-\sigma_\alpha^2}(2\gamma - \gamma 2) \tag{7-41}$$

Both quantities can be computed because σ_α can be computed. They are used to update the thresholds $T_n(0)$ and $T_s(0)$.

7-6-4 Threshold Update

The peak signal artifact threshold should be decreased by a factor proportional to the amount of target energy subtracted in the first iteration. The proportionality constant is B_p/N (see 7-29). Thus

$$T_s(1) = T_s(0) - \frac{B_p}{N}\, t_s(1) \tag{7-42}$$

where $t_s(1)$ is defined by Eq. 7-41.

The peak noise artifact threshold should be increased by B_p/N times the amount of noise energy added after the first iteration, that is,

$$T_n(1) = T_n(0) + \frac{B_p}{N}\, t_n(1) \tag{7-43}$$

where $t_n(1)$ is defined by Eq. 7-40.

The next image, denoted $\hat{s}_1(u)$ is then formed and the process is repeated:

$$\hat{s}_1(u) = \frac{1}{N} \sum_{n=1}^{N} e_n(1)\, e^{-jku\hat{x}_n} \tag{7-44}$$

The global maximum of $|\hat{s}_1(u)|$ is then found and canceled if its magnitude exceeds both $T_s(1)$ and $T_n(1)$, and so on. The CLEAN procedure is halted at that iteration for which the global maximum fails to exceed the larger threshold.

7-6-5 Summary of Coherent CLEAN Iterations

The CLEAN iterations start by computing the initial values of the signal and noise thresholds using Eqs. 7-30 and 7-31.

The ith iteration begins by determining whether the global maximum of $|\hat{s}_{i-1}(u)|$ is larger than both $T_s(i-1)$ and $T_n(i-1)$. If so, the target at the global maximum is canceled at the aperture using

$$e_n(i) = e_n(i-1) - \gamma \hat{s}_{i-1}(\hat{u}_i)\, e^{jk\hat{u}_i\hat{x}_n} \tag{7-45}$$

The phase-error variance is then updated using

$$\sigma_{\alpha_i}^2 = k^2(\hat{u}_i^2 \sigma_X^2 + \hat{v}_i^2 \sigma_Y^2) + \sigma_\varepsilon^2 \tag{7-46a}$$

The preceding expression represents the variance of the phase error introduced by the target subtraction procedure when adaptive beamforming has been used for phase and element-position error compensation. Without adaptive beamforming, a larger phase error variance arises:

$$\sigma_{\alpha_i}^2 = k^2(u_i'^2 \sigma_X^2 + (1 - u_i'^2)\sigma_Y^2) + \sigma_\beta^2 \tag{7-46b}$$

where u_i' is the actual target location and not the difference between the target and beamformer locations. In addition, σ_ε^2 is the variance of the actual phase errors at the array elements and not the variance of the residual phase errors after adaptive beamforming. Note that the second term on the right side of Eq. 7-46b is much larger than the corresponding term in Eq. 7-46a. This is because adaptive beamforming completely compensates for element-position errors in the beamforming direction; without adaptive beamforming, the phase errors introduced by the target subtraction procedures are larger.

$$t_s(i) = A_i^2\, e^{-\sigma_{\alpha i}^2}(2\gamma - \gamma^2) \tag{7-47}$$

$$t_n(i) = A_i^2(1 + \gamma^2 e^{-\sigma_{\alpha i}^2} - 2\gamma e^{-\sigma_{\alpha i}^2}) \tag{7-48}$$

$$T_s(i) = T_s(i-1) - \frac{B_p}{N}\, t_s(i) \tag{7-49}$$

$$T_n(i) = T_n(i-1) + \frac{B_p}{N}\, t_n(i) \tag{7-50}$$

Thus both thresholds are functions of σ_α, given by Eq. 7-46. In other words, the performance of CLEAN is limited by the variance of the element position and phase errors. The dependence is smaller when adaptive beamforming is used for compensating for these errors.

The image for the next iteration is then computed using

$$\hat{s}_i(u) = \frac{1}{N} \sum_{n=1}^{N} e_n(i)\, e^{-jku\hat{x}_n} \tag{7-51}$$

Equations 7-45 through 7-51 completely describe one iteration of CLEAN. The iterations continue until the global maximum of $|\hat{s}_i(u)|$ is smaller than either $T_s(i)$ or $T_n(i)$. At this point, the residual image $\hat{s}_i(u)$ is discarded and the targets that were subtracted in the first $(i-1)$ iterations are replaced using CLEAN

beams to form the cleaned image as follows:

$$s_c(u) = \sum_{k=0}^{i-1} \gamma \hat{s}_k(\hat{u}_k) H_c(u - u_k) \tag{7-52}$$

where $H_c(u)$ is equal to the array radiation pattern $H(u)$ within the mainbeam of $H(u)$ and is zero elsewhere.

7-6-6 Design of Random Arrays for Use with Coherent CLEAN

A useful array-design condition from which it is possible to calculate the required number of array elements such that the image is fully CLEANed, that is, all true targets are detected, has been derived in [121]. The derivation is not repeated here; only the basic principle involved behind the derivation and the results are stated. The derivation calculates a lower bound on the required number of array elements by ensuring that the intensity estimate of the largest target at each CLEAN iteration is larger than the intensity of the peak artifact in a probabilistic sense. The resulting lower bound on the number of receivers for a T target scene, denoted N_T, is [121]

$$N_T = TB_p\, e^{\sigma_\alpha^2}(2 - e^{-\sigma_\alpha^2}) \tag{7-53}$$

where σ_α^2 is the worst-case phase error variance at the array elements arising due to element position errors, electronic phase errors, and so on.

When adaptive beamforming is used for phase and element-position error compensation, σ_α^2 is given by Eq. 7-46a with $\hat{u}_i = (u_{\max} - u_0)$; $u_{\max}$ is the largest scan angle away from the beamforming direction u_0; $\hat{v}_i = (\sqrt{1 - u_{\max}^2} - \sqrt{1 - u_0^2})$; and σ_ε^2 is the variance of the residual phase errors at the array elements after adaptive beamforming.

Without adaptive beamforming, σ_α^2 is given by Eq. 7-46b with $u_i' = u_{\max}$, where σ_β^2 is the variance of the elemental phase errors without adaptive beamforming.

The number of array elements N should exceed N_T if all T targets are to be detected by CLEAN. Note that the value of N_T is larger when adaptive beamforming is not used because the phase error variance σ_α^2 is larger.

This completes the description and analysis of coherent CLEAN. The improvement in image quality obtained using CLEAN is the topic of the following section.

7-7 IMAGE QUALITY IMPROVEMENT

Two common image quality measures are *contrast* and *dynamic range*. A third measure, viz. reduction of speckle, is also important for coherent imaging applications. The improvements in contrast and dynamic range due to CLEAN are quantified in the following section. Reduction of speckle noise using CLEAN is discussed subsequently.

7-7-1 Effect of CLEAN on Contrast and Dynamic Range

Image contrast is defined as the ratio of the largest target intensity to the average image artifact level. Dynamic range is the ratio of the largest target intensity to the peak artifact level. The average and peak artifact levels of the cleaned image are determined by the thresholds after the final coherent CLEAN iteration. The signal threshold after the ith iteration can be obtained by combining Eqs. 7-30, 7-47, and 7-49:

$$\begin{aligned} T_s(i) &= T_s(0) - \frac{B_p}{N} \sum_{j=1}^{i} t_s(j) \\ &= \frac{B_p}{N} \left\{ \sigma_n^2 + \sum_{t=1}^{T} A_t^2 - \sum_{j=1}^{i} A_j^2 \, e^{-\sigma_{\alpha_j}^2} (2\gamma - \gamma^2) \right\} \end{aligned} \tag{7-54}$$

The noise threshold can be similarly obtained as

$$T_n(i) = \frac{B_p}{N} \left\{ \sigma_n^2 + \sum_{j=1}^{i} A_j^2 (1 - 2\gamma \exp(-\sigma_{\alpha_j}^2) + \gamma^2 \exp(-\sigma_{\alpha_j}^2)) \right\} \tag{7-55}$$

As the CLEAN iterations progress, the signal threshold decreases and the noise threshold increases. At the end of the CLEAN iterations, the noise threshold is generally larger than the signal threshold (they will be equal if all targets are subtracted out perfectly because no additional noise is introduced by the target subtraction procedure). Thus the peak artifact level of the cleaned image is equal to the noise threshold after all targets have been subtracted out, that is, $T_n(T)$ (the average artifact level is therefore $T_n(T)/B_p$). The peak artifact level of the raw image $\hat{s}_0(u)$ is the initial signal threshold $T_s(0)$ (average artifact level $= T_s(0)/B_p$). The ratio of the peak (average) artifact levels before and after application of CLEAN is equal to the improvement in dynamic range (contrast) due to CLEAN. The improvement in both dynamic range and contrast, denoted I, is therefore $T_s(0)/T_n(T)$, that is,

$$I = \frac{T_s(0)}{T_n(T)} \tag{7-56}$$

Under ideal conditions, that is, when the estimates of system and target parameters at each iteration of CLEAN are error-free, $T_n(T) = B_p \sigma_n^2 / N$ (a loop gain of unity is assumed) and

$$I = \frac{\sigma_n^2 + \sum_{t=1}^{T} A_t^2}{\sigma_n^2} \tag{7-57}$$

The improvement in contrast and dynamic range is clearly scene dependent. Equation 7-57 represents the asymptotic improvement ratio; it cannot be attained in practical systems. The asymptotic value of I for a single point target is the target SNR, that is, $I = A_1^2/\sigma_n^2$. This factor has been quoted as twice the SNR in [118] and [121] because the SNR was defined differently in those references.

The following computer simulations, taken from [121], illustrate the improvement in image quality obtained using CLEAN. Four targets were used in the first simulation. Their amplitudes, phases, and locations are shown in Table 7-1a. These targets were imaged using a 45-element random array 250 wavelengths long. The element locations were known to within one-tenth of a wavelength. The SNR per element was 13 dB with respect to the largest target.

TABLE 7-1

(a) Scene data: the target parameters

Direction ($u = \sin\theta$)	*Amplitude*	*Phase* (*deg*)
−0.39562	1.000	281.4
−0.18286	0.700	148.2
−0.03884	0.562	258.7
0.23211	0.350	2.6

(b) Image data: CLEAN process record (estimate of target parameters during process)

Direction	*Amplitude*	*Phase*
(1) −0.39554	0.84047	281.4
(2) −0.18281	0.75325	169.6
(3) −0.03900	0.51397	250.0
(4) 0.23220	0.25496	338.9
(5) −0.18276	0.21064	56.6
(6) −0.39585	0.13565	266.0

(c) Final estimate of target parameters after collecting targets within a beamwidth[a]

Direction	*Amplitude*	*Phase*
(1) −0.39555 (1.8)	0.97190 (2.8)	279.3 (2.1)
(2) −0.18281 (1.2)	0.69837 (0.2)	153.3 (5.3)
(3) −0.03900 (4.0)	0.51397 (8.5)	250.0 (8.7)
(4) 0.23220 (2.2)	0.25496 (27)	−21.1 (23.7)

[a] Error values in parentheses; direction—percent of one beamwidth; amplitude—percent of true amplitude; phase—phase error in degrees. (From ref. [121])

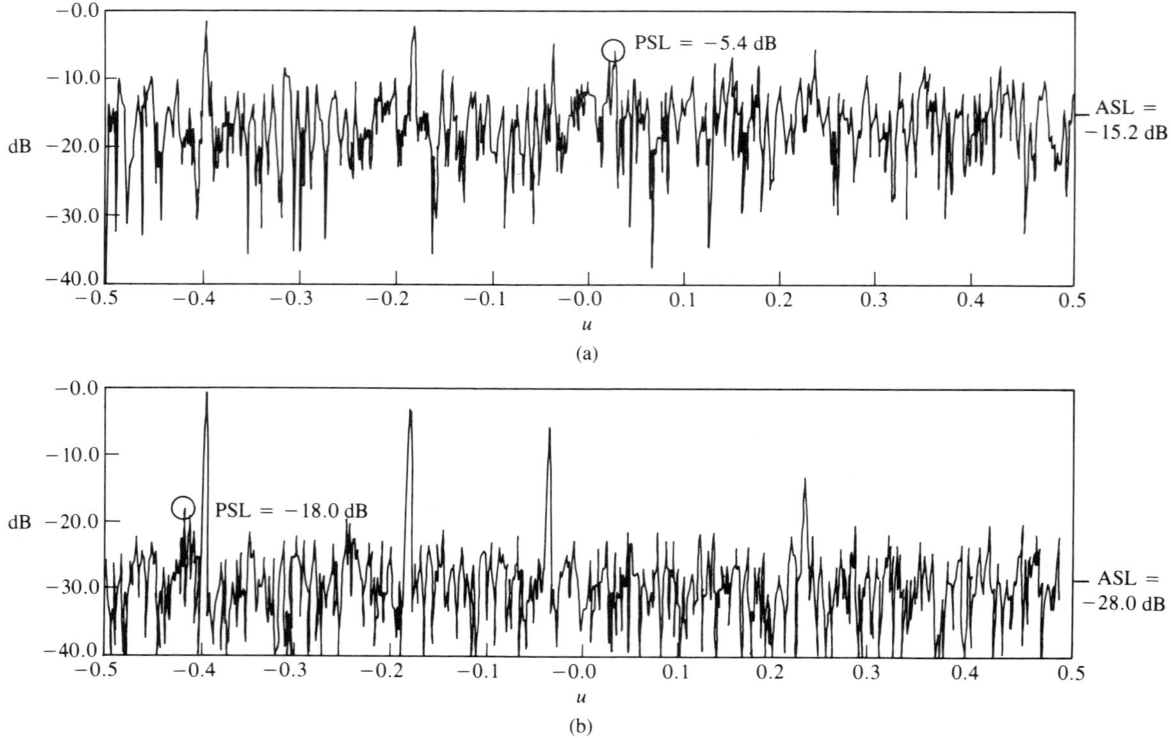

Figure 7-6 (a) Four-target nonCLEANed image. Negligible element-position error. (b) CLEANed image of (a). Imaging conditions: 10 dB SNR; four targets; 45 array elements; array size = 250λ; 0.1λ rms element-position error; $\pm 24^\circ$ field of view from broadside ($u_{\max} = 0.4$). (From ref. [121])

The raw image $\hat{s}_0(u)$ is shown in Fig. 7-6a. The average and peak sidelobe levels are -15.2 and -5.4 dB, consistent with random array theory [94]. It is evident that only the largest target can be identified with certainty; even accepting targets as strong as the next largest permits the possibility of false alarms arising from the sidelobes of the radiation pattern.

When the image of Fig. 7-6a is introduced as the primary image for the coherent CLEAN process, the improved image of Fig. 7-6b results. The average and peak sidelobe levels have dropped to -28.0 and -18.0 dB, an improvement in each of more than 12.5 dB. The theoretical asymptotic improvement in dynamic range and contrast for this target scenario is about 16 dB ($\approx (\sigma_n^2 + \sum A_i^2)/\sigma_n^2$). The actual improvement is smaller because of element-position errors. The theoretical value of I that includes element-position errors was computed, using Eqs. 7-54 through 7-56 with $\sigma_{\alpha_i} = ku_i(\lambda/10)$, to be 12 dB. Thus the experimental and theoretical values of the improvement in contrast and dynamic range agree to within 1 dB.

Table 7-1b shows the complex amplitudes and angular coordinates obtained at each CLEAN iteration. Table 7-1c shows the estimated target amplitudes and locations obtained using CLEAN. The errors between the actual and estimated target parameters are indeed small, as indicated by the numbers within the parentheses in Table 7-1c.

Figure 7-7 and Table 7-2 illustrate the application of CLEAN to a three-target scene. Here the situation is more stressful because the rms element position errors have been increased to 0.3 wavelengths, which is 3 to 10 times the normal tolerance in phased arrays. Again, the results are excellent, and agree closely with theory.

Reduction of speckle noise using CLEAN is discussed next.

7-7-2 Effect of CLEAN on Speckle

Coherent destructive interference in the sidelobe responses of strong scatterers within a target can cause the image of the target to break up into small pieces [33]. Figure 7-8 illustrates this phenomenon in one dimension. Figure 7-8a shows the reflectivity profile of a target that includes two strong scatterers. Figure 7-8b shows the mainlobe and sidelobe responses of the imaging system to the strong highlights. Because of the particular phasing of the echoes, cancellation occurs in the region between the strong scatterers, resulting in the one-dimensional intensity image shown in Fig. 7-8c. It is evident that such a target would appear as a pair of smaller targets.

CLEAN considerably reduces the speckle phenomenon by suppressing artifacts that arise due to leakage of target energy through the sidelobes. This is because removing the sidelobe artifacts also prevents coherent destructive interference in the sidelobe responses of the targets.

The use of CLEAN for eliminating speckle as well as improving contrast and dynamic range is illustrated by the following microwave experiment.

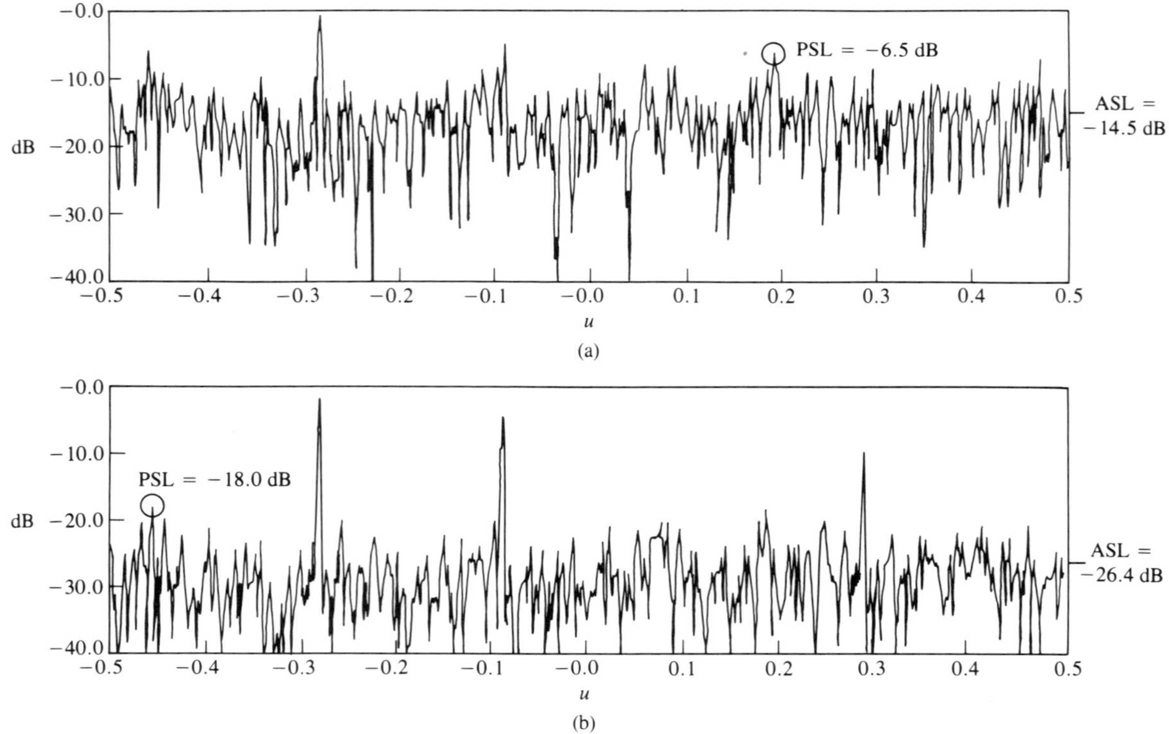

Figure 7-7 (a) Three-target nonCLEANed image. (b) CLEANed image of (a). Imaging conditions: 15 dB SNR; three targets; 40 array elements; array size = 250λ; 0.3λ rms element-position error; $\pm 17.5°$ field of view from broadside ($u_{max} = 0.3$). (From ref. [121])

TABLE 7-2
(a) Scene data: the target parameters

Direction ($u = \sin\theta$)	*Amplitude*	*Phase* (*deg*)
−0.28371	1.000	308.6
−0.08973	0.700	239.9
0.29112	0.350	292.2

(b) Image data: CLEAN process record (estimate of target parameters)[a]

Direction	*Amplitude*	*Phase*
(1) −0.28787 (4.0)	0.79623 (20)	302.5 (6.1)
(2) −0.09027 (13)	0.57991 (17)	252.8 (14)
(3) 0.29146 (8.5)	0.31450 (10)	306.9 (15)

[a] In this example each target was found only once. Error values in parentheses: direction—percent of one beamwidth; amplitude—percent of true amplitude; phase—phase error in degrees. (From ref. [121])

7-7-3 Application of Coherent CLEAN to Experimental Microwave Data

The following microwave experiment was reported in [121]. The data were obtained using a bistatic radar system operating at X-band ($\lambda = 3$ cm). A broad-beam (3.5°) pulsed transmitter illuminated a truck (2 m × 5.5 m) at a distance of 120 m. The transmitter was located slightly below the center of a 28.5-m long receiving array. The two-dimensional image of Fig. 7-9a was formed by laying a raster of successive one-dimensional intensity-modulated images from each range bin. The truck appears to be a collection of small targets. This is an example of the speckle phenomenon.

A one-dimensional deflection-modulated image taken through the cab of the truck (Fig. 7-9b) shows the problem more clearly because it does not suffer from the severe amplitude compression used in Fig. 7-9a. In this cut, it is seen that the target intensity drops severely at three locations. If one counted the number of targets at this radar range by setting a threshold to discriminate against clutter and sidelobe responses (as seen in this figure), one would mistakenly judge this portion of the truck to consist of three or more targets.

Figure 7-9c shows the same cut through the truck after CLEAN has been applied. The improvement is dramatic. CLEAN has removed the image artifacts and as a consequence the threshold has dropped by at least 5 dB. Now only a single target is observed, and this target is the width of the truck. Figure 7-9d is the two-dimensional clean image, showing a well-defined target of correct shape, size, and orientation.

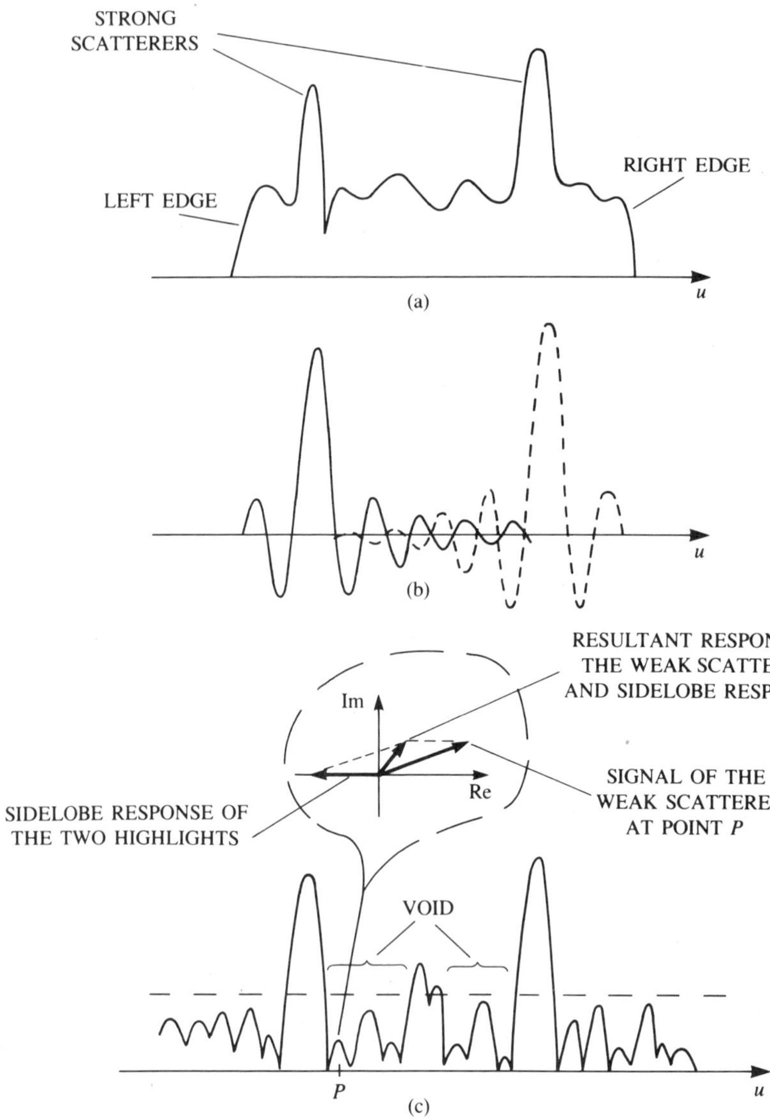

Figure 7-8 (a) Profile of target with two strong scatterers. (b) Main lobe and sidelobes of two strong scatterers. (c) Image of target. Phasor cancellation at point P is shown in detail.

To conduct a more demanding CLEAN experiment, the data were degraded significantly. Three-quarters of the elements were randomly removed from the 200-element array. The remaining elements formed an array whose average and peak sidelobe levels were -17 and -7 dB, respectively. The two-dimensional image of Fig. 7-10a results; it is the image of the same truck taken from a 50-element random array. Now the sidelobe responses are everywhere and it

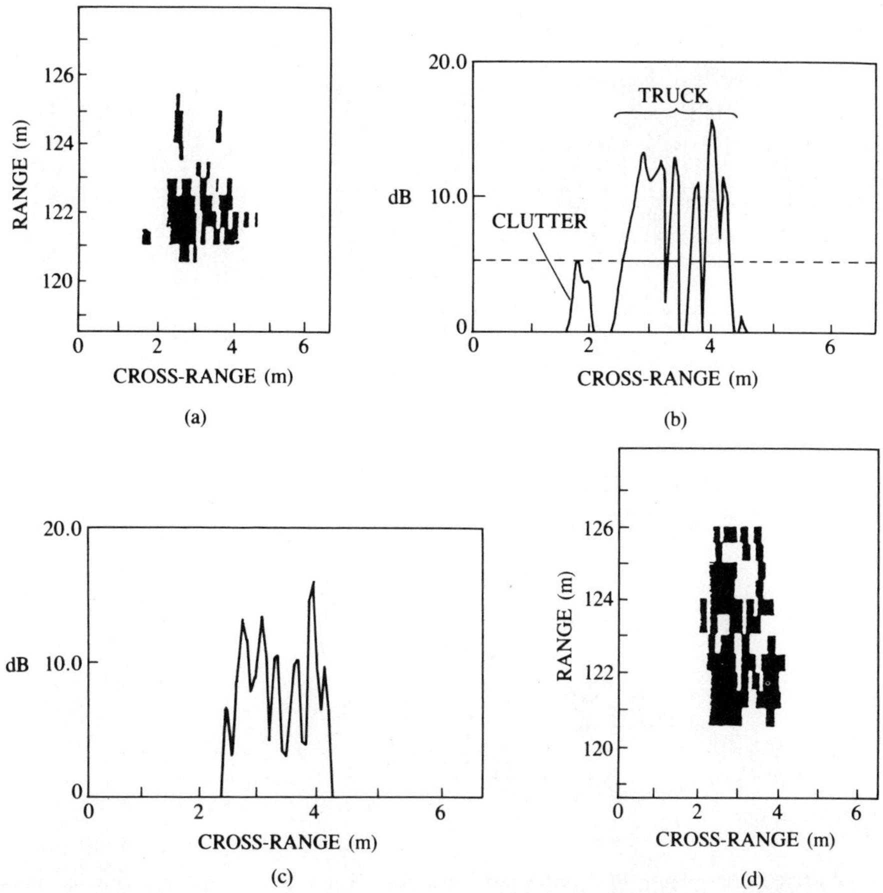

Figure 7-9 Performance of CLEAN technique for periodic array. (a) NonCLEANed two-dimensional image of truck by 200-element array. (b) NonCLEANed one-dimensional image of truck by 200-element array. (c) CLEANed one-dimensional image of truck by 200-element array. (d) CLEANed two-dimensional image of truck by 200-element array. (From ref. [121])

is unclear whether or where a target exists. Figure 7-10b is a one-dimensional image of the truck observed by the 50-element random array.

When the data of Fig. 7-10b initiate the CLEAN process, the one-dimensional image of Fig. 7-10c results. Here the target strength never drops to the residual clutter level and therefore no possibility exists of mistaking the target for a collection of smaller targets. Furthermore, the CLEAN process has reduced the sidelobe level (compare with Fig. 7-10b). The two-dimensional image of Fig. 7-10d is superior to the original image obtained from the high-quality array (Fig. 7-9a).

The comparison of Figs. 7-9d and 7-10d is even more striking. The former image is slightly superior to the latter. However, the 50-element array is much

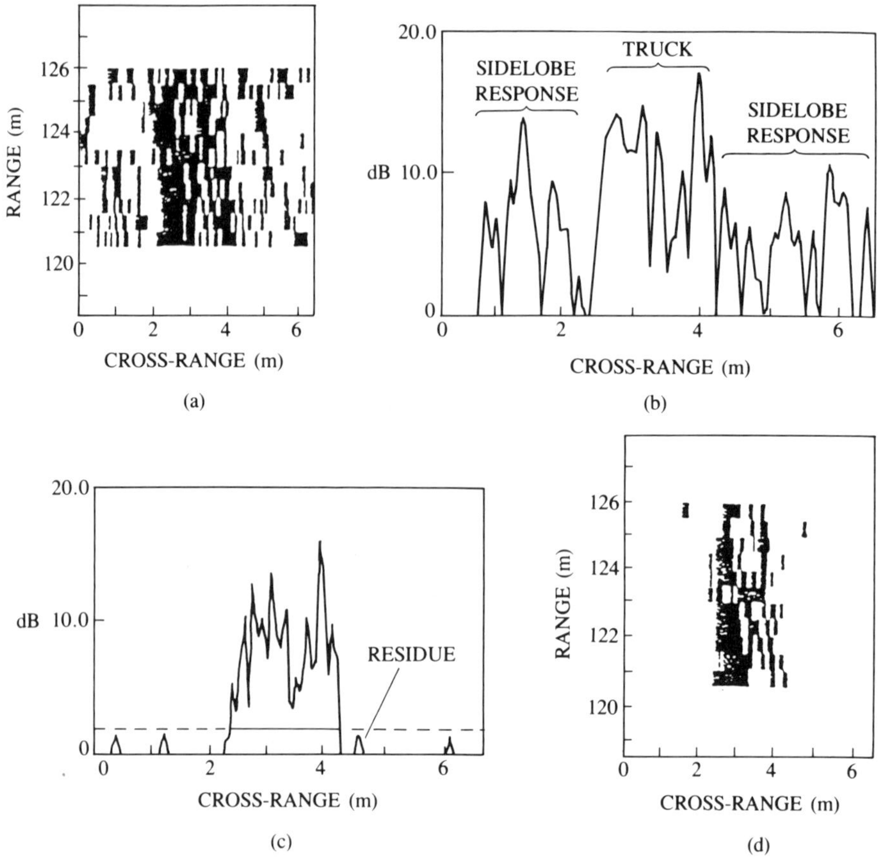

Figure 7-10 Performance of CLEAN technique for random array. (a) NonCLEANed two-dimensional image of truck by 50-element array. (b) NonCLEANed one-dimensional image of truck by 50-element array. (c) CLEANed one-dimensional image of truck by 50-element array. (d) CLEANed two-dimensional image of truck by 50-element array. (From ref. [121])

inferior to the 200-element array; its peak sidelobe level is -7 dB, which is 6 dB higher than that of the 200-element array. Furthermore, the sidelobes of the 200-element array fall off continuously from the main lobe, whereas the 50-element array, being random in structure, has a statistically uniform distribution of high sidelobes. Thus the sidelobe properties of the 50-element array are vastly inferior to those of the 200-element array. In addition, the SNR in the image of the 50-element array is 6 dB poorer than that of the 200-element array. Notwithstanding these differences, the CLEANed image of Fig. 7-10d is almost as high in quality as that of Fig. 7-9d. The differences are primarily due to the lower SNR of the 50-element example because CLEAN attacks the sidelobe energy irrespective of the distribution of the sidelobes.

Thus the application of CLEAN on the experimental data set greatly enhanced both dynamic range and contrast in addition to reducing speckle, as is evident from Figs. 7-9 and 7-10. In addition, the improvement in dynamic range and contrast closely followed theoretical predictions.

The preceding experiment was well designed in that the elemental SNRs were large and the array element positions were known with high accuracy. Therefore the global maxima of the images at each CLEAN iteration were larger than the signal and noise thresholds. This permitted all targets to be subtracted out. When the number of array elements is smaller than the lower bound N_T in Eq. 7-53, the CLEAN procedure will not be able to subtract all the targets out. Incomplete subtraction of target energy will leave gaps in the cleaned image; these gaps change the appearance of the target. Thus it is important to use at least as many elements as dictated by the lower bound of Eq. 7-53 while using CLEAN.

8

ADAPTIVE BEAMFORMING (ABF) I

8-1 DEFINITIONS AND PROPERTIES

Chapter 4 described many problems encountered in long-distance microwave imaging. One problem, that of aperture distortion, is in a category by itself: imaging with large apertures is not possible unless the problem is solved. The distortion is both geometric and electrical. These distortions are intrinsic to all systems sufficiently large to provide the resolution required of long-distance, long-wavelength imaging systems.

Two sets of self-calibration procedures are described in this and the following chapter. Adaptive beamforming (ABF) is the generic term. Other names are *self-cohering*, *self-calibrating*, *self-focusing*, *auto-focusing*, *self-phasing*, and *self-synchronizing*.

Adaptive control of the weight vector of a large phased array is in experimental use today in terrestrial high-resolution microwave imaging as well as in radio astronomy. The radio camera was first described in a concept paper in 1973 [93]. The first radio camera experiment demonstrating ABF was performed two years later [79]. The algorithm upon which most subsequent radio camera experiments has been based was published in 1981 [95], with further details in 1983 [97]. The work reported in these and other references demonstrates that diffraction-limited imagery can be obtained from very large antenna arrays even when they are highly distorted.

Since the early part of the last decade the same principle has been used in radio astronomy to correct for uncertainties in the locations of the phase centers of the dishes that form large radio astronomy arrays, as well as to compensate for wavefront distortion induced by spatial variations in the refractive index of the propagation medium. Review articles by Cornwell for the use of these procedures in radio astronomy are found in [21] and [22].

An interesting property of the radio camera is that it is a near-field instrument. In actuality its means of operation and its performance are independent of whether the target is in the near field or the far field. In practice, however, high-resolution imaging implies near-field operation. The transition between the near field and the far field occurs at that distance at which the wavefront from a source, when viewed at the antenna array, changes from a quadratically curved front to an approximately planar front. This distance is in the neighborhood of L^2/λ, where L is the size of the aperture. At this distance the cross-range dimension of the resolution cell equals L. At greater distances the cross-range dimension is proportionally larger than L and at shorter distances it is proportionally smaller, assuming that the array is properly focused.

An easy test of the near-field vs. far-field operation of the radio camera is whether or not the cross-range resolution is smaller than or exceeds the antenna size. It is easy to see that for practical radar usage the imaging system must automatically and dynamically accommodate to near-field operation. The minimum available beamwidth is λ/L and the minimum available cross-range resolution for a system focused to range R is $\Delta S = \lambda R/L$. A typical value of ΔS is 1 m. The typical range of radar target distances is a few to hundreds of kilometers. The required aperture size is $L = \lambda R/\Delta S$, which at 3 cm wavelength, a target distance of 3 km, and for a required resolution of 1 m, evaluates to 30 m. In this example the resolution cell is 30 times smaller than the aperture; hence the target is very deep in the near field (i.e., close to the antenna).

The practical significance of near-field operation is that the imaging system must focus to the target distance. The radio camera automatically performs this function in the process of adaptive beamforming. ABF, being designed to self-calibrate the system in the presence of arbitrary phase errors across the array, easily accommodates for the quadratic phase variation induced across the aperture by a near-field target. The two groups of ABF techniques that have been developed at the Valley Forge Research Center (VFRC) compensate for aperture distortion and also focus the array at the target distance.

A distinction can be made between self-calibrating procedures according to whether they are local or global. A *local procedure* applies adaptive control to portions within the entire system to correct their errors, but fails to compensate for all the errors in the system. An example of local adaptivity is the use of a pilot signal, that is, a known waveform at a known time, injected into the input of the RF section of a phased array. By measuring the output signal it is possible to deduce the distortions in the receiving path and to transmit feedback control signals to the phase shifters and amplifiers for the purpose of compensating for these errors. This procedure will self-calibrate the receiver but will not correct for element position errors or the distortions caused by weather or propagation conditions.

Global adaptivity applies feedback control over the entire system, simultaneously correcting for propagation medium effects, geometric distortion in the aperture, and phase errors in the distribution network.

Adaptive beamforming as applied to the radio camera is a globally adaptive procedure. Instead of introducing an artificial pilot signal into the receiver, as is the practice when local adaptivity is to be achieved, echoes from the target are used to derive the control for the phase shifters or the weight vector of the array. These echoes experience all the distortions mentioned above. They interact with medium turbulence and weather. Their phases are distorted by element-position error. And they experience all the phase and amplitude errors due to circuit mistunings, component aging, oscillator drift, and so on. It is these distorted signals that are converted into diffraction-limited images by ABF. To do so, the system must correct all errors encountered by the echoes, not merely the ones induced in the receiver. Consequently, the ABF control procedure is global.

A companion procedure to ABF is called *self-survey*. The function of a self-survey procedure is to determine the locations of the parts of a large antenna system, for once the element positions are known, it is then possible to calculate the phase shifts required to form and point a beam in an arbitrary direction. A self-survey system employs a minimum of three, and typically four to six, beacons or transmitters. These sources are distributed in azimuth throughout the sector that the array is designed to scan. Either time of arrival or phase measurements of the beacon signals are made at each of the antenna elements. The number of measurements is sufficient to calculate the entire geometry of the system, that is, the positions of all beacons and antenna elements.

Thus a two-step operation is involved: first is the survey of the system, which is followed by beamforming and scanning. ABF is a more efficient procedure. It recognizes that the function of the antenna array is only the latter of the two operations performed by the self-survey technique. It bypasses the survey stage completely. It ignores where the elements are located and therefore eliminates the requirement to calculate their locations. It makes only those measurements that permit the system to adjust the phase shifts at each element to compensate for the position error. Furthermore, the requirement of the self-survey method that several beacons be deployed over the scanning sector, which limits the use of the procedure to fixed installations in which the user has control of the nearby terrain, is avoided completely. Instead, in the ABF procedures described in this chapter, only a single source in the general direction in which the system is scanning is needed. And experience shows that this source does not have to be a beacon or transmitter; in general, some scatterer in the field of view acts as a suitable replacement. Such a scatterer is called a *beamforming source*. Even that modest requirement is eliminated in the algorithms given in the next chapter.

There are several desirable properties of an adaptive beamforming algorithm:

1. *Independence of array geometry and distortion.* The algorithm should be unconstrained by the locations of the elements or the errors in their positions. In simpler terms, the self-calibration procedure should operate successfully without regard to array design or the nature of the errors in the system. This is accomplished to a large extent in the algorithms described in this and the following chapter.

2. *Independence of source distribution.* This highly desirable property is difficult to achieve. The algorithms described in this chapter impose conditions upon the source distribution while those in the next chapter do not.

3. *Passive.* Adaptive beamforming is most desirable when applied in the passive mode, which means in the receiving system only. When self-calibration is accomplished in this manner, the transmitter and receiver functions are independent and the equipment need not be collocated. This permits bistatic operation (receiver and transmitter do not share a common aperture) and possibly the use of available on-the-air radiation to illuminate the target scene. All the algorithms have this desirable property.

4. *Minimum intra-array communication.* All algorithms require calculations made from complex field measurements at the elements. To perform a calculation based upon all the measurements, there must exist within the system an intra-array communication subsystem to deliver the data to a common point for computing. The preferred algorithms are those that minimize this additional burden upon an already complex system.

5. *Noniterative.* The calculations referred to above can be made at the element level (also called the aperture plane of the array), which is prior to the image processor. Or they can be made in the output or image plane following the image processor. Corrections for the errors are generally applied to the signals entering the image processor. Thus if the calculations are made in the aperture plane, they can be applied directly to the distorted signals. On the other hand, if they are derived from the output of the image processor and delivered to the input signals for correction, a new image must be formed before the corrections can be updated or a new weight vector formed. Such an iterative procedure across the image processor is costly in time and in the total amount of computation required. In general, a noniterative procedure is preferred.

6. *Amplitude only.* It would be highly desirable if amplitude measurements would suffice, for then the imaging system would simplify enormously. Detection would be by an envelope detector at each element. Without phase detectors there would be no need for frequency- and phase-locked local oscillators (LOs). And without such LOs there would be no need for an LO distribution system within the array. (See [97], Chapters 5 and 8, for the system complexities that result from the need to preserve phase.)

No procedure has yet been discovered that either works without phase information or that reconstructs phase from amplitude measurements made in a very sparse array. However, because long-distance imaging is a near-field procedure, there is, in principle, information in the modulus of the complex field measurement that might compensate for the missing phase information. The reason for this conjecture is as follows: When the target is in the far field the relative amplitudes of the element voltages in the array are independent of target distance. Not so in the near field, however, for there the phase and amplitude of the radiation field couple. This means that the set of amplitudes across the array due to radiation or reflection from a target scene that is located in the near field is a function of distance to the target. Consequently, amplitude measurements made at two or more distances from the target convey information about the scene that is normally carried only in the arguments of the measurements.

The second possibility—the reconstruction of phase from amplitude measurements—is a subject of current research [80], [123], [124]. As this field matures, solutions may arise that are useful to the large-array problem. One promising technique is reported by Tsao. Only 10% of the receivers in the array require phase measurement; 90% of the array consists of amplitude-only detectors [120].

8-2 ADAPTIVE BEAMFORMING ON A SINGLE DOMINANT SCATTERER

The primary procedure used at VFRC today assumes that there exists somewhere in the field of view (FOV) of the imaging system a point-like scatterer or source having large radar cross-section or source strength. The theory governing the requirement on cross-section or source strength is given in [96]. Such a target or source radiates a nearly spherical wavefront, thereby inducing a simple phase variation across the array. Provided that the array is linear or planar, the phase variation is linear if the target or the source is in the far field, or approximately quadratic when the source is in the near field. Deviations from such simple behavior indicate geometric distortion in the array, electrical mistunings, or wavefront distortion due to turbulence in the propagation medium.

Consider a one-dimensional, far-field source distribution $s = s(u)$. Its radiation field in the axis of the array is

$$e = e(x) = \mathscr{F}^{-1}\{s(u)\}$$

where the right-hand side is the inverse Fourier transform of the source distribution $s(u)$, and $u = \sin\theta$, where θ is the angle from the normal to the array. For simplicity, operator notation is used in the following development. $\mathscr{F}$ indicates the Fourier transform and $\mathscr{F}^{-1}$ its inverse. The electric field in the x-axis is then written

$$e = e(x) = \mathscr{F}^{-1}s \tag{8-1}$$

Now imagine that the system could measure e and form a weight vector w equal to its conjugate:

$$w = w(x) = e^* = \mathscr{F}s^* \tag{8-2}$$

The radiation pattern of the system is the inverse transform of w, which is

$$f = f(u) = \mathscr{F}^{-1}w = \mathscr{F}^{-1}\mathscr{F}s^* = s^* \tag{8-3}$$

Thus the radiation pattern is the complex conjugate of the source distribution.* The most interesting aspect of this procedure is that no account was taken of the shape of the antenna, the placement of its parts, or geometric or electrical errors in the antenna. To utilize this relationship in an ABF algorithm, whereby the shape of the radiation pattern is the modulus of the angular profile of the source, it is necessary that there be at least one scatterer of small physical size that has a radar cross section (RCS) so large that its reflection dominates all the other scatterers in its range cell.† A corner reflector is an example; its RCS is approximately T^4/λ^2, where T is the length of one side and its reflection profile approximates a δ-function. Let this reflector be at distance $r = r_0$. The source distribution at this range is

$$s_0 = s_0(u) \approx \delta(u - u_0) \tag{8-4}$$

where u_0 is the bearing to the dominant scatterer.

Next, let the adaptive system recognize the presence of the single, dominant scatterer at $r = r_0$ (by the procedure described in Section 8-3) and form its weight vector from measurements of the radiation from that range. Then from Eqs. 8-3 and 8-4,

$$f = f(u) \approx \delta(u - u_0) \tag{8-5}$$

is the radiation pattern of the antenna. When this radiation pattern scans the source distribution $s_i = s_i(u)$ at range R_i, the image delivered by the system is

$$\begin{aligned}\hat{s}_i = \hat{s}_i(u) &= \int f(x)s_i(u - x)\,dx \triangleq f(u) * s_i(u)\\ &\approx \delta(u - u_0) * s_i(u)\\ &\approx s_i(u - u_0)\end{aligned} \tag{8-6}$$

Thus the output is an excellent replica of the target scene except for the unknown angular displacement u_0.

* The weight vector is chosen as the complex conjugate of the electric field to ensure that the radiation pattern, when formed from the e-field of a point source, points to that source. This is called *retrodirective beamforming*. If the conjugate is not employed, the beam points in the mirror or Snells-law direction.

† A range cell or range bin is an interval in range such that all scatterers within it return echoes that overlap in time in the receiver. Its nominal extent is $c\tau/2$, where c is the speed of light and τ is the radar pulse duration.

We now introduce an antenna into the equations. Let the antenna be a phased array that is designed as a sampled aperture of finite length. Let the design location of the nth element be x_{0n} and the weight w_{0n}. The aperture, as designed, can be described by the weighting function

$$w_0 = w_0(x) = \sum^{N} w_{0n}\delta(x - x_{0n}) \tag{8-7}$$

where the weights, as a rule, are real, and are selected to provide shaping to the side radiation pattern. Under the conditions described in Chapter 4 that lead to gain errors and aperture distortion, the actual weight vector differs from Eq. 8-7 and can be written

$$w(x) = \sum w_n \delta(x - x_n) \tag{8-8}$$

where

$$w_n = a_n e^{j\phi_n} \tag{8-9}$$

Now the weights have become complex. The w_n differ from the design values w_{0n} primarily due to the phase errors in the system and the x_n differ from the x_{0n} due to the geometric errors in the locations of the elements of the antenna. Let us use the echoes from r_0 to self-calibrate the array. Then for the most part the moduli of the w_n do not differ significantly from the design values because the wavefront of the radiation from a point source is spherical; hence $a_n \approx w_{0n}$.

The weight vector can be viewed as a transducer that converts the electric field across the aperture to a current $i = i(x)$ that drives the electronic system; i is given by

$$i = we = w\mathscr{F}^{-1}s \tag{8-10}$$

It is the current in the distorted aperture. Let the adaptive beamforming subsystem measure i and form a compensation or correction weight vector

$$c = c(x) = i^* = w^*e^* = w^*\mathscr{F}s^* \tag{8-11}$$

The composite weight vector is the cascade or product

$$wc = ww^*\mathscr{F}s^* \tag{8-12}$$

The term

$$\begin{aligned} ww^* &= \sum_n \sum_m w_n w_m^* \delta(x - x_n)\delta(x - x_m) \\ &= \sum_n w_n w_n^* \delta(x - x_n) \\ &= \sum_n a_n^2 \delta(x - x_n) \triangleq w_1 \end{aligned} \tag{8-13}$$

is a sampling function having sample points at the actual element location x_n, and real weights a_n^2, implying that the phase errors in the aperture have been eliminated. The composite weight vector is

$$wc = w_1 \mathscr{F} s^* \tag{8-14}$$

and the radiation pattern is

$$\begin{aligned} f = f(u) &= \mathscr{F}^{-1} wc = \mathscr{F}^{-1} w_1 \mathscr{F} s^* = \mathscr{F}^{-1} w_1 * \mathscr{F}^{-1} \mathscr{F} s^* \\ &= \mathscr{F}^{-1} w_1 * s^* \end{aligned} \tag{8-15}$$

where $\mathscr{F}^{-1} w_1 \triangleq f_1(u)$ is the radiation pattern of the sampled aperture defined by Eq. 8-13. Assuming, as before, a source distribution at $r = r_0$ given by Eq. 8-4 and the composite weight vector (8-12) formed from measurements of the echoes from that range, the adaptively formed radiation pattern is

$$f(u) \approx f_1(u) * \delta(u - u_0) = f_1(u - u_0) \tag{8-16}$$

The key feature of Eq. 8-16 is that all phase errors have been eliminated by the adaptively-formed compensation weight vector. The aperture weights a_n^2 are real, as they should be. They differ from the design values w_{0n}. However, because of the approximate relationship $a_n \approx w_{0n}$, that difference can be easily corrected by modifying the initial w_{0n} to $\sqrt{w_{0n}}$ to account for the fact that the final weights are a_n^2. Alternately the correction can be introduced in the signal processing computer. The simplest procedure (used for all VFRC imaging) is to set $w_{0n} = 1$ for all n, and to perform the desired aperture tapering upon the stored data records in the computer. With this procedure the square-root relationship between the weights of w_1 and w_0 is avoided and an arbitrary taper most suitable to the particular imaging problem can be applied.

8-3 THE RADIO CAMERA

The adaptive, retrodirective beamforming procedure described above is the key to the design of the radio camera. The radio camera is a large imaging system that operates at wavelengths very large compared to optics, and its large size implies that it suffers from distortions in its shape and in errors in the locations of its parts. Propagation and circuit errors are assumed as well. The means for applying the procedure given in the preceding section are described in detail in Chapter 5 of [97] and in [95].

Figure 8-1 illustrates the logic by which calibration information is extracted from the radiation field from a target. A point source or target, identified as a phase synchronizing source in the sketch, reflects a transmitted pulse toward the array. Because of the small size of the target, it reradiates a spherical

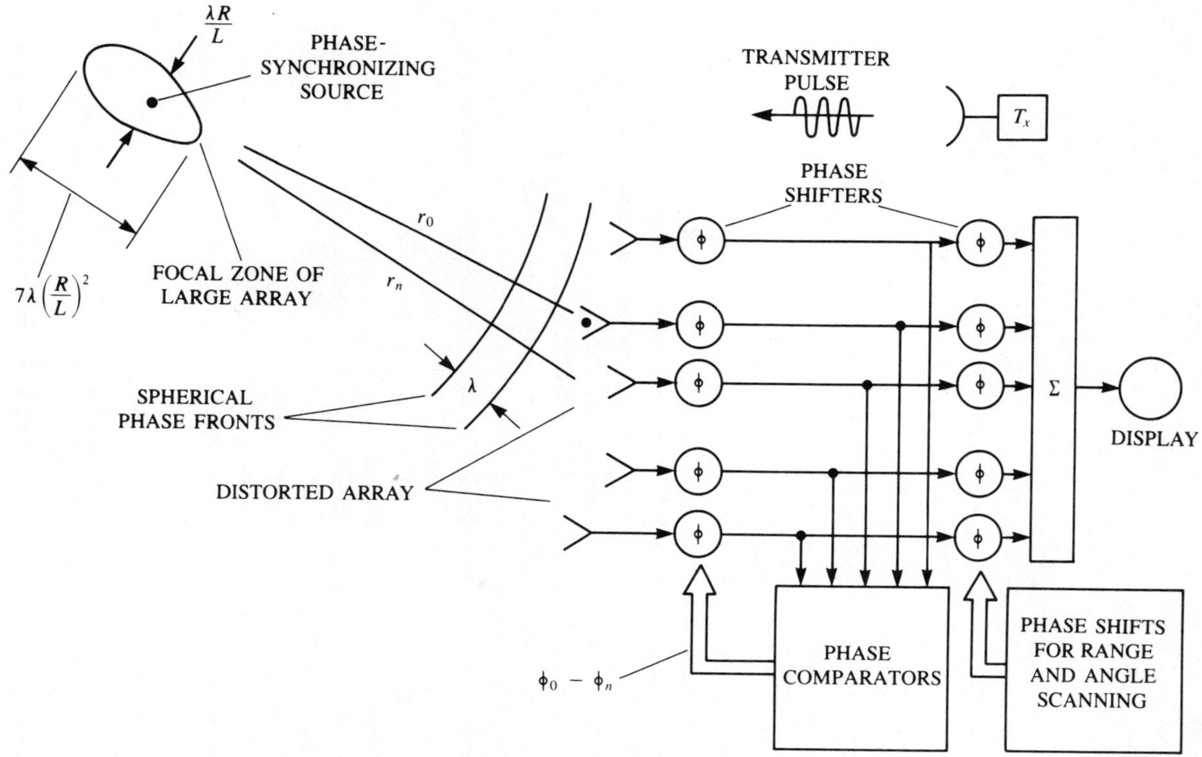

Figure 8-1 Distorted array self-coheres on a phase-synchronizing source: > = antenna elements; ·> = reference element; L = size of array; R = nominal distance to array; r_n = distance to nth element. (From ref. [97])

radiation field, meaning that the amplitude and phase are constant on any sphere centered on the synchronizing source. Assuming free-space radiation, the complex field amplitudes measured by the elements of, say, a linear or planar array, would have equal moduli. Their phases, however, would be random and meaningless because of the geometrical distortion, even with element position errors as little as several tenths of a wavelength. Any circuit mistunings or mismatches would further distort the measured phases. The system compensates for the phase distortion across the aperture by comparing the phases measured at the various elements and deriving feedback-control signals from them to drive the first bank of phase shifters so that the signals passing through the phase shifters are all cophased (described below). Doing so phase shifts each receiver channel by the negative of the received signal. Because the amplitudes are equal and the phase shifts are the negatives of the measured signal phases, the weight vector is the complex conjugate of the field measurements. The first bank of phase shifters and the circuit that controls it is the calibration or compensation subsystem. The sum of the cophased signals is that of an array focused on the phase synchronizing target. The two-dimensional radiation pattern of the array is centered at this point. The beam cross section in the focal zone is $\lambda R/L$ in the lateral or cross-range dimension and approximately $7\lambda(\mathrm{R/L})^2$ in the range or longitudinal dimension.

Once the beam is formed, a second set of phase shifters can scan it to form a map or an image of a target. In essence, the inputs to the second bank of phase shifters can be viewed as the outputs of antenna elements of a properly designed, error-free array. Scanning is an open-loop procedure, exactly as in a conventional phased array. The extent to which scanning can be accomplished is limited, however, by the amount of geometrical distortion in the array. This is because only a single beamforming source is used to calibrate the system. The calibration is correct in the direction of the beamforming source and at its distance. Residual distortions develop as the beam is scanned away from the location of the beamformer. Errors in the array in the direction of the beamforming source and at its range are fully compensated. The errors that dominate under scanning are the distortions in the axis of the array [97, Section 7.3]. Errors normal to the array direction and to the direction to the beamforming source have a relatively minor effect.

The tolerance theory for scanning in the near field of the array is given in [97, Section 7.5]. There it is shown that the maximum scan angle from the focal point is

$$\theta_{\max} = \frac{\lambda}{4\pi\sigma_x} \tag{8-17}$$

where σ_x is the standard deviation of a zero-mean random variable in the x-position of each of the antenna elements in the array. Equation 8-17 is derived on the assumption that the maximum nominal tolerance on the loss of array gain is 1 dB. The angular field of view is twice $\theta_{\max}$ given by Eq. 8-17 and the cross-range dimension of the sector that may be scanned by the adaptively formed beam is R times this value, as given in Eq. 8-18:

$$\Delta S = 2\theta_{\max} R = \frac{\lambda R}{2\pi\sigma_x} \tag{8-18}$$

The extent in range ΔR over which the beam may be scanned is much larger. It is shown in [97, Section 7.5] that the ratio of the dimensions of the scanning sector is given by

$$\frac{\Delta R}{\Delta S} = \frac{\sqrt{12}R}{L} \tag{8-19}$$

where R is the range to the focal zone and L is the size of the array. In general, $\Delta R \gg \Delta S$ because normally $R \gg L$. Thus the position-error limitation to scanning is most heavily felt in the cross-range or angle direction.

The two banks of phase shifters in Fig. 8-1 indicate only that two distinct functions must be performed. These functions can be performed in hardware or within a computer. The procedure used at VFRC is illustrated in Fig. 8-2. A search procedure is indicated by which the imaging system discovers the phase synchronizing source, which is a critical operation for the success of the system.

The search procedure is based upon a minimum amplitude variance (MAV) test upon the echoes streaming across the array. Referring to Fig. 8-2, a transmitter launches a pulse toward a target area, in which is embedded a small target of large radar cross section. A corner reflector is illustrated in the figure, although experience shows that it is rarely needed because targets of opportunity having appropriate scattering properties are plentiful (Section 8-5). The reflected echo from the range bin containing the reference target is dominated by the reflection from that target. The distorted array consists of antenna elements followed by receivers that coherently detect and demodulate the signals. The data are sampled and stored as shown in the lower right of the figure. The sampling interval is approximately the pulse duration. Each row of data is the echo trace received by a single element. Successive rows correspond to successive elements in the array. Each column of data is called a *range bin.* The data represent a snapshot of the radiation field from sources at a distance corresponding to the time of arrival of those signals. For simplicity, it is assumed that range walk or range-bin slippage has been corrected.

To locate the range bin in which the target most closely resembles a point source, the signal processor calculates the normalized variance of the measured amplitudes of the signals received across the array in each range bin. The range bin with the smallest variance is chosen, for that echo sequence most closely approximates the expected field from a point source. Designate that range by R_0. The weight vector adopted by the system is the complex conjugate of the complex column vector of signals received at that range. It is shown in Chapter 6

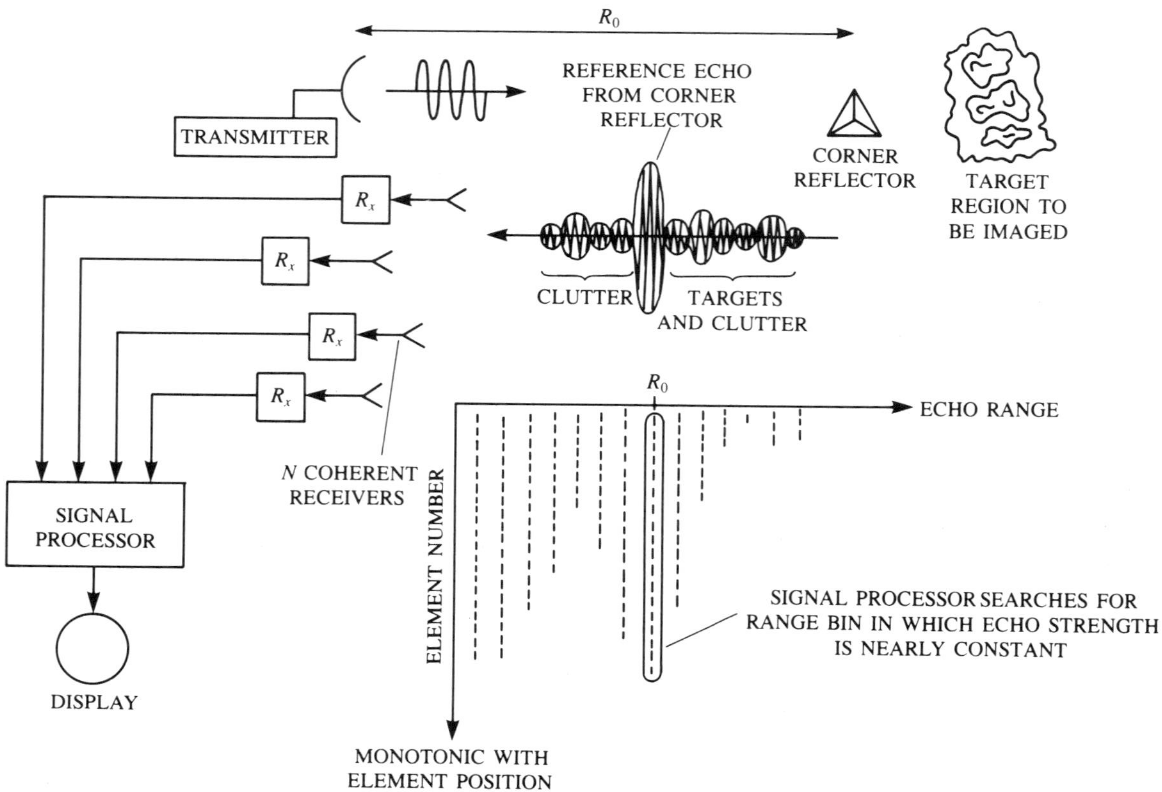

Figure 8-2 Radio camera data format and procedures. (From ref. [95], [97])

TABLE 8-1 Steps in radio camera imaging [95], [97]

1	Measure and store complex envelopes of echo samples	$V_{in}e^{j\psi_{in}}$ (range bin ↗ i; ↖ element number n)
2	Correct amplitudes by dividing by element pattern estimate $\hat{f}_n$	$A_{in}e^{j\psi_{in}}$
3	Find R_0 such that $A_{0n} \approx A$, all n	$Ae^{j\psi_{0n}}$
4	Phase rotate at R_0 by phase conjugate in relation to reference element, $\exp j(\psi_{00} - \psi_{0n})$	$Ae^{j\psi_{00}}$
5	Phase rotate at all range elements	$A_{in}e^{j(\psi_{in} - \psi_{0n} + \psi_{00})}$
6	Focus at each range R_i	$A_{in}e^{j[\psi_{in} - \psi_{0n} + \psi_{00} + (kx_n^2/2)(1/R_i - 1/R_0)]} \triangleq B_{in}$
7	Phase shift linearly with angle	$B_{in}e^{-jkx_nu}$
8	Sum at each range element	$\hat{s}_i(u) = \sum_{n=1}^{N} B_{in}e^{-jkx_nu}$

of [97], and also in [96], that the reference target echo need be no more than 4 dB greater than noise and clutter in that range bin for successful phase synchronization of the array to be achieved.

Table 8-1 describes the algorithm. The measured signals from the ith range bin at the nth element, corrected by the shape of the element pattern, are $A_{in} \exp(j\psi_{in})$. Step 3 finds the reference range R_0 in which the amplitudes A_{0n} are nearly equal. Because the A_{0n} are nearly equal, the measured value at the nth element can be written approximately as $A \exp(j\psi_{0n})$. Its conjugate $A \exp(-j\psi_{0n})$ is the weight vector. Step 4 gives the phase-corrected signals in the zeroth range bin. The corrected inth term is proportional to $A_{in} \exp[j(\psi_{0n} - \psi_{0n})]$ (step 5).

This term has now been corrected for array distortion. It also has been corrected for the near-field curvature of the phase-front associated with the distance R_0 to the phase synchronizing source. Step 6 alters the phase curvature by replacing the quadratic phase variation of the reference range by the proper curvature for the ith range bin, and similarly for all other bins. Following this step, the beam may be scanned in angle, for it is now in focus at all ranges. Steps 7 and 8 complete this process.

High-resolution imaging implies that the resolution cell in the transverse dimension is smaller than the array. The target is always in the near field and the array must therefore be focused. Adaptive beamforming automatically focuses the array to R_0 (step 5). However, it is rarely necessary to perform step 6, which refocuses the array simultaneously to all range bins. At most it is necessary to refocus in steps of hundreds of meters. This is because the depth of field of an array is usually larger than the extent in the range dimension of the sector being imaged.

The depth of field of an aperture (array, lens, dish, etc.) is the extent in range over which the aperture is nominally in focus and is approximately $7\lambda(R/L)^2$ [94].

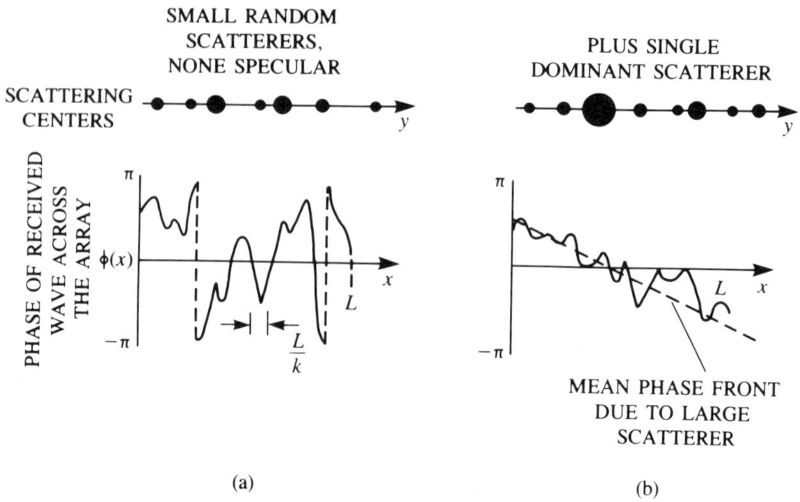

Figure 8-3 Phase distributions across array. (From ref. [94])

The cross-range resolution is approximately $\Delta S = \lambda R/L$. Combining,

$$\text{D.O.F.} = 7\lambda\left(\frac{R}{L}\right)^2 = 7\lambda\left(\frac{\Delta S}{\lambda}\right)^2 = \frac{7}{\lambda}(\Delta S)^2$$

Given a 3-cm system focused to a 1-m beam cross section, the depth of field is about 200 m.

Figure 8-3 (top) illustrates the spatial distribution of scatterers of various sizes located at some range R. The illustration to the left shows a group of small scatterers of random sizes and locations. This distribution models statistically homogeneous clutter, uniformly distributed throughout the illuminated sector. On the right, one large target or scatterer has been added. The bottom sketches picture the difference between the phasefronts of the fields received at the array for the two cases. The phasefront on the left is a sample of a zero-mean random process whereas the phasefront on the right has a well-defined slope determined by the bearing from the array to the large scatterer. Superimposed on the average phasefront is a perturbation field caused by the radiation from all the remaining, small scatterers in the range bin.

The argument that follows is based upon plane wave propagation, but it is equally valid in the near field. The sum of the waves emanating from the scatterers is the radiation field at the receiving array. Only those scatterers illuminated by the transmitting beam contribute to the reradiation field. Call the source function of this field $\rho(u)$. We can write this source function as

$$\rho = \rho(u) = f_T(u)s'(u) \tag{8-20}$$

$f_T = f_T(u)$ is the radiation pattern of the transmitter and

$$s' = s'(u) = \sum_i s'_i \delta(u - u_i) \tag{8-21}$$

is the scatterer distribution, s'_i and u_i being the complex strength and bearing of the ith scatterer. The electric field in the receiving array is

$$\begin{aligned} e = e(x) &= \mathscr{F}^{-1}\rho \\ &= \mathscr{F}^{-1} \sum_i f_T(u) s'_i \delta(u - u_i) \\ &= \sum_i s_i \exp(jkxu_i) \end{aligned} \tag{8-22}$$

where $s_i = s'_i f_T(u_i)$ and $\mathscr{F}^{-1}$ means inverse Fourier transform. The electric field is the sum of randomly weighted (by the s_i) plane waves arriving from random directions u_i, each convolved with w_T.

Let d be the size of the transmitting antenna and L the size of the receiving antenna. The transmitted beamwidth $\Delta\theta_T \approx \lambda/d$ and the width of the clutter patch illuminated at range R by the beam is $\Delta s = \lambda R/d$. The reradiation is lobular and the typical lobe width is $\lambda/\Delta s = d/R$. At the receiver, also a distance R from the clutter patch, the lobe cross section is $(d/R)R = d$, which is approximately the separation between the lobes. These L/d lobes across the receiving array are statistically independent because they are derived from the Fourier transform of a band-limited random process. Thus the phase of the e-field across the receiving array is a band-limited random variable characterized by the order of L/d random values in the interval $[-\pi, \pi]$. Such a field is useless for phase synchronizing the receiving array.

The situation changes dramatically when a large radar cross section (RCS) reflector is added (upper right). The sum of the scattered fields from the clutter does not change, but added to it is the strong plane wave from the prominent scatterer. The e-field can be written

$$e = e(x) = s_0 \exp(jkxu_0) + \sum_i s_i \exp(jkxu_i) \tag{8-23}$$

in which the first term dominates. This field can be viewed as a perturbed plane wave, that is, its direction of arrival is approximately u_0, meaning that its phase slope is nearly but not exactly ku_0 at every point. The phase of the e-field (8-23) is

$$\begin{aligned} \phi(x) &= \tan^{-1} \frac{s_0 \sin kxu_0 + \sum_i s_i \sin kxu_i}{s_0 \cos kxu_0 + \sum_i s_i \cos kxu_i} \\ &= kxu_0 + \tan^{-1}\left(\frac{\sum_i s_i \sin kx(u_i - u_0)}{s_0 + \sum_i s_i \cos kx(u_i - u_0)}\right) \end{aligned} \tag{8-24}$$

The sum field from the random scatterers can be written as

$$A(x) \exp j\alpha(x) = \sum_i s_i \exp(jkxu_i) \tag{8-25}$$

With $A \ll s_0$ throughout the array when s_0 is large compared to the sum of all the other scatterers,

$$\begin{aligned}\phi(x) &\approx kxu_0 + \tan^{-1}\left(\frac{A(x)\sin\alpha(x)}{s_0}\right)\\ &\approx kxu_0 + \frac{A(x)}{s_0}\sin\alpha(x)\\ &= kxu_0 + \delta\phi(x)\end{aligned} \tag{8-26}$$

where $\delta\phi(x)$ is the deviation of $\phi(x)$ from kxu_0.

Equation 8-26 is the phase available for measurement by the system for self-calibration of the array. The first term is the linear phase term with constant slope, which is the desired term. The second term is the error phase. The dominant field due to s_0 need exceed the scattered field (8-25) by no more than 6 dB to enable the dominant scatterer algorithm (DSA) to extract the phase information it needs with adequate accuracy to form and scan a main beam [97, Section 7.6].

The procedure is summarized in the mathematics that follows. Consider a linear array with N elements located at (x_n, y_n), $1 \leqslant n \leqslant N$. The y_n are the deviations in the element positions in the array that are to be corrected by the adaptive beamforming procedure. Let β'_n, $1 \leqslant n \leqslant N$, $0 \leqslant |\beta'_n| \leqslant \pi$, denote the electronic phase error at each of these elements. These errors are caused by mistunings, impedance mismatches, oscillator drifts, and so on. Assume that somewhere in the FOV of the imaging system there exists a range bin containing a single dominant scatterer plus cutter. Let R_0 denote the distance of this range bin from the origin, which is assumed to be located at the first element $(x_1 = y_1 = 0)$ (see Fig. 8-9). Let the bearing to the scatterer be $u_0 = \sin\theta_0$, where θ_0 is the angle measured clockwise from broadside to the array.

The phase of the echo from the R_0 range bin measured at the nth receiver is

$$\phi'_{0,n} = -kr_{0,n} + \beta'_n + \delta\phi'_{0,n} \tag{8-27}$$

where $k = 2\pi/\lambda$, $r_{0,n}$ is the distance from the beamforming scatterer (also called adaptive beamforming source, beamformer, or reference target) to the nth receiver, and is equal to

$$r_{0,n} = [(x_n - R_0\sin\theta_0)^2 + (y_n - R_0\cos\theta_0)^2]^{1/2} \tag{8-28}$$

β'_n is the electronic phase error at the nth receiver and $\delta\phi'_{0,n}$ is the phase perturbation in the spherical wavefront of the beamforming scatterer due to clutter and noise.

The Fresnel approximation for the distance $r_{0,n}$ can be used in most microwave imaging scenarios. Further, the perturbation of the nth element in a direction perpendicular to the array, y_n, will be much smaller than its location along the array, that is, x_n. Expanding Eq. 8-28 in a Taylor's series and ignoring all terms other than the linear and quadratic terms in x and y yields

$$r_{0,n} = R_0 - x_n u_0 - y_n(1 - u_0^2)^{1/2} + x_n^2(1 - u_0^2)/2R_0 \tag{8-29}$$

A similar expression pertains to any other range bin. Thus, $r_{m,n}$, the distance from a scatterer in the mth range bin to the nth element, is found from Eq. 8-29 with all "0" subscripts replaced by "m."

The signal phase $\phi'_{m,1}$ at the array origin is treated as a reference phase for the mth bin and is subtracted from all the phases measured in the array, thereby removing the target distance R_m. In the reference range bin R_0, the difference between the phases at the nth element and the reference element is (substituting Eq. 8-29 into Eq. 8-27)

$$\begin{aligned}\phi_{0,n} = \phi'_{0,n} - \phi'_{0,1} &= k[x_n u_0 + y_n(1 - u_0^2)^{1/2}] \\ &\quad - \frac{kx_n^2(1 - u_0^2)}{2R_0} + \beta'_n - \beta'_1 + \delta\phi_{0,n}\end{aligned} \tag{8-30}$$

where $\delta\phi_{0,n} = \delta\phi'_{0,n} - \delta\phi'_{0,1}$. The negative (or conjugate phase) of Eq. 8-30 is the phase shift introduced into the nth channel to compensate for array distortion and electrical phase errors in the receivers. This is the ABF process.

It is desirable to combine the second, fourth, and fifth terms to form a phase error $\beta_n = \beta'_n - \beta'_1 + ky_n(1 - u_0^2)^{1/2}$ and to treat it as being nearly independent of the viewing angle of the imaging system. The phase errors β'_n and β'_1 are electronic errors and therefore are independent of viewing angle. The last term, on the other hand, is not; it is an explicit function of the angle, which for any scanning problem can vary only over the width of the imaging sector. However, the FOV of most microwave imaging systems is narrow because the transmitter beamwidth is typically the order of a few degrees. Hence this error may be assumed to be independent of this small variation. Consequently, summing the electronic phase error $\beta'_n - \beta'_1$ and the perturbation error $ky_n(1 - u_0^2)^{1/2}$ gives a phase error β_n that is almost independent of the beamformer direction u_0. Using $\beta_n = \beta'_n - \beta'_1 + ky_n(1 - u_0^2)^{1/2}$ in Eq. 8-30 yields

$$\phi_{0,n} = kx_n u_0 - \frac{kx_n^2(1 - u_0^2)}{2R_0} + \beta_n + \delta\phi_{0,n} \tag{8-31}$$

The first two terms point and focus the array to (u_0, R_0). The third term, β_n, is the phase error from the R_0 bin caused by electrical phase distortion and element position errors. Except for the small variation across the FOV, this error is constant for all range bins; hence its inclusion in the weight vector of the array eliminates the primary source of distortion. The fourth term is caused by clutter and noise. Because the clutter-induced phase perturbation is not constant from range bin to bin, the $\delta\phi_{0,n}$ term does not compensate for this source of phase error. Let $\delta\phi_{m,n}$ be the clutter-induced phase perturbation in the mth range bin; then $\delta\phi_{m,n} - \delta\phi_{0,n}$ is the residual phase error after adaptive beamforming and is responsible for the loss in mainbeam gain and the rise in the sidelobe level that result from the ABF process.

The nth component of the phase-compensating weight vector is

$$w_{n,\mathrm{DSA}} = \exp(-j\phi_{0,n}) \tag{8-32}$$

Thus the DSA phase corrections compensate for both electronic phase errors and perturbation of the receivers along the y-axis; however, they are contaminated by clutter and focus the array to u_0 and not to the center of the FOV, which we may call $\bar{u}$. Focusing the array to u_0 leads to a pointing error $\bar{u} - u_0$ in the target location estimates. The multiple scatterer algorithm (MSA), described in Section 8-6, reduces the clutter noise in the phase corrections and yields a better estimate of the target location.

8-4 DSA EXPERIMENTS

Radio camera images have been made with the DSA applied to data from three experimental facilities. The first is a cable-supported synthetic receiving array 83 m in length. This array is 2500 wavelengths at the 3-cm wavelength at which it was operating. Its resolving power is comparable to human vision. The experimental procedure is shown schematically in Fig. 8-4. A single receiver was moved along a cable 10 m above the ground. The cable was 83 m long and supported only at the ends and center. The radar transmitter was directed approximately normal to the array. The receiver with 1-ft dish was pulled from end to end in approximately two minutes. It was turned on 330 times at predetermined locations as it moved over the 83-m course. The estimated uncertainty in antenna location at each of the sample points was of the order of 1 wavelength rms.

Figure 4-8 shows images made with this array. The scene is of rows of houses on streets in the town of Phoenixville with and without adaptive beamforming. The same microwave data were used to form these images. The dominant scatterer adaptive beamforming algorithm was used to form the upper image. Careful examination of the image data shows that the "corrected" array was operating diffraction-limited. The lower picture, without adaptive beamforming, is a useless array of pixels. The image of the housing development in Figs. 2-12 and 5-14 were also made with this array.

The second facility was a physical aperture consisting of 32 X-band (3-cm) receivers each with 12-inch horns (Figs. 8-5a and b). The receivers were spaced

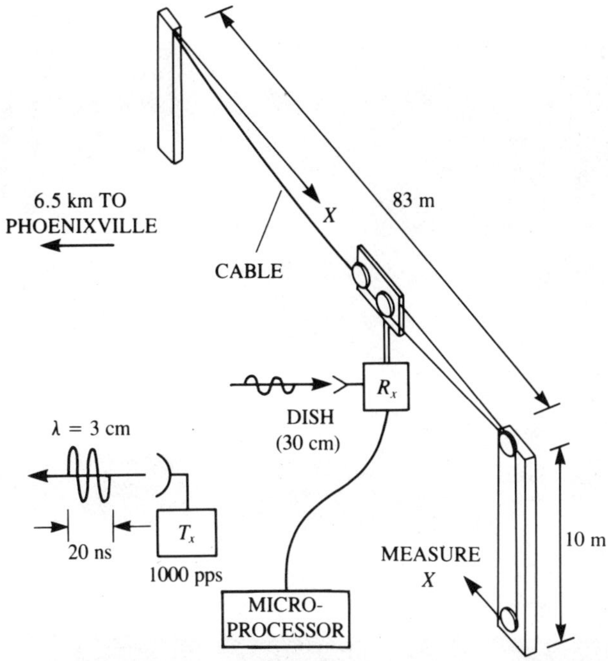

Figure 8-4 Cable-supported synthetic receiving aperture at the Valley Forge Research Center.

(a)

Figure 8-5 (a) 32-Element microwave array. Each module is a microwave receiver at 3-cm wavelength. The mean spacing is 1 m. The elevation look angle of each module is approximately

(legend to Fig. 8-5 continued on next page)

(b)

(legend continued from previous page)
30°. Placement and pointing were done by eye. This array is highly distorted relative to the nominal 0.1λ standard tolerance in array design. This distortion is responsible for the poor imaging performance in Fig. 8-6c; it is electrically corrected by adaptive beamforming, as is demonstrated in Fig. 8-6a. (b) Front and rear views of array modules. (From ref. [102])

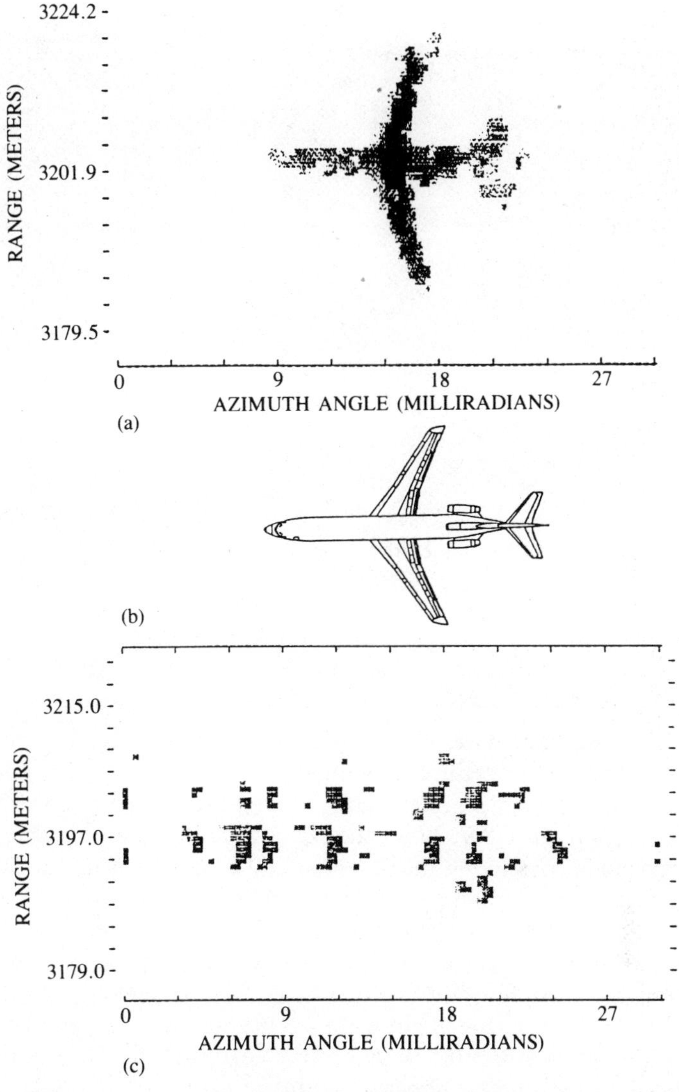

Figure 8-6 Microwave image of Boeing 727 obtained with a 32-m X-band radio camera. The airplane is flying at a distance of 3.2 km from Valley Forge Research Center. (a) With adaptive beamforming. (b) Plan view drawing of aircraft. (c) Without adaptive beamforming. (From ref. [102])

by approximately 1 m, the element placements being inaccurate over the 32-m course of the array. The receivers were placed on a laboratory roof and were not bolted to a common platform. The upper portion of Fig. 8-6 is a picture of a Boeing 727 flying at a distance of 3.2 km from the laboratory made with this array. It is the same aircraft pictured in Figs. 5-6 and 5-13. The lower

Figure 8-7 Monostatic X-band radar used for ISAR imaging, showing pedestal and 1.2-m dish.

portion shows the same data without adaptive beamforming. The importance of the self-calibration procedure is again evident.

A third facility is a single, monostatic X-band radar used in an ISAR mode (see Section 5-7). Figure 8-7 shows the pedestal and 1.2-m dish antenna. The flexible waveguide connecting the antenna to the equipment in the laboratory is visible. Figure 4-9 shows a pair of images made from a common set of data with and without adaptive beamforming. The data are from the same experiment used to produce the image of Fig. 5-15. The location of the antenna did not change from pulse to pulse in this experiment. The distortion instead was due to the perturbations in the flight path of the airplane target due to wind turbulence (Fig. 8-8). Adaptive beamforming corrected the distortion; without it, the echo data produced a hopeless jumble of pixels.

8-5 BEAMFORMER STATISTICS

Dependence upon implanted corner reflectors as adaptive beamforming sources is unrealistic. Practical system design requires that echoes from targets themselves serve for the self-calibration process. Three studies have been conducted of the frequency of occurrence of targets of opportunity that can serve as beamforming sources. The first, by Yadin, examined airborne radar data obtained by the Naval Research Laboratory in the upper UHF band (850 MHz) [128].

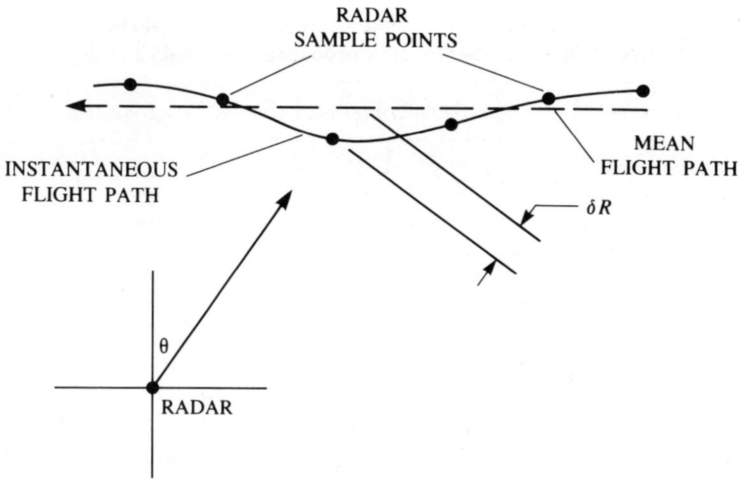

Figure 8-8 Flight path distortion in ISAR.

The data were radar ground clutter obtained from an 8-element receiving system flown at 200 knots over the southeastern portion of the United States at an altitude of 15 000 to 20 000 ft. Because the antenna array was a rigid structure with known characteristics, it was possible to compare the adaptively formed radiation pattern based upon the ground clutter with the intrinsic pattern of the antenna itself. This comparison was made from echoes from a large number of range bins. The criterion of whether or not a particular range bin contained a suitable adaptive beamforming source was determined by the rise in the sidelobe level; an acceptable range bin was one for which the sidelobe level rose no more than 3 dB. Based upon this arbitrary criterion, it was found that approximately 10% of the range bins were acceptable candidates for adaptive beamforming. In view of the fact that at most one range bin per radar trace is needed for the ABF process and that at the most one trace per beamwidth is necessary, the observed frequency of occurrence is encouraging.

A second study also examined the frequency of occurrence of beamforming sources amidst echoes from ground targets. The experiments are in ground-to-ground imaging at very low grazing angles [37]. The array sketched in Fig. 8-4 was employed. Radiation was in the X-band. Data were obtained at distances of 4 to 7 km from farmland, and residential and industrial areas from a town. The results are shown in Table 8-2. These results are not as optimistic as the observed frequency of occurrence in the airborne case but are nevertheless very satisfactory.

The primary differences between the experiments were the grazing angles and the criteria of acceptability of ABF range bin data. The airborne grazing angle was a few degrees while the ground-to-ground grazing angle was nearly zero. The criteria of acceptability were also different. In the airborne experiment,

TABLE 8-2 Variance test effectiveness [37]. All values relative to total number of bins processed. $\sigma_{Ni}^2 \leqslant 0.12$

Run type	*Beamformers available*	*Detected*
Industrial	16.9%	5.8%
Residential	11.6%	1.3%
Undeveloped	2.7%	1.8%

a range bin was accepted when the radiation pattern adaptively formed from its echoes had satisfactory sidelobe properties. A more sophisticated measure was chosen for the ground-to-ground experiments. First, a corner reflector was used to form the weight vector at the array, after which an image of a target sector was obtained. This image is the reference image. Second, the echoes from each range bin were used to obtain a set of trial weight vectors, and an image was formed from each one. Next, the correlation coefficient between each trial image and the reference image was calculated. The criterion for acceptability was a correlation coefficient in excess of 0.88, a value obtained by a subjective viewing test using 63 subjects (see [101] and Chapter 10).

Both tests lead to optimistic inferences. Typically there are hundreds of range bins in a radar trace. Thus, when the frequency of occurrence of acceptable range bins is few percent, the probability is nearly unity that a suitable target of opportunity will be found.

A third study of a different nature involves the use of the dominant scatterer algorithm for adaptive beamforming and imaging on aircraft, which are isolated targets. Here the question is the fraction of aircraft snapshots obtained during the passage of a target through the transmitter beam for which a single range bin proves suitable, that is, for which imaging can be performed. Table 8-3 shows the results of observations made from approximately 100 data sets obtained from about 30 experiments at the Valley Forge Research Center.

TABLE 8-3 Percentage of aircraft imaging experiments in which adaptive beamforming, with the dominant scatterer algorithm, was successful

	Range cell	
	1 m	3 m
Percent success	87	28
Number of experiments	18	11
Number of snapshots	68	36

The results are very satisfactory for the shorter pulse duration. The reason that they were less satisfactory for the longer pulse duration is that the probability that a single scatterer exists in any single range bin decreases rapidly with the size of the range cell.

While these results are satisfactory on the average, they do not guarantee the successful use of this algorithm at all times. A more robust procedure that increases the probability of success still further is described in the next section. Moreover, the dominant scatterer algorithm with the minimum amplitude variance test is sure to fail under those circumstances in which no dominant scatterer exists. An example is the backscatter from the surface of the sea. Fields of vegetation, crowns of trees, and large extents of sand are in a similar category. In these cases, the backscatter is statistically homogeneous and a different type of algorithm is required. Chapter 9 addresses this problem.

8-6 MULTIPLE SCATTERER ALGORITHM (MSA)

When no high-quality adaptive beamforming source is found, a form of range-bin diversity sometimes can be employed to synthesize a useful ABF source. The echo data from several range bins are combined to produce a weight vector superior to what would be available from any of the individual range bins. A modification to the DSA was invented by Attia to deal with this situation [4]. Experiments were reported in [3]. The method is called by its inventor the *robust minimum variance algorithm* (RMVA); here it is called the *multiple scatterer algorithm* (MSA).

Conceptually, the procedure works as follows. In Fig. 8-9 there is a transmitter of beamwidth $\Delta\theta$ illuminating a set of isolated scatterers in different range bins. These scatterers are called *candidate* beamformers, meaning that none is suitable for adaptive beamforming by itself, but that using them together a satisfactory virtual beamformer can be synthesized. Amplitude and phase data are recorded at each element in the array for each of these range bins. The algorithm proceeds as follows. The range bin for which the echo amplitudes at the various receivers in the array have the smallest variance is selected as the reference bin R_0, as in the earlier algorithm. Second, the phases $\phi_{0,n}$ of the received signals from R_0 are subtracted from the phases $\phi_{m,n}$ of the signals of M candidate bins, thereby removing the random phase errors β_n, again as in the former algorithm. With the β_n deleted, the difference phases form quadratic sequences, modulo 2π, across the array. These phase sequences are then "unwrapped" by removing the modulo 2π discontinuities. While the unwrapped phases are corrected for the aperture distortion, there remain residual phase errors caused by the clutter. These clutter-induced residual phase errors are unbiased and therefore can be reduced by averaging. Averaging yields

$$\frac{1}{M}\left\{\sum_{m}^{M}(\phi_{m,n}-\phi_{0,n})^{u}\right\}$$

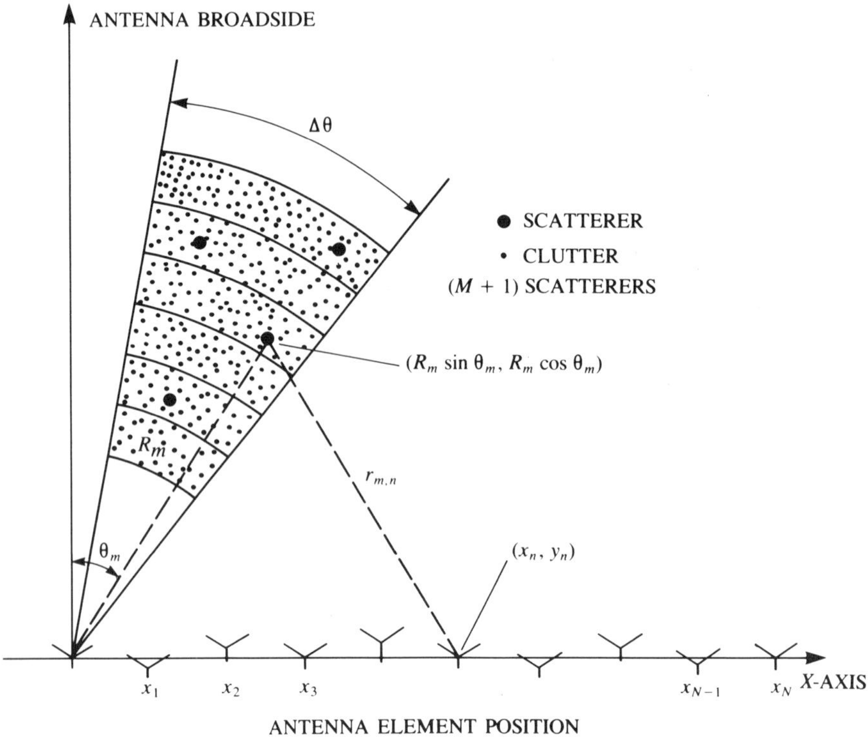

Figure 8-9 Antenna array and target geometry. (From ref. [47])

which is the averaged unwrapped phase of the difference phases of the M candidate beamformers from the reference ABF bin at the nth element. The superscript "u" means unwrapping.

The steps can be summarized as follows:

1. Find R_0.
2. Form $\phi_{m,n} - \phi_{0,n}$ to remove β_n.
3. Unwrap $\phi_{m,n} - \phi_{0,n} = (\phi_{m,n} - \phi_{0,n})^u$ to remove modulo 2π ambiguities.
4. Take the average $(1/M)\{\sum_m^M (\phi_{m,n} - \phi_{0,n})^u$ of the unwrapped phases to reduce the variance of the clutter-induced residual phase errors.

Each of the range bins is assumed to contain a single scatterer embedded in clutter where the scatterer is not sufficiently large to operate the DSA. We call such a scatterer *subdominant*. In none is the target-to-clutter ratio adequate for the DSA to operate satisfactorily. There are $M + 1$ such scatterers. Let R_m, $1 \leqslant m \leqslant M + 1$, be their distances from the origin and $u_m = \sin \theta_m$ their directions relative to the normal to the array. The geometry is shown in Fig. 8-9. The mathematics of Section 8-3 can be carried over directly from the DSA to the MSA by replacing all subscripts "0", denoting range bin R_0, by "m", denoting range bin R_m. $\phi_{m,n}$ is found in this manner from Eq. 8-31. The difference phases are obtained by

$$\phi_{m,n} - \phi_{0,n} = \left[kx_n(u_m - u_0) - kx_n^2\left(\frac{1}{2R_m} - \frac{1}{2R_0}\right) + kx_n^2\left(\frac{u_m^2}{2R_m} - \frac{u_0^2}{2R_0}\right) + \delta\phi_{m,n} - \delta\phi_{0,n} \right]_{\text{mod } 2\pi} \qquad (8\text{-}33)$$

which is then unwrapped:

$$(\phi_{m,n} - \phi_{0,n})^u = kx_n(u_m - u_0) - kx_n^2\left(\frac{1}{2R_m} - \frac{1}{2R_0}\right) + kx_n^2\left(\frac{u_m^2}{2R_m} - \frac{u_0^2}{2R_0}\right) + \delta\phi_{m,n} - \delta\phi_{0,n} \qquad (8\text{-}34)$$

For many microwave imaging systems and target geometries the quantity

$$kx_n^2\left(\frac{u_0^2}{2R_0} - \frac{u_m^2}{2R_m}\right) \ll \pi$$

even in very near-field imaging situations. Further, $1/2R_0 \approx 1/2R_m$; therefore, Eq. 8-34 is nearly linear in x (first term) and is perturbed by the distortions in the phasefronts caused by clutter in range bins R_0 and R_m (last two terms). The average of M such quantities will also be a straight line with a smaller perturbation provided that the clutter in the various candidate beamformer range bins are uncorrelated. The unwrapping is illustrated in Fig. 8-10a, where the dots represent $\phi_{m,n} - \phi_{0,n}$ versus element position n. Each of the M curves $\phi_{m,n} - \phi_{0,n}$ versus n exhibits discontinuities at different values of n. Unwrapping removes these discontinuities and permits averaging the M curves to obtain a linear regression line with smaller perturbations. The symbols "*" at the left and right edges of Fig. 8-10a represent the unwrapped values of the symbols "•". The residual clutter-induced phase errors $\delta\phi_{m,n} - \delta\phi_{0,n}$ in Eq. 8-34 are depicted in Fig. 8-10b.

The success of the procedure depends upon the quality of the phase unwrapping. It is clear that phase unwrapping will be a relatively easy task if the slope of $\phi_{m,n} - \phi_{0,n}$ is small, that is, $k(u_0 - u_m)\Delta x$ is very much less than 2π where Δx is the nominal interelement spacing. The residual phase errors are easily obtained by calculating the differential phases between the unwrapped phases and the linear regression line. The dots in Fig. 8-10b represent $\delta\phi_{m,n} - \delta\phi_{0,n}$ versus n. The slope of a linear periodic array with $\lambda/2$ spacing is $\pi(u_m - u_0)$; $|u_m - u_0|$ cannot exceed the transmitter beamwidth, which is rarely

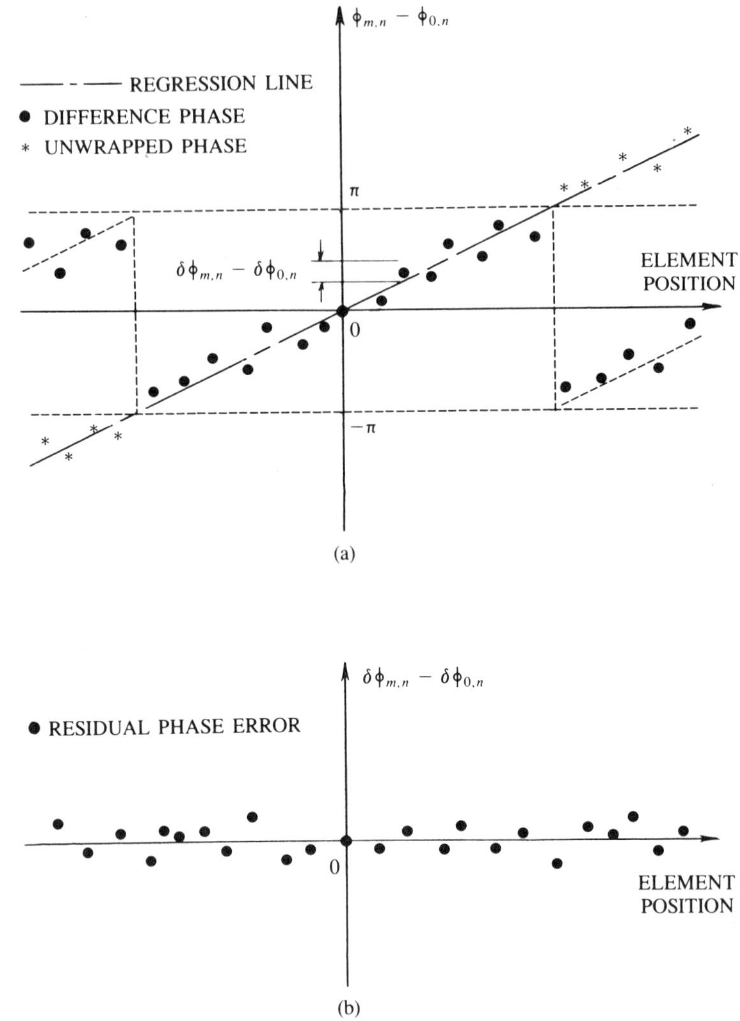

Figure 8-10 (a) Phase unwrapping. (b) Residual phase errors between the regression line and the unwrapped phases. (From ref. [48])

more than a few degrees. Hence in most imaging scenarios there will be many data points between the discontinuities of $\phi_{m,n} - \phi_{0,n}$. However, for an array with large interelement spacing the number of data points between 2π jumps can be small and the phase unwrapping problem becomes more difficult, in which case MSA may prove less applicable. The second requirement for successful unwrapping is that the perturbations in the straight line $kx_n(u_m - u_0)$ due to clutter should be much smaller than π. It is for this reason that the range bin with the smallest normalized amplitude variance (and therefore the smallest clutter-induced phase perturbation) is chosen to unwrap the phases of the candidate beamformers.

The unwrapped value of $\phi_{m,n} - \phi_{0,n}$, denoted $(\phi_{m,n} - \phi_{0,n})^u$, is given in Eq. 8-34. The average of M unwrapped phase sequence curves is

$$\mu_n = \frac{1}{M} \sum_{m=1}^{M} (\phi_{m,n} - \phi_{0,n})^u$$
$$= kx_n\left[\left(\frac{1}{M}\sum u_m\right) - u_0\right] - \frac{kx_n^2}{2}\left[\left(\frac{1}{M}\sum \frac{1-u_m^2}{R_m}\right) - \frac{1-u_0^2}{R_0}\right]$$
$$+ \frac{1}{M}\sum \delta\phi_{m,n} - \delta\phi_{0,n} \tag{8-35}$$

The second, fourth, and sixth terms sum to $\beta_n - \phi_{0,n}$ (see Eq. 8-31). Adding Eq. 8-31 to Eq. 8-35 replaces these terms by β_n, which reduces the bias and random errors. This sum is called ϕ_n^*:

$$\phi_n^* = \mu_n + \phi_{0,n}$$
$$= \frac{kx_n}{M} \sum_{m=1}^{M} u_m - \frac{kx_n^2}{M} \sum_{m=1}^{M} \frac{1-u_m^2}{2R_m} + \beta_n + \frac{1}{M} \sum_{m=1}^{M} \delta\phi_{m,n} \tag{8-36}$$

The negative of Eq. 8-36 is the MSA error-compensating phase shift introduced into the nth channel. Therefore, the nth component of the MSA weight vector is

$$w_{n,\text{MSA}} = \exp(-j\phi_n^*) \tag{8-37}$$

The first two terms in Eq. 8-36 point and focus the array to map coordinates (u, R):

$$\left\{\frac{1}{M}\sum_{m=1}^{M} u_m, \left(\frac{1}{M}\sum_{m=1}^{M}\frac{1}{R_m}\right)^{-1}\right\}$$

The third term is the phase error caused by electrical phase distortion and element-position error. The fourth term is the averaged clutter-induced phase error of the M candidate beamformers. This term is smaller than the corresponding DSA term ($\delta\phi_{0,n}$ in Eq. 8-31) because of the averaging.

8-7 COMPARISON OF DSA AND MSA

8-7-1 Angular Accuracy

In this section, the angular biases and errors in the DSA and MSA are compared. Equations 8-31 and 8-36 define the error-compensating phase shifters. The first term in Eq. 8-31 is $kx_n u_0$ and in Eq. 8-36 it is $(kx_n \sum_m^M u_m)/M$.

These terms steer the beams. The DSA centers the phase-corrected image on u_0, whereas the MSA centers the phase-corrected image on $(\sum_m^M u_m)/M$, which is the average of the locations of the M candidate beamformers. The average is more likely to be closer to the center of the FOV than the location of the reference beamformer u_0. Therefore, one advantage of the MSA over the DSA is that it focuses the array closer to the center of the FOV than the DSA, thereby reducing the uncertainty in target location.

An estimate of the improvement in angular accuracy is obtained using the theorem in statistics that the standard deviation of the sample mean of M independent random variables varies as $M^{-1/2}$, and the realistic assumption that the angular locations of the candidate beamformers are random and independent. The expected location of u_0 is within the main lobe (of width $\Delta\theta$) of the transmitting beam, for otherwise the target would be too weakly illuminated to become the reference reflector. For the same reason the directions to all the other candidate beamformers also fall within the main beam. These bearings to the targets may be assumed to be approximately uniformally distributed within the main beam because there is no physical reason for their distribution to be biassed. Thus, after averaging, the standard deviation of the sample-mean pointing direction is $\sigma_\theta/M^{1/2}$. Hence the accuracy in the angular coordinate of the microwave image is improved by the factor $M^{1/2}$.

The second terms in Eqs. 8-31 and 8-36 focus the array to R_0 and $[(1/M)\sum R_m^{-1})]^{-1}$, respectively. There is no advantage of one over the other because the quadratic phase curvature from either distance is deterministic and therefore can be removed to permit refocusing the array for range bins lying outside the depth of field of the array.

8-7-2 Residual Phase-Error Variance

The third terms in Eqs. 8-31 and 8-36 are the phase errors caused by electrical phase distortion and element position error. This error is constant for all range bins and is corrected by both ABF procedures.

The fourth terms in Eqs. 8-31 and 8-36 are the clutter-induced phase errors. The MSA term is smaller because of the M-fold averaging. Let σ_{DSA} be the standard deviation of $\delta\phi_{0,n}$ of Eq. 8-31 and σ_{MSA} be the standard deviation of

$$\frac{1}{M}\sum_{m=1}^{M} \delta\phi_{m,n}$$

of Eq. 8-36. Given the realistic assumption that all phase errors are independent, and under the assumption that the $\delta\phi_{m,n}$ and $\delta\phi_{0,n}$ have common variances, the relation between the phase-error variances is

$$\sigma_{\text{MSA}}^2 \approx \frac{1}{M}\sigma_{\text{DSA}}^2 \tag{8-38}$$

The significance of Eq. 8-38 lies in its influence upon the radiation pattern. σ^2_{MSA} and σ^2_{DSA} are the variances of the phase errors in the phase corrections across the array after ABF. It is these residual errors that determine the qualities of the radiation patterns formed by these algorithms. The effects of random, uncorrelated phase errors were discussed in Section 3-6-2. Equation 3-32 describes the expected power pattern Eff^* when phase errors are present and amplitude errors are not. It is convenient to normalize Eq. 3-32 so that the on-axis ($u = 0$) gain in the absence of phase errors is unity. These changes lead to

$$Eff^* \approx e^{-\sigma^2} f_0 f_0^* + \frac{\overline{a^2}}{\bar{a}^2} \frac{\sigma^2}{N} \tag{8-39}$$

For the problem being discussed σ^2 is either σ^2_{DSA} or σ^2_{MSA}. $f = f(u)$ is the radiation pattern of the array with phase error, $f_0 = f_0(u)$ is the design pattern, $\overline{a^2}/\bar{a}^2$ is the square of the ratio of the rms to mean value of the aperture weights and is normally close to and slightly greater than unity, and N is the number of elements. Clearly $\sigma^2 > 0$ reduces mainlobe gain and increases sidelobes. Only when σ_{DSA} is not sufficiently small is there reason to reduce the residual phase-error variance with the more complicated algorithm. For example, assume that $\sigma^2_{\text{DSA}} = 0.1(\text{radian})^2$, $N = 300$, the allowed loss in mainlobe gain is 0.5 dB and the sidelobe specification, after ABF, is -30 dB. The expected mainlobe gain, relative to the design gain, is $e^{-0.1} \approx 0.9$ or -0.43 dB, which is acceptable. However, the sidelobe level is too high. The average sidelobe is approximately 0.1/300 or about -35 dB and the peak sidelobe is typically 10 dB higher, which exceeds the specification by 5 dB. Equation 8-38 indicates that MSA will meet the specifications for a value of $M = 3$.

However, Eq. 8-38 is somewhat optimistic because the second assumption, equality of variances, is not often satisfied. The "best" range bin is selected for the DSA, implying that the rms of the phase perturbations $\delta\phi_{m,n}$ of the MSA exceeds that of the $\delta\phi_{0,n}$ of the DSA. Let $\alpha \geqslant 1$ be the average of the variances in the M candidate beamforming bins relative to the DSA reference ABF bin. Then the proper relation between the phase-error variances of the two algorithms becomes

$$\sigma^2_{\text{MSA}} = \frac{\alpha}{M} \sigma^2_{\text{DSA}} \tag{8-40}$$

Because MSA is a more complicated algorithm, achievement of the improvement given by Eq. 8-40 requires additional computation. Experience with the algorithm indicates that the computation is 2 to 3 times larger than the DSA process.

MSA plays the role of a threshold extender to DSA. The lowest-amplitude variance bin is selected for the DSA reference. Only if it proves inadequate is the more complicated MSA employed. The reference bin data are used to unwrap the phases of a small set of the M candidate beamformers. A few dB reduction in the phase-error variance has been observed in ground clutter experiments. This advantage is demonstrated in the next section.

8-7-3 Image Correlation

The DSA and MSA can also be compared by correlating the DSA and MSA phase-corrected images with a perfect image. The correlation measure has proved useful in comparing microwave images [101]. The correlation between two images $I_1(u)$ and $I_2(u)$ is

$$\rho \triangleq \frac{\displaystyle\int_{-1}^{1} I_1(u) I_2(u)\, du}{\left(\displaystyle\int_{-1}^{1} I_1^2(u)\, du \int_{-1}^{1} I_2^2(u)\, du\right)^{1/2}} \tag{8-41}$$

where $I_1(u)$ is the error-free image of the target scene and $I_2(u)$ is the erroneous image. Assume that the perturbations in the phase corrections caused by clutter are independent and identically distributed Gaussian random variables with zero mean and variance σ^2. A lower bound on the correlation ρ between the error-free and erroneous images has been derived in [46] in terms of the residual phase-error variance; it is given by

$$\rho_0 = \tfrac{1}{2}(1 + e^{-\sigma^2}) \approx e^{-\sigma^2/2} \tag{8-42}$$

so that $\rho \geqslant \rho_0$. By comparing Eqs. 8-42 and 8-39 we see that the expected value of the mainbeam gain relative to the design gain G_0 is the square of ρ_0:

$$E\left[\frac{G}{G_0}\right] = \exp(-\sigma_{\delta\phi}^2) \approx \rho_0^2 \tag{8-43}$$

8-8 DSA AND MSA EXPERIMENTS

The 83-m, 330-element quasilinear array pictured in Fig. 8-4 was used to image the nuclear power plant in Limerick, Pa, located 17.6 km from the array. The array was highly distorted; the locations of the receivers differed from their assumed locations by one or more wavelengths in each dimension.

Beamforming bins having normalized amplitude variances not larger than 0.12 have been shown to yield satisfactory images when the DSA is used for phase cohering an array [37]. A survey of the normalized amplitude variances of the 75 range bins in the FOV yielded 6 bins that could possibly contain a single dominant reflector embedded in clutter. These range bins were ordered in accordance with their variances and numbered 0 through 5; their normalized amplitude variances were 0.06, 0.07, 0.09, 0.10, 0.10, and 0.12. Bin 0 contains the highest-quality beamformer. A comparison was made between the DSA and the MSA both using bin 0 as the reference bin. Bins 1 and 2, the next higher-quality bins, were used as the candidate bins for the MSA test. These results, shown in parts (a) and (b) of Figs. 8-11 to 8-13, compare the DSA and the MSA under low-variance conditions, that is, highly favorable for imaging.

A second comparison was made using high-variance range bins, none of which was favorable for imaging. These range bins had normalized amplitude variances of 0.20, 0.21, and 0.23 and were numbered 6 through 8. The results of these experiments are shown in parts (c) and (d) of Figs. 8-11 to 8-13.

8-8-1 1-D Cross-Range Images of Bin 2

Figure 8-11 shows the cross-range images of bin 2 obtained from the four experiments. The dashed line, which is 25 dB below the peak, is a typical threshold value below which signals are not displayed in range–azimuth imagery. It is close to the average sidelobe level of $10 \log(1/N)$, where N is

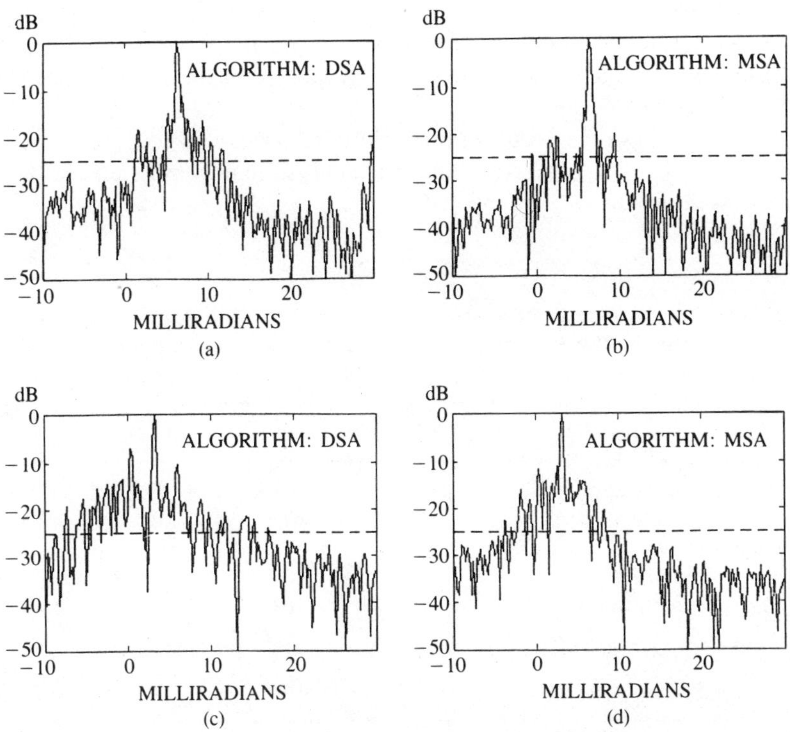

Figure 8-11 Cross-range image of bin 2 of Limerick nuclear power plant. *Low-variance case:* (a) DSA with bin 0 as the reference bin. (b) MSA with bin 0 as the reference bin and bins 1 and 2 as the candidate bins. *High-variance case:* (c) DSA with bin 8 as the reference bin. (d) MSA with bin 8 as the reference bin and bins 6 and 7 as the candidate bins. (From ref. [48])

the number of antenna array elements. The upper two parts compare the low-variance experiments and the lower two parts compare the high-variance experiments.

Figure 8.11a shows that the DSA peak sidelobe level (PSL) is about −13 dB. Figure 8-11b shows the improvement due to the MSA: the PSL is decreased to about −16 dB. Furthermore, the sidelobe artifacts around the main beam, evident in the DSA image, have been largely removed.

The high-variance comparison is made in the lower two figures. Figure 8-11c shows the performance of the DSA with bin 8 as the reference bin. The PSL has risen to about −7 dB and many sidelobes rise above the threshold. This experiment demonstrates that bin 8, whose normalized amplitude variance is 0.23, is a very poor beamformer for the DSA. However, when additional candidate beamformer bins of equally poor quality are added, the MSA performs reasonably satisfactorily. Figure 8-11d shows this case. The PSL has decreased to about −11 dB. This MSA image is almost as good as the DSA image of Fig. 8-11a formed from the highest-quality beamformer.*

In addition to a comparison based upon the PSL, a second useful comparison is the number of apparent targets appearing above the display threshold. Both values are listed in Table 8-4 for the four cases, along with the normalized amplitude variance information. From the table it is evident that the MSA high-variance case approaches the quality of the low-variance DSA case.

Figure 8-12 shows the details of the residual phase errors across the aperture for the four cases. The rms errors and the losses in mainbeam gain (calculated from Eq. 8-43) are listed in Table 8-5.

TABLE 8-4 Comparison of DSA and MSA self-calibrating algorithms in low- and high-variance cases. The MSA high-variance case is almost as good as the DSA low-variance case

	Low variance		*High variance*	
Self-calibration algorithm	DSA	MSA	DSA	MSA
Experiment in Fig. 3-11	a	b	d	d
Amplitude variance				
Reference beamformer	0.06	0.06	0.23	0.23
Candidate beamformers		0.07		0.20
		0.09		0.21
Peak sidelobe (dB)	−13	−16	−7	−11
Number of targets > −25 dB	15	10	26	15

* It is interesting to observe the shift in the optical axes of the lower images relative to the upper ones. The radio camera points in the direction of the reference reflector. The shift reflects the difference in directions of the reference reflectors in bins 0 and the virtual MSA source.

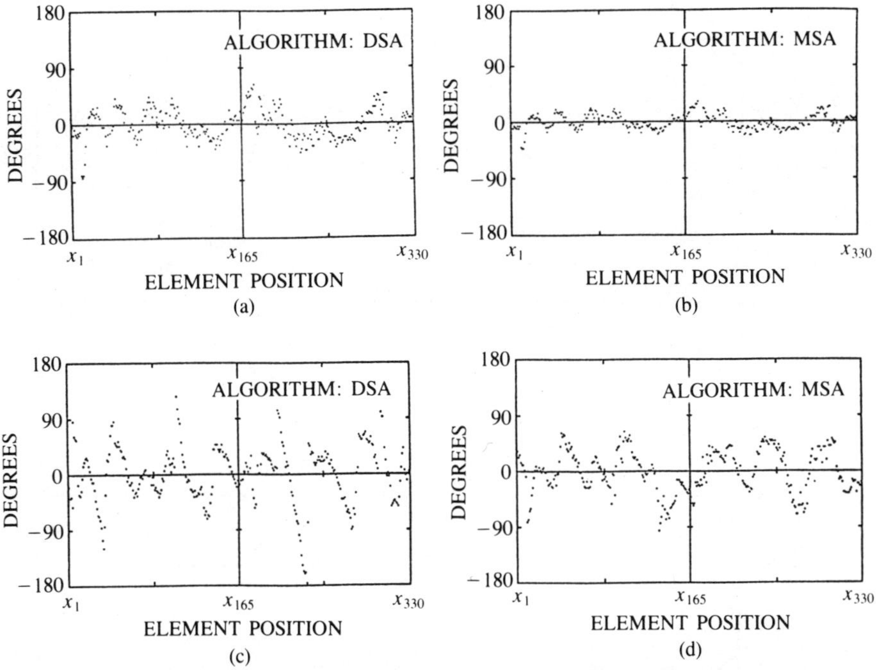

Figure 8-12 Phase errors of bin 2 of Limerick nuclear power plant. *Low-variance case:* (a) DSA with bin 0 as the reference bin. (b) MSA with bin 0 as the reference bin and bins 1 and 2 as the candidate bin. *High-variance case:* (c) DSA with bin 8 as the reference bin. (d) MSA with bin 8 as the reference bin and bins 6 and 7 as the candidate bins. (From ref. [48])

TABLE 8-5 Residual rms phase errors for the four cases of Fig. 8-11 and the losses in mainbeam gain that result from these errors

	Low variance		*High variance*	
Self-calibration algorithm	DSA	MSA	DSA	MSA
Experiment in Fig. 3-12	a	b	c	d
Residual rms phase error (deg)	23.57	10.07	46.74	33.35
ΔG (dB)	−0.73	−0.13	−2.89	−1.47

8-8-2 2-D Range–Azimuth Images

Figure 8-13 shows range–azimuth images of buildings at the site of the Limerick nuclear power plant. The upper two images, both obtained from low-variance beamformers, are very similar and both are satisfactory. Although the MSA reduces the phase errors and thereby intrinsically produces a higher-quality image, the DSA image of Fig. 8-13a is of sufficient quality that use of the MSA is not warranted.

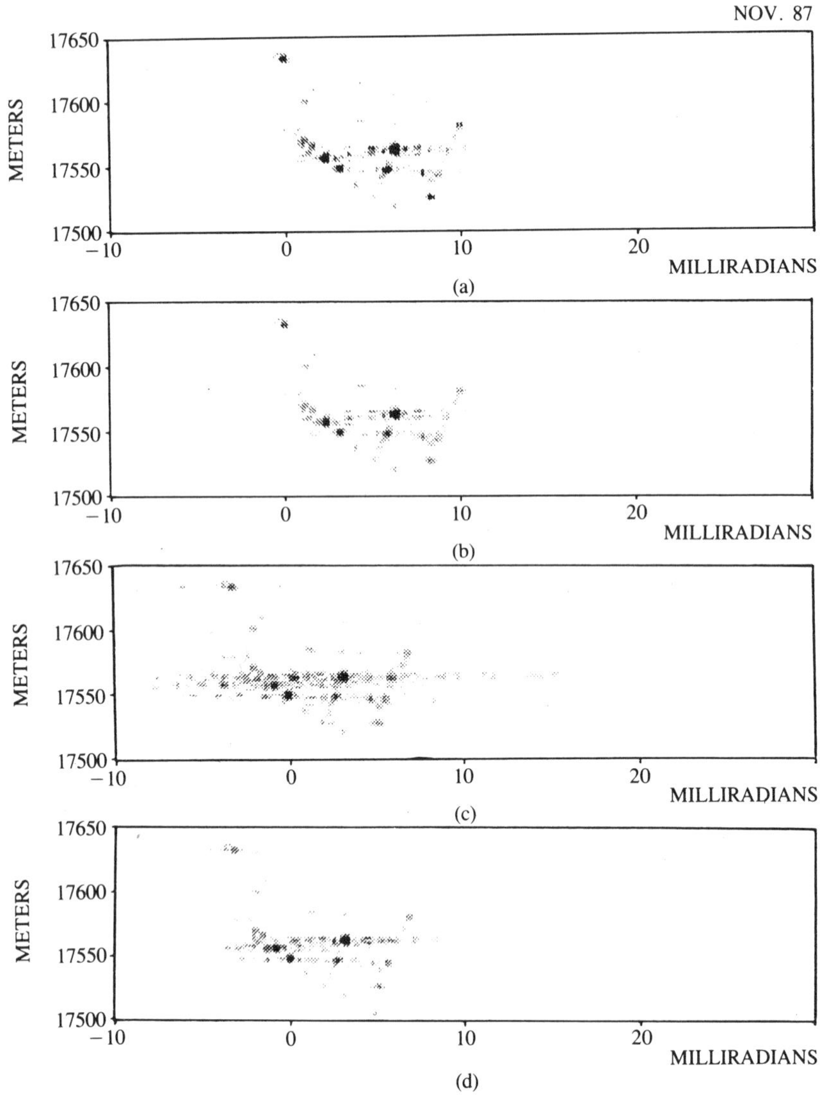

Figure 8-13 Range–azimuth images at site of Limerick nuclear power plant. *Low-variance case:* (a) DSA with bin 0 as reference bin. (b) MSA with bin 0 as the reference bin and bins 1 and 2 as the candidate bins. *High-variance case:* (c) DSA with bin 8 as the reference bin. (d) MSA with bin 8 as the reference bin and bins 6 and 7 as the candidate bins. (From ref. [48])

Not so with the imagery from the low-quality beamformers. Figure 8-13c shows the DSA image driven by a range bin with a normalized amplitude variance of 0.23. This image is badly degraded by high sidelobes; it is not a satisfactory image. The same low-quality reference beamformer plus two equally low-quality candidate bins produce a considerably improved MSA image

(Fig. 8-13d). Here the sidelobe artifacts have been largely removed and the image is not significantly inferior to the high-quality images in Figs. 8-13a and 8-13b.

8-8-3 Image Correlation

It is interesting to form an image correlation comparison based on the definition given in Eq. 8-41. The best of the four images is Fig. 8-13b, as determined by the residual rms phase data. Using it as the reference image, its correlations with Figs. 8-13a, c, and d are 0.98, 0.87, and 0.92, respectively. In [101] it was shown that a correlation coefficient of 0.9 or greater indicates satisfactory image formation. Based on that criterion, experiment (c) fails while (a) and (d) are satisfactory. Thus we again see that three range bins having normalized amplitude variances 0.20 or greater can be used with the MSA to obtain a useful image, one nearly as good as the image obtained using the DSA with the best beamformer. Moreover, if only one of the three range bins is used for beamforming purposes, the resulting image will have large sidelobe artifacts, as is demonstrated in Figs. 8-11c and 8-13c. In addition, the loss in mainlobe gain relative to experiment (b) can be calculated using Eq. 8-43. Table 8-6 summarizes these findings.

TABLE 8-6 Image correlations and losses of mainbeam gain relative to experiment (b), again demonstrating that the MSA high-variance case (Fig. 8-13d) is nearly as good as the DSA low-variance case (Fig. 8-13a)

	Low variance		*High variance*	
Self-calibration algorithm	DSA	MSA	DSA	MSA
Experiment in Fig. 8-13	a	b (reference)	c	d
Image correlation	0.98	1.0	0.87	0.92
ΔG (dB)	−0.18	0	−1.21	−0.72

The last experiment illustrates the capability of a dominant scatterer algorithm to image a distant target with human-optical resolution. The VFRC 83-m array was used to image an industrial site in Phoenixville, Pa, located at a distance of 8.2 km from the array. Figure 8-14a is an aerial photo of the Cromby Power Plant which shows two high stacks, a water tower, the power plant building, and two connected storage silos. The microwave image of the white blocked area in the upper figure is shown below. The radar line of sight is indicated by the white vertical line in the center of the upper figure. The radio camera image was obtained by applying the MSA to the best set of three range bins; it reveals the two high stacks (A and B), the water tower (C), the two connected silos (D), and the power plant building (E). The resolution cell is 3 m

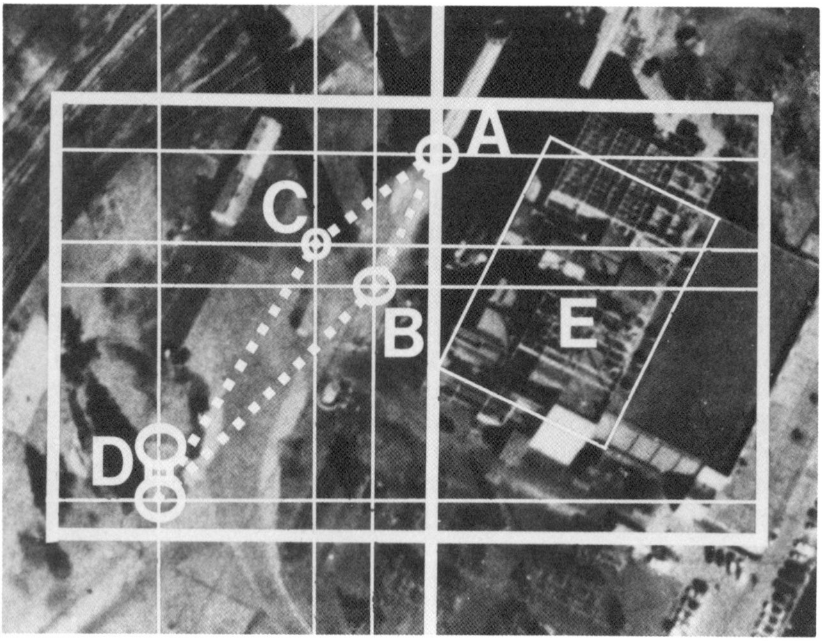

(a)

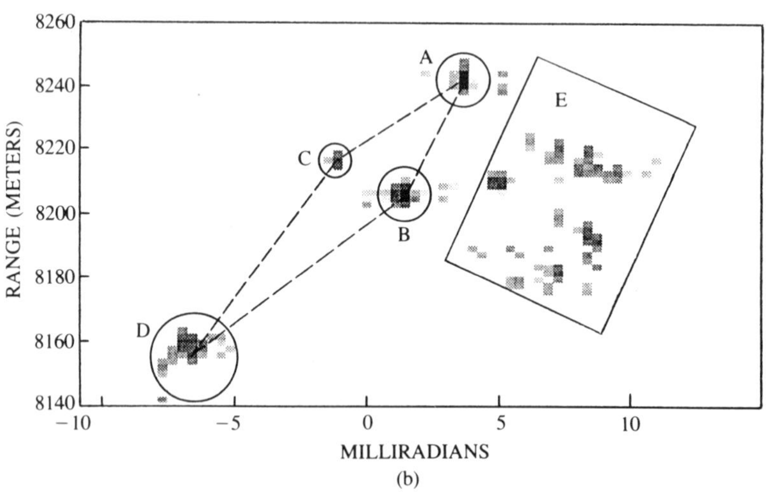

(b)

Figure 8-14 Optical photograph and radio camera image of Cromby power plant. (a) Aerial photograph, Phoenixville, Pa. (1975): A, B, two high stacks; C, water tower, D, two storage silos (erected after photograph); E, power plant building. (b) Radio camera image (3-cm wavelength). (From ref. [48])

in both range and cross-range. The MSA phase corrections compensated for the large geometric distortion of the phased array antenna, and very little sidelobe energy is evident in the microwave image.

The variance range from 0.12 to 0.20 is the DSA transition zone. The algorithm always works at the low end and never works at the high end. The MSA, on the other hand, provides a useful image from three range bins having normalized amplitude variances as large as the order of 0.2. In this sense MSA acts like a threshold extender, permitting the radio camera to function under adverse clutter conditions.

The performance of each algorithm can be improved by subarray processing.

8-9 USE OF SUBARRAYS

Other variations of the basic algorithm improve its operation. One modification is the division of the large array into sections or subarrays to enhance the adaptive beamforming process when the beamforming source is not an ideal point reflector. Figure 8-15 shows the problem. A target of size T reradiates energy toward the array. The radiation is always lobular and the nominal lobe width is λ/T almost independent of the distribution of the scatterers on the target. The phase across the central portion of a lobe is fairly constant, while it changes abruptly and often randomly from lobe to lobe. For the ABF process to proceed smoothly it is necessary for the entire array to be illuminated by the central portion of a single lobe. A typical lobe cross section at the array, which is a distance R from the target, is $\lambda R/T$. Unless L, the array size, is smaller than $\lambda R/T$, this condition is not likely to be satisfied. It is shown in [97] and

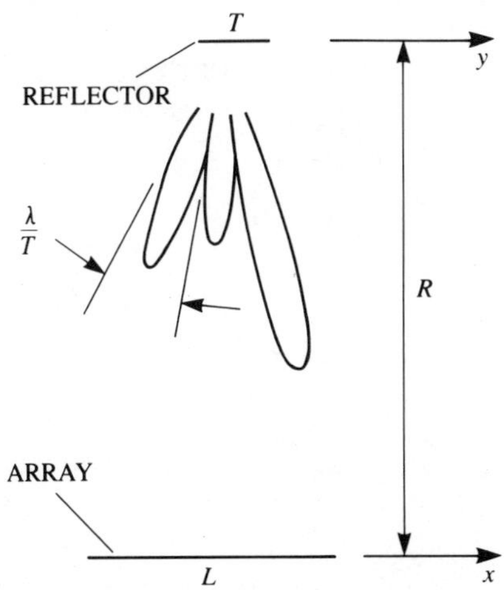

Figure 8-15 The reradiation pattern of the target is lobular. (From ref. [97])

[96] that the minimum target size for satisfactory beamforming is

$$T < \frac{\lambda R}{2L} \tag{8-44}$$

Experience with man-made reflectors indicates that the condition (8-44) is violated sometimes and imaging quality with the DSA or the MSA degrades. It is often possible to correct this situation by breaking the array into sections or subarrays each small enough to satisfy Eq. 3-44. Let $L_1 < L$ be a subarray such that

$$T < \frac{\lambda R}{2L_1} \tag{8-45}$$

The phase of the reradiated wavefront from the target across the subarray is nearly constant and phase corrections for element-position errors, electrical mistunings, and so on, can be accomplished. Let $L_2, L_3, \ldots,$ be other such subarrays. Phase compensation is performed across each subarray. Thus each subarray has its aberrations corrected. However, the phase reference for a particular subarray is the phase of the lobe of the reradiation pattern of the target that illuminates that subarray. Because the lobe phases must be assumed to be random with respect to each other, the array segments, each internally phase synchronized, are also randomly phased with respect to each other. Thus the remaining problem is to link the array segments by phase correcting for the random phase differences between segments of the array. Figure 8-16 illustrates this situation. The abscissa is range and the ordinate is element

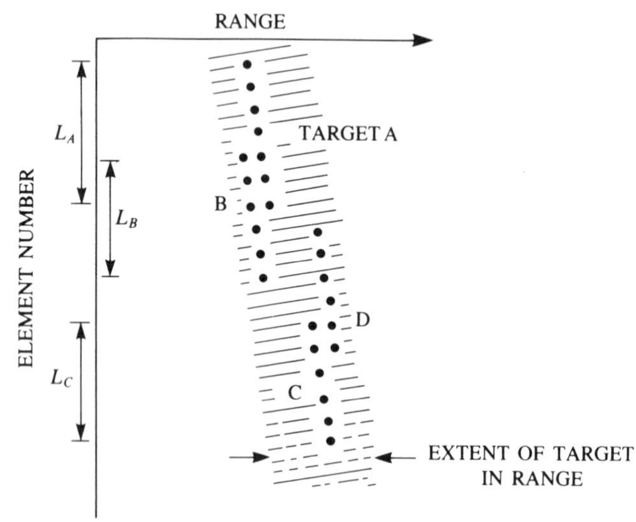

Figure 8-16 Subarrays L_A, L_B, and L_C self-cohere on targets A, B, and C. L_A and L_B are overlapping subarrays; L_B and L_C are nonoverlapping.

number. Target echoes fall in the shaded area. Let A, B, and C be prominent scatterers. None persists throughout the array, however. A provides a strong, stable echo over that portion of the array labeled L_A, and similarly for B and C. Subarrays L_A and L_B overlap; L_B and L_C do not.

Array elements within L_A are self-calibrated by the echoes from A using, say, the DSA. Echoes from B self-calibrate subarray L_B. The self-calibrated subarrays do not share a common phase reference and therefore exhibit a random offset phase between them. The offset between L_A and L_B can easily be measured because targets A and B appear in both data sets. Let ϕ_{AA} be the phase of target A as observed by subarray L_A and ϕ_{AB} its phase as observed by subarray L_B. The difference $\phi_{AA} - \phi_{AB}$ is the phase offset between the self-calibrated subarrays. Adding this difference to all the phases in the L_B data set eliminates the random phase offset and extends the calibration over the two subarrays.

An additional step is required when the subarrays do not overlap. In that case a target common to both subarrays is found after self-calibration. Call this target D. The phase difference $\phi_{DB} - \phi_{DC}$ is added to all the phases in the L_C data set to eliminate the random phase offset between subarrays L_B and L_C.

The technique is more complex when applied to the MSA. Two useful procedures for solving the problem are now discussed. The first deals with nonoverlapping subarrays, the second with overlapping array segments. Both techniques are described in the context of the MSA.

The phase sequence $\phi_{0,n}$ from the reference range bin R_0, given in Eq. 8-31, is

$$\phi_{0,n} = kx_n u_0 - \frac{kx_n^2(1-u_0^2)}{2R_0} + \delta\phi_{0,n} + \beta_n \tag{8-46}$$

where $\delta\phi_{0,n}$ are the perturbations in the phasefront caused by clutter and β_n are the phase errors due to errors in array geometry and mistunings in the receivers. The averaged unwrapped phase differences of the sequences $\phi_{m,n}$ from M candidate beamformers, given in Eq. 8-35, are

$$\begin{aligned}\mu_n &= \frac{1}{M}\sum_{m=1}^{M}(\phi_{m,n} - \phi_{0,n})^u \\ &= kx_n\left(\frac{1}{M}\sum u_m - u_0\right) - \frac{kx_n^2}{2}\left(\frac{1}{M}\sum\frac{1-u_n^2}{R_m} - \frac{1-u_0^2}{R_0}\right) \\ &\quad + \left(\frac{1}{M}\sum\delta\phi_{m,n}\right) - \delta\phi_{0,n}\end{aligned} \tag{8-47}$$

The nth component of the weight vector formed by the MSA is (from Eq. 8-37)

$$w_{n,\mathrm{MSA}} = \exp[-j(\mu_n + \delta_{0,n})] \tag{8-48}$$

Equation 8-47 is nearly a straight line (because the quadratic term is very small) perturbed by zero-mean clutter-induced phase errors. Such a sequence is illustrated in Fig. 8-10a. This figure portrays the case in which the main reflector is physically small, that is, satisfies condition (8-44), and overwhelmingly dominates the backscatter from the clutter. In this case the data points cluster about a linear regression line.

Sometimes, however, condition (8-44) is not satisfied. Then, due to the lobular nature of the reradiation, the data sequence exhibits the phenomenon shown in Fig. 8-16. The low spatial frequency variation shown there induces a phase error that adds to an independent, random perturbation from element to element. This phenomenon is evident in the residual phase errors shown in Fig. 8-17 and indicates a lobular reradiation across the array. The residual phase-error variance is increased because it is the sum of the variances of the random component and the low-frequency undulation. It is often possible to remove much of the latter contribution by estimating, from the data, the phase change from lobe to lobe. These phase jumps are then deleted from the data.

Figure 8-17 illustrates one procedure. A μ_n sequence is illustrated in the upper figure. The array is broken into three subarrays and a linear regression line (least-mean-square fit to the data) is found for each subarray (a). The result is a lower residual phase-error variance (b). The discontinuities in this piecewise linear fit to the data are assumed to be due to phase changes between lobes in the reradiation field; they are removed by adding fixed phases to each section to make the piecewise linear fit continuous (c). The modified sequence of averaged unwrapped differential phases is called μ_n^* and the nth component of the modified MSA weight becomes

$$w_{n,\mathrm{MSA}}^* = \exp \sum [-j(\mu_n^* + \phi_{0,n})] \tag{8-49}$$

Figure 8-18 illustrates another subarray procedure for estimating the correction phases. The arrays are overlapped in this case. Exactly as described before, regression lines are fitted to the subarray μ_n data (a). Each overlap region now has two regression lines. Each pair is averaged to form a new regression line in its limited interval. The result is a discontinuous piecewise linear fit to the data, which are then adjusted exactly as described before to obtain a continuous piecewise linear fit (b). The phase-corrected data are the μ_n^* and the modified MSA weight vector is given by Eq. 8-49.

8-10 SUBARRAY EXPERIMENTS

Figure 8-19 shows four 1-D images of a corner reflector for which beamforming was accomplished with a relatively unsatisfactory scatterer, identified in the figure as the echo from bin 61. The target area was a residential street in Phoenixville, Pa. The amplitude variance from that range bin was 0.21. Figure 8-19a shows the DSA performance. The maximum amplitude reached is

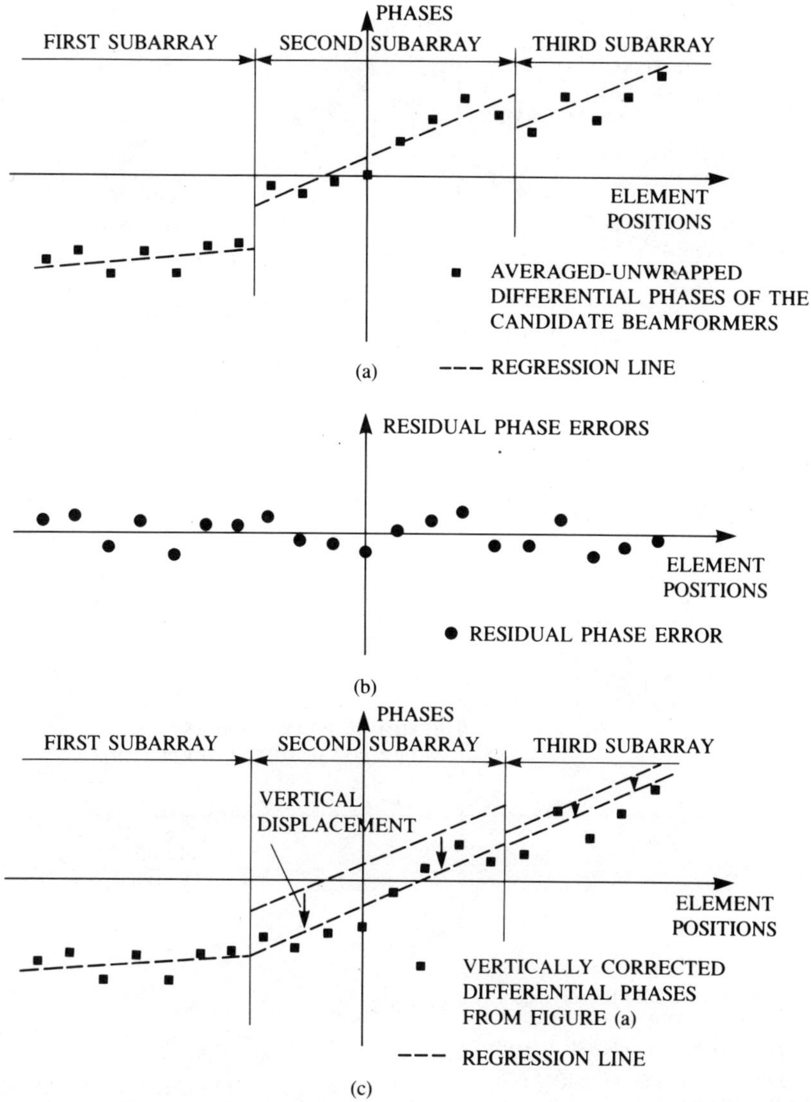

Figure 8-17 Nonoverlapping subarray processing. (a) Regression line in each subarray. (b) Residual phase errors after removing the regression line from the differential phases. (c) Vertically corrected differential phases from (a). (From ref. [47])

53.26 units and the peak sidelobe is -6.9 dB. The image extends over nearly 30 mradians at the -25 dB level and does not look as expected from a corner reflector.

Figures 8-19b, c, and d show the performances of MSA, MSA with nonoverlapping subarrays, and MSA with overlapping subarrays, respectively.

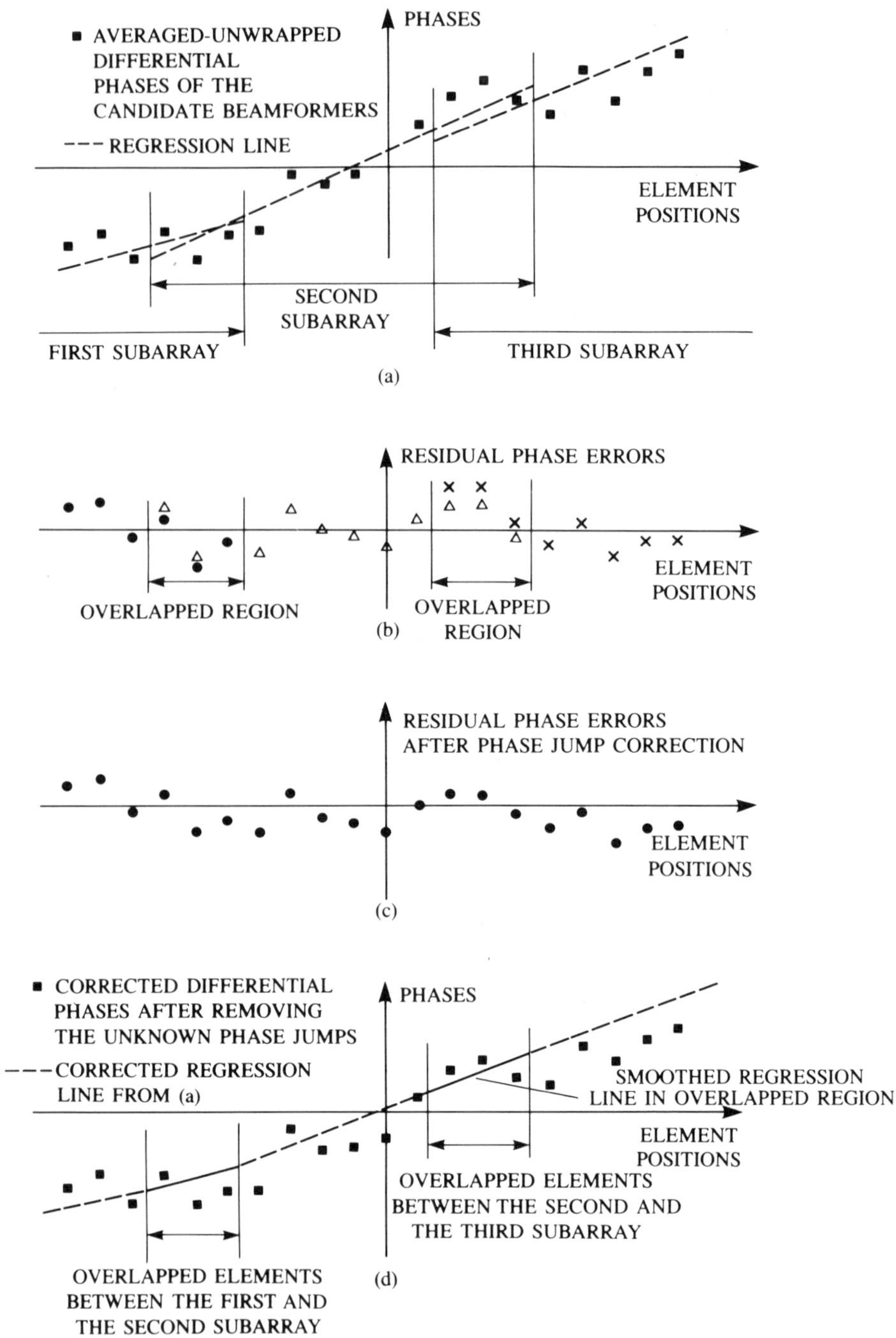

Figure 8-18 Overlapping subarray processing. (a) Regression line in each subarray. (b) Residual phase errors after removing the regression line from the differential phases. (c) Corrected residual phase errors. (d) Corrected regression line and corrected phases. (From ref. [47])

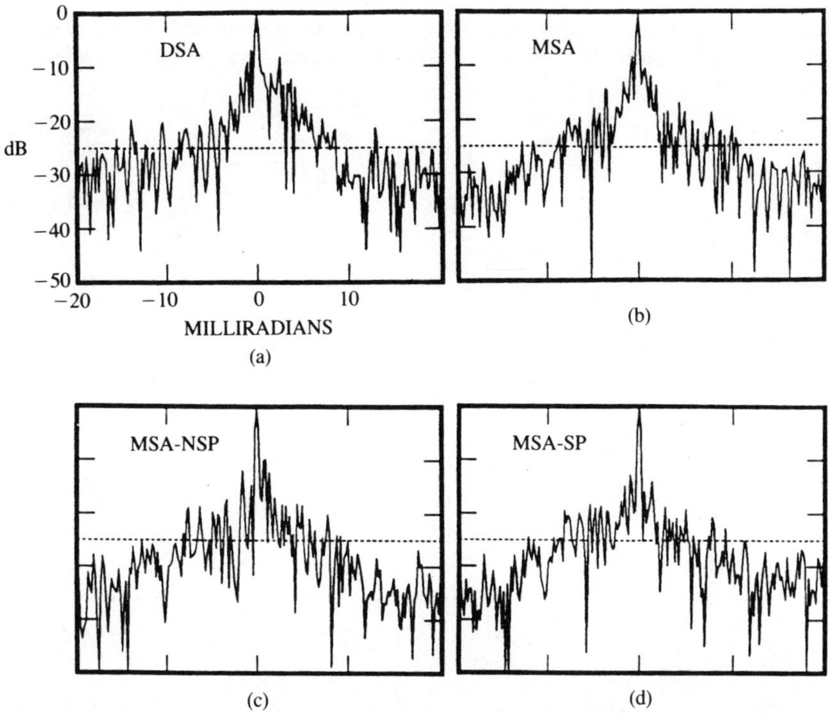

Figure 8-19 Cross-range images of corner reflector. (a) DSA with bin 61. (b) MSA with bins 61, 67, and 68. (c) MSA with nonoverlapping subarrays, using bins 61, 67, 68, and three subarrays, each containing 110 array elements. (d) MSA with overlapping subarrays, using bins 61, 67, 68, and three subarrays, each containing 113 array elements with 8 elements overlapped. (From ref. [47])

The superiority of Fig. 8-19d is evident. The peak amplitude has increased to 69.45 units, indicating an increase of 2.3 dB in mainlobe gain, and the peak sidelobe has dropped to -14.4 dB. The extent of the image above the -25 dB level is reduced to 17 mradians. Table 8-7 lists the results for the four cases.

Another example is shown in Fig. 8-20. The 1-D images are of a target in the Limerick nuclear power plant. The target is located in bin 32. The beamformer is located in bin 26. The amplitude variance of its echoes across the array is 0.20. Table 8-8 lists the performances of the four procedures.

The results in these two experiments are similar. Comparing the top rows (DSA) in Tables 8-7 and 8-8 with the bottom rows (MSA with overlapping subarrays), we see an improvement in mainlobe gain of about 2 dB, a reduction in sidelobe level of about 6 dB and a reduction in the angular size of the target of nearly 2:1.

TABLE 8-7 Comparison of DSA and MSA, with and without subarray processing. Data set 309 (High Street). Target is a corner reflector. The amplitude variance of the beamformer is 0.21

	Max. amplitude	*Relative mainlobe gain* (dB)	*PSL* (dB)	*Extent of image above* −25 dB (mradians)
DSA	53.26	−2.30	−6.9	28
MSA	61.34	−1.08	−8.4	23
With nonoverlapping subarrays	66.68	−0.35	−10.0	18
With overlapping subarrays	69.43	0	−14.4	17

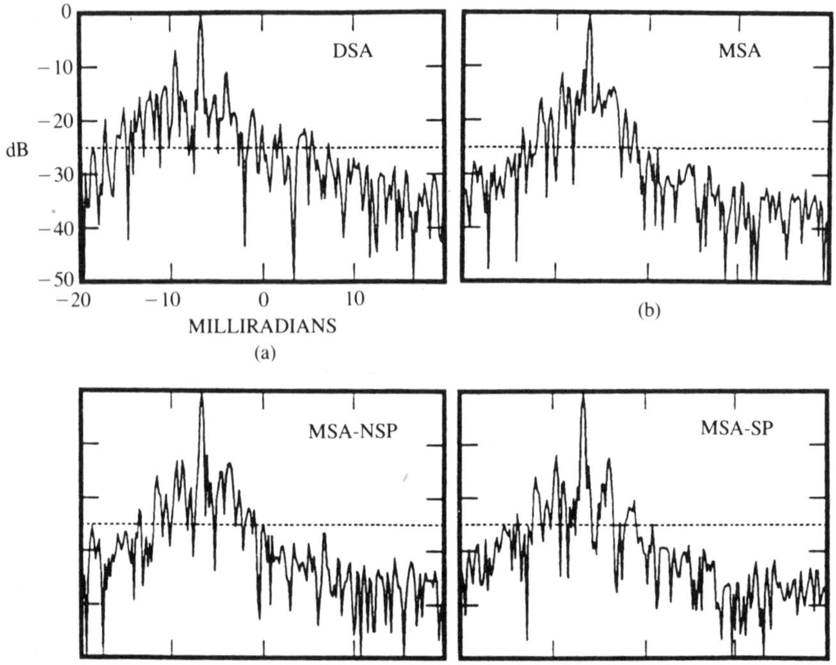

(a) (b) (c) (d)

Figure 8-20 Cross-range images of bin 32 of Limerick nuclear power plant. (a) DSA with bin 26. (b) MSA with bins 17, 25, and 26. (c) MSA with nonoverlapping subarrays, using bins 17, 25, 26, and three subarrays, each containing 110 array elements. (d) MSA with overlapping subarrays, using 17, 25, 26, and three subarrays, each containing 113 array elements with 7 elements overlapped.

TABLE 8-8 Comparison of DSA and MSA, with and without subarray processing. Data set 301 (Limerick nuclear power plant). The amplitude variance of the beamformer is 0.20

	Max. amplitude	*Relative mainlobe gain* (dB)	*PSL* (dB)	*Extent of image above* −25 dB (mradians)
DSA	51.08	−1.43	−6.9	24
MSA	56.89	−0.49	−10.7	13
With nonoverlapping subarrays	58.34	−0.28	−11.9	13
With overlapping subarrays	60.22	0	−12.1	13

8-11 OBSERVATIONS

A dominant scatterer algorithm will successfully self-calibrate a distorted phased array when a single suitable reflector (large RCS and small physical size) is embedded in clutter. The target-to-clutter ratio must be large enough to ensure that the reradiation wavefront is dominated by the prominent reflector. A test is the normalized amplitude variance of the echo across the array. A value of 0.12 or less ensures that the DSA will provide a useful image.

Under the same condition, the MSA will provide a slightly superior image but the improvement is often not sufficient to warrant the extra complexity. The strength of MSA is that it extends the useful variance range to 0.2. Subarray processing further improves its performance.

9

ADAPTIVE BEAMFORMING II

The previous chapter discussed adaptive beamforming algorithms that use data from range bins, each containing a single prominent scatterer imbedded in clutter, to infer the phase errors at the array elements. This chapter discusses adaptive beamforming techniques that use data from other types of range bins.

The spatial correlation algorithms (SCA) use data from a set of range bins whose scattering functions $s(u)$ are sample functions of the same random process (e.g., sea clutter echoes). Sections 9-1 and 9-2 discuss the unit lag SCA [2], [5]. Sections 9-3 through 9-5 discuss the multiple lag SCA [109] and compare the performance of unit-lag and multiple-lag SCAs with the DSA and MSA.

The DSA uses data from a single range bin consisting of a single prominent scatterer imbedded in clutter to self-calibrate the array. The MSA uses a few such range bins. The spatial correlation algorithms operate on a different principle. They require data from several range bins whose angular spectra are functions of the same random process. An algorithm, developed by Tsao [119], uses an energy conservation principle to self-cohere the array. A version of this algorithm, called the *energy conservation algorithm* (ECA), is described in Section 9-6. Computer simulations are used to compare ECA performance with the unit- and multiple-lag SCAs. The computational complexities of the various algorithms are also estimated and compared.

9-1 UNIT-LAG SPATIAL CORRELATION ALGORITHM

9-1-1 Concept

Figure 8-3b showed a prominent scatterer imbedded in a range bin containing clutter. The radiation field at the array was dominated by the large scatterer.

As a consequence, the DSA was able to capture a relatively smooth, linear phase front and self-calibration was successful. The situation pictured in Fig. 8-3a is different. There no single scatterer is prominent. Instead, the field at the array (Eq. 8-22) is the sum of randomly weighted plane waves arriving from random directions within the illuminated sector. Both situations are common. Corners figure prominently in man-made targets. Natural terrain such as farmland, crowns of trees, and forests is less structured. Still more statistically homogenous is sea clutter. The unit-lag spatial correlation algorithm (SCA) was developed to handle these cases [2], [5]. This algorithm, like the MSA, is also due to Attia.

Let $s(u)$ represent the complex reflectivity function of one of the range bins in Fig. 8-3a. $s(u)$ can be considered a random process. For statistically homogeneous clutter, both the real and imaginary parts of $s(u)$ have probability distribution functions (pdfs) that are independent of u. The source strength is the random process $s(u)$ modulated by the transmitted beam $f_T(u)$. Let $S = E\{|s(u)|^2\}$ be the mean squared value of the random process $s(u)$. The mean squared scattered power as a function of u is then $S|f_T(u)|^2$.

The complex scatterer strengths in the range bin shown in Fig. 8-3a represent a sample function of the random process $s(u)$. It is reasonable to assume that the scattering functions $s(u)$ in many other range bins are also derived from the same random process, especially for target scenarios such as sea clutter. Consider the correlation between the data measured at two array elements at x_1 and x_2, $R(x_1, x_2) = E\{e(x_1)e^*(x_2)\}$, where $e(x)$ is the data sample measured at aperture location x. The Van Cittert–Zernicke theorem states that $R(x_1, x_2)$ is spatially stationary when the underlying source distribution is incoherent, that is, $R(x_1, x_2) = R(x_1 - x_2)$. Sea clutter echoes possess this property because the source distribution $s(u)$ is incoherent. The averaging required to estimate the correlation $R(x_1, x_2)$ can be performed over range because the source distributions $s(u)$, at different range bins, are different realizations of the same random process.

Under far-field conditions, the autocorrelation function of the measured aperture data $R(x)$, and the mean angular power density function $I(u) = S|f_T(u)|^2$, are Fourier transform pairs, as described by the Van Cittert–Zernicke theorem. The theoretical value of this correlation function is known because the transmitted power pattern is known. The measured correlation function will differ from the known theoretical value because of phase errors at the receiving aperture. The SCA uses the difference between the known and measured correlation phases to estimate the unknown phase errors at the array elements.

The concept is introduced in simplified form by the following example. Assume that $|f_T(u)|$ is an even function of u, which is typical of most antenna patterns. Then $R(x)$ is real, by the property of the Fourier transform. Further, assume that far-field conditions prevail and that the receiving array is periodic.

These three assumptions, viz. an even transmitted pattern, far-field geometries, and periodic receiving arrays are not necessary for the SCA; they are used for simplifying this discussion. The implications of the three assumptions are discussed later.

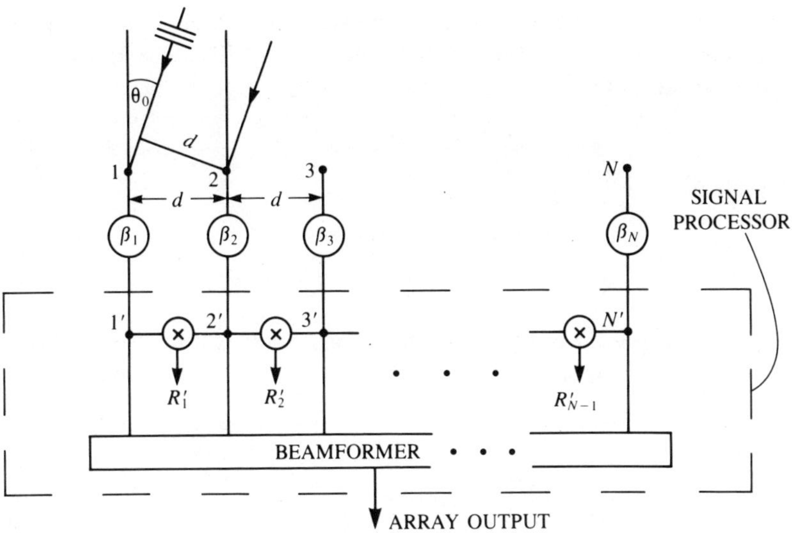

Figure 9-1 Array model and signal processor configuration for self-synchronization by spatial correlation algorithm. (After ref. [5])

Figure 9-1 sketches a periodic array of N elements steered to θ_0, the angle from broadside. The phase shift ϕ_n in the nth channel, introduced by the beamformer, is $nkdu_0$ where $k = 2\pi/\lambda$ is the wavenumber, d is the element spacing, and $u_0 = \sin\theta_0$. There is a phase error β_n in the nth channel, resulting from element position errors and electrical mistuning. The left channel is considered the reference. Its phase error is arbitrarily assigned the value of zero. In each channel of the beamformer but the reference channel is a phase-error correcting weight. The sum is the array output.

Information for determining the correct values of the weights is obtained from the products of signals in adjacent channels. The first product is $e'_2 e'^*_1$, the second product is $e'_3 e'^*_2$, and so on. Each product is integrated or averaged over K range bins.

$$\hat{R}(2, 1) = \frac{1}{K} \sum_{k=1}^{K} e'_{2,k} e'^*_{1,k} \tag{9-1}$$

is an estimate of the correlation coefficient of the radiation field between the first and second elements modified by the phase error difference $\beta_2 - \beta_1$, which, because β_1 is assigned the value zero, equals β_2. $\hat{R}(2, 1)$ is a random variable having an expected value

$$E\{\hat{R}(2, 1)\} = R(2, 1)e^{j\beta_2} \tag{9-2}$$

where $R(2, 1)$ is the error-free correlation coefficient. It is real because the transmit pattern $f_T(u)$ is assumed to be an even function of u and was pointed to broadside. An estimate of the correlation coefficient $\hat{R}(2, 1)$ is approximately equal to its mean, that is

$$\hat{R}(2, 1) \approx R(2, 1)e^{j\beta_2} \tag{9-3a}$$

In a similar fashion

$$\hat{R}(3, 2) \approx R(3, 2)e^{j(\beta_3 - \beta_2)} \tag{9-3b}$$

$$\hat{R}(N, N-1) \approx R(N, N-1)e^{j(\beta_N - \beta_{N-1})} \tag{9-3c}$$

The measured phases of the correlation coefficients can be used to estimate the elemental phase errors. Setting the first error to zero, we obtain

$$\hat{\beta}_1 = 0, \qquad \hat{\beta}_2 = \text{Phase}\ \{\hat{R}(2, 1)\}, \qquad \hat{\beta}_3 = \text{Phase}\ \{\hat{R}(3, 2)\} + \hat{\beta}_2, \ldots \tag{9-4}$$

or

$$\hat{\beta}_n = \sum_{m=1}^{n-1} \text{Phase}\ \{\hat{R}(m+1, m)\}, \qquad n > 1,\ \hat{\beta}_1 = 0 \tag{9-5}$$

The phase compensation vector for the array is the conjugate of the estimated phase-error vector, whose components are $\hat{\beta}_n$.

9-1-2 Unit-Lag SCA for Near-Field Geometries and Off-Broadside Steering

Three constraints were implied in the discussion above. First, the transmitter pattern $f_T(u)$ was assumed to be symmetric about its axis, which was implicitly oriented broadside to the receiving array. Second, the description of the algorithm was given in far-field terms. Third, the array was assumed to be periodic. The implications of these assumptions on the ability of the SCA to calibrate the array is now discussed.

It is now shown that neither the near-field assumption nor the broadside steering condition is necessary for effective implementation of the SCA because the SCA automatically introduces the phase shift for steering and focusing the array into the phase corrections $\hat{\beta}_n$. With near-field curvature, the measured correlations, denoted $\hat{R}_1(m, n)$, can be expressed as

$$\hat{R}_1(m, n) \approx \hat{R}(m, n) \exp(-jk(x_m^2 - x_n^2)/2r) \tag{9-6}$$

where x_m is the location of the mth element, r represents range, $\hat{R}(m, n)$ is the far-field correlation coefficient given by Eq. 9-1 and the Fresnel kernel $\exp(-jkx^2/2r)$ arises due to near-field curvature. The approximation sign indicates that the relationship Eq. 9-6 holds in the region of validity of the Fresnel transform.

Equation 9-6 is further modified when the array is steered away from broadside. Assume that u_0 is the steering angle. The function $S|f_T(u - u_0)|^2$ is the scattered angular power density spectrum with the axis of symmetry generalized from $u = 0$ to $u = u_0$. From Fourier theory, it is easy to show that the measured phases of the correlation coefficients $\hat{R}(m, n)$ are multiplied by $\exp[jku_0(x_m - x_n)]$ to account for off-broadside steering. With near-field geometry and off-broadside steering included, the measured correlation coefficients, denoted $\hat{R}_2(m, n)$, can be expressed as

$$\hat{R}_2(m, n) \approx \hat{R}(m, n) \exp\{jk[(x_m - x_n)u_0 - (x_m^2 - x_n^2)(1 - u_0^2)/2r]\} \quad (9\text{-}7)$$

where u_0 is the steering angle.

The implication of the assumption that the transmitter radiation pattern is symmetrical is that the correlation function $R(m, n)$ is real. This assumption is violated, however, when the reradiation used to self-calibrate is from a few adjacent range bins of an asymmetrical target. Another case is when only a single range bin is available and its source distribution is unsymmetrical, and a technique such as transmitter location diversity is used to estimate the correlation function.

In what follows, the assumption that the phase of the error-free correlation coefficient is zero is removed. Call that phase $\psi(x_m - x_n)$. The spatial correlation algorithm uses the correlation phases to determine the phase errors. Spatial stationarity of the correlation function demands that the phase of the measured far-field correlation coefficient $\hat{R}(m, n)$ depends only on $(x_m - x_n)$ in the absence of elemental phase errors (all $\beta_n = 0$). With phase errors included, the phase of $\hat{R}(m, n)$ equals $\beta_m - \beta_n + \psi(x_m - x_n)$. Therefore, the phase of $\hat{R}_2(m, n)$ in Eq. 9-7, can be expressed as

$$\begin{aligned} \text{Phase}\,\{\hat{R}_2(m, n)\} = {} & \{\beta_m + k[x_m u_0 - x_m^2(1 - u_0^2)/2r]\} \\ & - \{\beta_n + k[x_n u_0 - x_n^2(1 - u_0^2)/2r]\} \\ & + \psi(x_m - x_n) + \delta(m, n) \end{aligned} \quad (9\text{-}8)$$

The first (second) term on the right side of Eq. 9-8 is the sum of the phase error at the mth (nth) element and the phase shift necessary to steer the mth (nth) element towards look direction u_0 and range r. The third term is the phase of the error-free correlation coefficient $\hat{R}(m, n)$ and depends only on $(x_m - x_n)$. The fourth term $\delta(m, n)$ represents noise.

The unit lag SCA uses the phases of the $(N - 1)$ correlation coefficients $\hat{R}_2(m, n)$ to determine the phase errors $\beta_2, \ldots, \beta_N$ relative to β_1. Using Eqs. 9-5 and 9-8, the unit lag SCA phase error estimates can be expressed as the sum of three terms:

$$\hat{\beta}_n = \sum_{m=1}^{n-1} \text{Phase}\,\{\hat{R}_2(m+1, m)\}, \qquad n > 1$$

$$= \{\beta_n + k[x_n u_0 - x_n^2(1 - u_0^2)/2r]\} + \sum_{m=1}^{n-1} \delta(m+1, m) + \sum_{m=1}^{n-1} \psi(x_{m+1} - x_m) \quad (9\text{-}9)$$

The components of the weight vector are $\exp(-j\hat{\beta}_n)$.

The first term on the right side of Eq. 9-9 is the sum of the phase error at the nth element and the phase shift necessary to steer the nth element towards look direction u_0 and range r. The second term represents noise.

Whether or not Eq. 9-9 produces a useful weight vector depends upon the third term, which is the sum of the phases of the error-free correlation coefficient $\hat{R}(m+1, m)$, $1 \leqslant m \leqslant n-1$. The phase-error estimates $\hat{\beta}_n$ become meaningless if this term varies randomly as a function of n. It is clear that the performance of the SCA is unaffected when $\psi(m+1, m) = 0$ for all m, that is, the error-free unit-lag correlation coefficients $R(m+1, m)$ are real and positive. What is less obvious is that performance is unaffected irrespective of ψ when the array is periodic (see Section 9-1-3).

$R(m+1, m)$ will be real when the product of the transmitted beam pattern and the scatterer distribution is symmetrical about the beamsteering axis of the transmitter. Further, $R(m+1, m)$ will be positive when all the spacings between adjacent elements $|x_{m+1} - x_m|$ are smaller than the null-to-null width of the correlation function $R(x)$. As shown in Chapter 8 (see the discussion following Eq. 8-20 and Fig. 8-3), the width of the main lobe of $R(x)$ is approximately equal to the length of the transmitting antenna L_t. Thus the error-free correlation phases $\psi(m+1, m) = 0$ when (a) the product of the transmitted beam pattern and the scatterer distribution is symmetrical about the beamsteering axis and (b) $|x_{m+1} - x_m| \leqslant L_t$, all m. Under these conditions, the phase corrections can be rewritten as

$$\hat{\beta}_n = \{\beta_n + k[x_n u_0 - x_n^2(1 - u_0^2)/2r]\} + \sum_{m=1}^{n-1} \delta(m+1, m); \qquad n > 1 \quad (9\text{-}10)$$

As mentioned at the beginning of this section, the phase correction $\hat{\beta}_n$ in Eq. 9-10 is the sum of the phase error β_n (the phase required to steer the nth element towards look direction u_0 and range r) and noise. That is, the SCA not only estimates the phase errors β_n but also steers the receiving array towards the transmit direction.

No assumptions about array periodicity were made while deriving Eq. 9-10. Thus the unit-lag SCA is capable of calibrating either periodic or aperiodic arrays when (a) the product of the transmitted beam power pattern and the angular power spectrum of the scatterer distribution is symmetrical about the optical axis $u = u_0$, and (b) $|x_m - x_{m+1}|$ is smaller than half the null-to-null width of $R(x)$, that is, $|x_m - x_{m+1}| < L_t$, all m, where L_t is the length of the transmitting aperture.

9-1-3 Unit Lag SCA for Complex Correlation Functions

Sometimes the correlation is complex. This occurs when the source distribution is not symmetric. When the correlation function is not real and its phase varies linearly with element position, the algorithm still works. It fails, however, when the phase variation is nonlinear. Under this condition, there is no hope of using the correlation measurements to phase an *aperiodic* array. This can be seen by observing that the second term in Eq. 9-9, $\sum_{m=1}^{n-1} \psi(x_{m+1} - x_m)$, varies randomly with n. Hence different random numbers are added on to each of the phase corrections β_n, thereby rendering them useless for phase correcting the array.

However, a periodic array can be calibrated even under these circumstances. For a periodic array, the second term in Eq. 9-9, $\sum_{m=1}^{n-1} \psi(x_{m+1} - x_m) = (n-1)\psi(d)$, where d is the interelement spacing and $\psi(d)$ is the phase of the correlation function $R(x)$ for $x = d$. In this case the phase error estimate in Eq. 9-9 can be expressed as

$$\hat{\beta}_n = \{\beta_n + k[x_n u_0 - x_n^2(1 - u_0^2)/2R]\} + (n-1)\psi(d) + \sum_{m=1}^{n-1} \delta(m+1, m); \qquad n > 1 \quad (9\text{-}11)$$

In addition to the true phase error, the steering phase, and noise, the phase correction contains a linear phase term $(n-1)\psi(d)$. This linear phase term has the effect of pointing the array towards a direction different from the look direction u_0. The result is a beampointing error but not a distortion of the beam. The beampointing error Δu is found by equating the linear phase slope $\psi(d)$ to $kd\,\Delta u$. The beampointing error, therefore, is $\Delta u = \psi(d)/kd$.

Thus periodic arrays can be phase-cohered using the unit-lag SCA even when the phase of the correlation function varies in a nonlinear fashion with interelement spacing. As shown above, this phase variation introduces a beampointing error. Beampointing errors are acceptable for some applications such as imaging where the pointing error leads only to an image shift but not to a degradation in beam quality or image quality. It is important to note that if the phase of the correlation function can be approximated as a linear function for small lags, an *aperiodic* array can also be cohered with a beampointing error $\Delta u = k/\alpha$, where α is the slope of the phase of the correlation function.

9-1-4 Effect of Noise on the SCA

From Eqs. 9-10 or 9-11, we see that the phase corrections steer the array toward look direction u_0 (or $u_0 + \Delta u$) and range r, in addition to estimating the phase errors. From Eq. 9-9, the noise term in the phase correction for the nth element, denoted $\delta\phi_n$, can be expressed as

$$\delta\phi_n = \sum_{m=1}^{n-1} \delta(m+1, m), \qquad n > 1 \qquad (9\text{-}12)$$

Clearly, the number of noise terms $\delta(m+1, m)$ increases linearly with element number n. Thus one would expect the variance of the residual errors in the phase corrections to increase linearly with element number. This result was obtained in [5].

The increase in the residual phase error variance with element number results in the following relationship between the array mainbeam gain (after adaptive beamforming using the SCA) and the number of array elements. The gain of the array is expected to increase directly with the number of elements N. However, because the phase error variance grows linearly with element number, the average phase error variance grows with the number of array elements. As a consequence, the loss in mainlobe gain due to the residual phase error also grows with N. It is shown in [5] that the expected value of the mainbeam gain of a periodic array can be expressed as

$$E\left\{\frac{G}{G_0}\right\} = \frac{1}{N} + \frac{2}{N^2}\frac{r^{1/2}[(N-1) - Nr + r^N]}{(1-r)^2} \tag{9-13}$$

where $r = \exp(-\sigma_\delta^2)$ and σ_δ^2 is the variance of the noise terms $\delta(m+1, m)$ in Eq. 9-8. Figure 9-2 plots Eq. 9-13 for different values of N.

9-2 UNIT-LAG SCA EXPERIMENTS

9-2-1 Computer Simulations

In [5], a clutter patch was simulated on the computer at X-band ($\lambda = 3$ cm) by uniformly distributing a large number of scatterers (1000) within its boundary

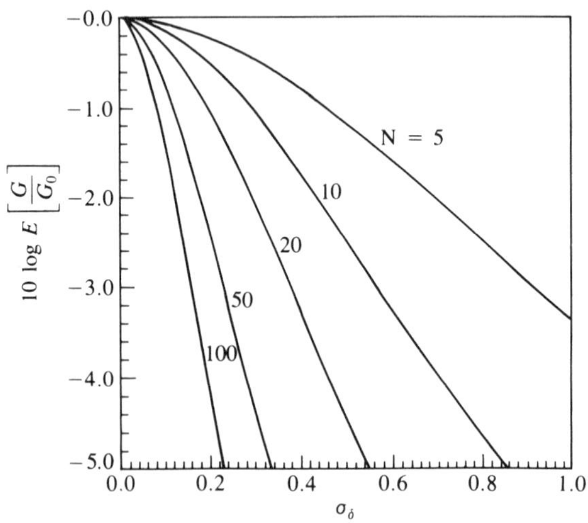

Figure 9-2 Normalized mainbeam gain (dB) versus σ_δ (radians) for the spatial correlation algorithm. (From ref. [5])

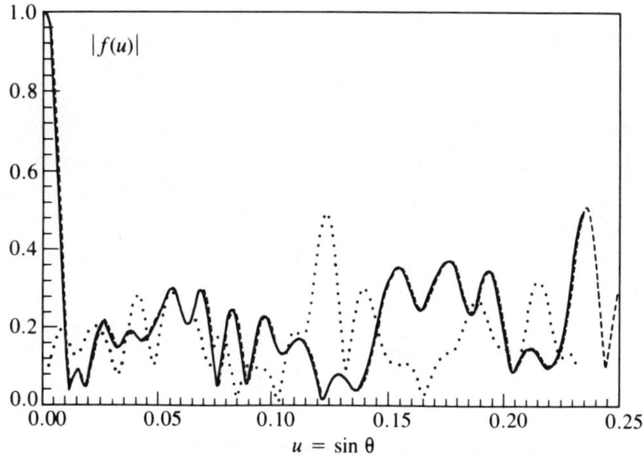

Figure 9-3 Computer simulation experiment of the unit-lag spatial correlation algorithm. ——, Error-free pattern; ····, distorted due to phase errors; ----, recovered by the algorithm ($N = 20$, $L = 100\lambda$, clutter to noise ratio (CNR) = 20 dB, $M = 100$). (From ref. [5])

($\Delta R = 15$ m, $\Delta\theta = 0.06$ radians). The phase of each scatterer was taken as an independent random variable uniformly distributed over the interval $[-\pi, \pi]$, while its amplitude was taken to be proportional to both the transmitter radiation pattern and a uniformly distributed random quantity. The SCA was tested through the simulation of periodic and random arrays. Random unknown phase errors uniformly distributed over the interval $[-\pi, \pi]$ were inserted in each receiving channel. Also, an independent additive narrowband Gaussian noise component was injected into each channel on reception. It is reported in [5] that the algorithm was highly successful, always enabling the recovery of the error-free radiation pattern almost completely. Figure 9-3 shows the results of one such simulation experiment. It corresponds to an $N = 20$ element random array with a nominal length $L = 100\lambda$. The clutter-to-noise ratio (CNR) is 20 dB and the sample size $K = 100$. The solid line plots the error-free radiation pattern while the dotted line represents the distorted pattern (due to the phase errors). The dashed line plots the pattern recovered by applying the SCA. The agreement is excellent.

9-2-2 A Sea Clutter Experiment

In the Naval Research Laboratory experiment referred to earlier in Section 8-5 the radar data-taking equipment was also flown over the North Atlantic Ocean. Eight channels of sea clutter echoes were recorded, one for each dipole antenna in the array, in response to each pulse transmission. The receiving antenna was a rigid, 8-element equal-gain phased receiving array operating at 850 MHz. Its

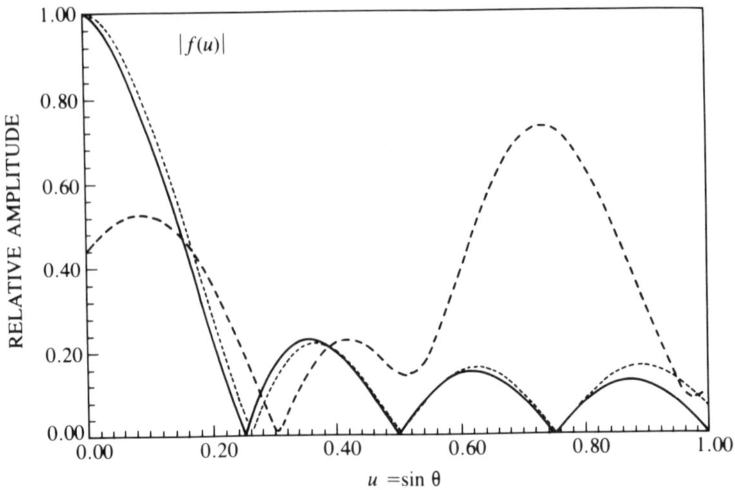

Figure 9-4 Sea clutter experiment with unit-lag spatial correlation algorithm. ——, Ideal radiation pattern; ----, destroyed by phase errors; ····, recovered by algorithm. (From ref. [5])

radiation pattern was of the form $\sin(Nx)/N \sin x$, where $N = 8$ is the number of elements and x is proportional to the sine of the angle from boresite. Its peak sidelobe is approximately -12 dB. The pattern is shown as the solid curve in Fig. 9.4.

To conduct the experiment, random phases representing element-position errors and electrical system errors were added in each channel. The distribution of these phase errors was uniform in the interval $[-\pi, \pi]$. Phase errors destroyed the radiation pattern, as expected; it is shown as the dotted curve. Sea clutter echoes measured in each of the eight receiving channels were then phase-shifted by these random phases and processed by the SCA. One hundred range bins of data were used to smooth the correlation estimates. Phase corrections were calculated from the measured correlation coefficients according to Eq. 9-5 and cascaded with the random phase errors. The radiation pattern following phase compensation is shown dotted in the figure. It is nearly indistinguishable from the original, error-free pattern down to the third sidelobe, where the difference is 1.8 dB at the level of -16 dB. There would be no way to distinguish these patterns in a realistic operating system. This experiment demonstrates the ability of the SCA to phase-compensate on sea clutter; it is typical of many similar tests.

Two-dimensional SCA experiments on ground clutter and on complicated echoes from an airplane also were performed. These results are shown in Section 9-4, where the performances of several algorithms are compared.

9-3 MULTIPLE-LAG SPATIAL CORRELATION ALGORITHM (MLSCA)

9-3-1 Muller's Theorem

The SCA described in Section 9-1 employed only the unit-lag correlation coefficient of the correlation function.* The phase errors were derived directly from the arguments of the measured values of $R(n + 1, n)$ at each point in the array. By the same procedure that led to Eq. 9-5, it is easy to see that other correlation coefficients contain phase error information as well. Consider $R(n + 2, n)$, for example. It provides the phase error between two elements, spaced by twice the interelement spacing, at each point in the array at which it is measured. Similarly, $R(n + 3, n)$ provides the phase error between two elements spaced by three times the interelement spacing. Thus measurement of all the available coefficients from lags 1 through $(N - 1)$ provides phase error information as directly as the unit-lag correlation coefficients alone. The obvious question to be answered is "If $R(n + 1, n)$, measured everywhere in the array, solves the phase-error compensation problem, why utilize any other, more complicated correlation algorithm?"

The answer is that two factors limit the quality of the unit-lag correlation measurements $\hat{R}(n + 1, n)$. The first is noise. The quality of the estimate is a function of the signal-to-noise ratio (SNR) of the measured samples of the electric field, e_n and e_{n+1}. Second, the number of range bins used for smoothing the correlation coefficients may be insufficient to produce suitably accurate estimates of the phase errors.

When any or both of these conditions prevail, the estimated phase errors, deduced from the arguments of $\hat{R}(n + 1, n)$, can be in error. The solution is to estimate the phase errors using the correlation coefficients for several lags. In essence, an additional domain of averaging or smoothing is introduced.

Historically, the concept was first introduced by Muller as an important theorem in optical telescopy [69]. The problem of poor seeing conditions has plagued astronomers since the invention of the telescope. The problem is due to the nonhomogeneity of the refractive index of the atmosphere. Muller discovered that it is possible to fully phase-compensate a multiple-mirror telescope (analogous to a phased array with very large subapertures for elements) by feedback-controlling optical time delays placed in front of each optical aperture so as to maximize the integral of certain functions of the square of the image intensity in the composite image plane. Because of the Fourier relations between the e-field in the aperture and the image, and between an intensity image and the autocorrelation in the aperture, Muller's image-plane algorithm can be replaced by an equivalent aperture-plane algorithm that uses

* The following discussion assumes periodic arrays and far-field geometries. However, the results obtained below hold for near-field geometries as well. Further, random arrays can be calibrated by the MLSCA when the phases of the true correlation coefficients for all the lags being considered are known.

the measured correlation coefficients $\hat{R}(m, n)$ [39], [109]. Both Muller's image plane algorithm and its equivalent in the aperture domain are iterative. A noniterative aperture-domain algorithm is described later in this section. Muller's image-plane algorithm is described first.

Of the class of solutions discovered by Muller the simplest one to describe is as follows. Let the image formed by the ith mirror be I_i. Let all the images be brought together onto a common focal plane and registered. Form an objective function J, called the *sharpness function*, that is the integral of the square of the intensity distribution in the image plane. Let the time delays referred to above be represented more generally by a set of weights w_i. The optimization or search procedure adjusts the w_i until

$$\frac{\partial J}{\partial w_i} = 0, \qquad \text{all } i \tag{9-14}$$

The theorem demonstrates that when this condition is achieved the wavefront distortion has been fully compensated and that the image is without error, except for an arbitrary and unknown displacement of the optical axis.

Figure 9-5, taken from [69], is a schematic illustration of Muller's algorithm. A device capable of measuring the image sharpness J is placed at the focal plane of the telescope. The disturbing atmosphere degrades image sharpness by randomly perturbing the phase of the incoming light. A phase corrector placed in the optical path is capable of restoring the proper phase. Individual elements of the phase corrector are adjusted in turn to maximize the sharpness of the image, as measured by the device placed in the image plane. If the entire telescope objective can be corrected within the time that atmospheric distortion holds constant, the resulting image will be nearly diffraction limited.

The multiple-mirror telescope can be viewed as an optical phase array. While

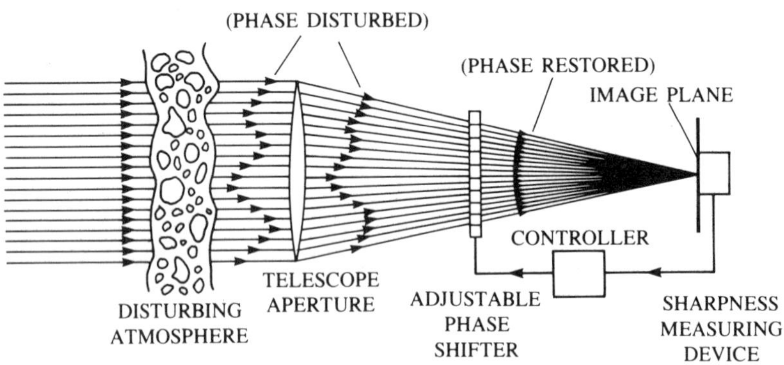

Figure 9-5 Schematic diagram of Muller's image-restoring system. Arrow heads indicate the relative phase of the wave. The adjustable phase shifter corrects the phase of the incoming wave. (Adapted from ref. [69])

Muller was concerned with wavefront distortion caused by atmospheric turbulence, his theorem is applicable to any source of apparent distortion of the wavefront, as caused, for example, by errors in the positions of the parts of the telescope.

In mathematical terms, let $s = s(u)$ be the source distribution, $f_0 = f_0(u)$ the design radiation pattern of the multiple-mirror telescope array, and $f = f(u) = f_0 + \delta f$ the actual pattern. The optimum complex image is

$$\hat{s}_0 = s * f_0 \tag{9-15}$$

while the actual distortion image is

$$\begin{aligned} \hat{s} &= s * f = s * (f_0 + \delta f) \\ &= \hat{s}_0 + \delta s \end{aligned} \tag{9-16}$$

The display $I = I(u)$ is either the magnitude or the squared magnitude of Eq. 9-16 and I_0 is displayed when the radiation pattern is error free. The statement of Muller's theorem is that

$$I(u) = I_0(u + \Delta u) \tag{9-17}$$

when Eq. 9-14 is satisfied, where Δu is the shift in the optical axis. That is, the phase-compensated image will be a replica of the error-free image (except for a shift Δu in the image plane) after optimizing Muller's objective function J.

A limitation of the theorem is the requirement that the radiation field upon which it operates be derived from an incoherent distribution of sources. Stars, being independent energy generators, satisfy this requirement (see Fig. 1-3). The phase relations between the radiations from each source within the passbands of the receivers are continuously changing. The result is a radiation field at the receiving antenna that is continuously changing but whose statistics are invariant with time.

9-3-2 The Microwave Case

Not so in microwave radar, however. There the phase relations between the echoes from targets are dictated by the geometry of the locations of the transmitter, the targets, and the receiver. Each transmitted pulse results in the same set of echoes and hence the same radiation field at the receiving array.

Figure 9-6 schematically portrays a surface J as a function of two weights, indicated by the phase shifts ϕ_1 and ϕ_2. The global maximum of J occurs at the coordinates of the correct values of ϕ_1 and ϕ_2 when the source distribution is incoherent. For coherent or partially coherent source distributions, the secondary peaks could be as large, and the search procedure can mistakenly settle for a nonglobal peak. Further, there is no guarantee that the global maximum occurs at the coordinates of the optimal weights.

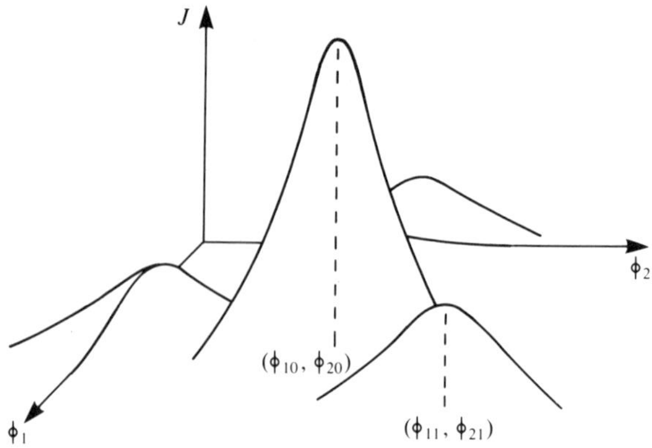

Figure 9-6 Sharpness function is globally maximized by correct phase compensators ϕ_{10} and ϕ_{20}.

The primary potential benefit of the procedure, in microwave radar, is less its ability to compensate for medium turbulence than for distortion in the aperture. Large antenna systems will be intrinsically distorted due to nonrigidity of their structures, differences in gravitational forces across the antennas, differential windloading, differential weather effects, and so on. To achieve the benefits of self-calibration based upon Muller's theorem, means must be developed for altering the phase relations of the electric fields reflected from the targets so as to emulate an incoherent distribution of radio sources.

Two methods for doing so have been discussed earlier. Each method builds up a diversity image suitable for initiation of the search for the weight vector. The first procedure uses transmitter-location diversity as its tool (Chapter 6). The second procedure uses range-bin diversity (Section 9-1). In transmitter-location diversity, successive radar pulses are emitted from different array elements. The distance between the transmitter and targets changes from pulse to pulse; this also changes the phase of the target echoes from pulse to pulse. Therefore, transmitter-location diversity has the ability of making coherent targets appear incoherent to the receiving array. Transmitter-location diversity solves the problem completely. Its limitation is its practicality, however, for it is not a solution applicable to all system designs. In general, the diversity image I_D is the weighted sum of images I_m, $1 \leqslant m \leqslant M$, obtained from a set of K snapshots of the radiation field (Section 5-1). These images are called *trial* or *constituent images*. The trial images differ either because the diversity procedure alters the radiation pattern between successive transmissions or because the procedure alters the phase relations among scatterers in the source function, or both. Thus the diversity image to which Muller's search procedure is applied can be written

$$I_D = \sum_{k=1}^{K} a_k I_k \tag{9-18}$$

The aperture correlation $R(x)$ is the inverse Fourier transform of I_D. The following section describes a method that makes use of several correlation lags to estimate the phase errors.

9-3-3 Multiple-Lag SCA

From Eq. 9-8, observe that the phase of the correlation between the mth and nth element contains information about the elemental phase errors β_m and β_n. This section describes a noniterative procedure for estimating the phase errors using the autocorrelation phases from lag one through L. This procedure is computationally more efficient than Muller's image plane iterative algorithm, and its equivalent in the aperture plane. The method was developed in [109]. It uses only the arguments of the data correlations, and estimates the phase corrections $\hat{\beta}_n$ by solving a set of linear equations. The algorithm is then modified to make use of the correlation moduli also. First, the technique is described for the unit-lag case. This is equivalent to the spatial correlation algorithm described in Section 9-1. The technique is then generalized to make use of the correlation coefficients for larger lags.

9-3-3-1 Unit-Lag SCA Revisited

Let $\hat{\phi}(n+1, n)$ denote the phase of the unit-lag correlation estimate $\hat{R}(n+1, n)$. Using Eq. 9-8, it is easy to show that this phase is

$$\begin{aligned}\hat{\phi}(n+1, n) &= \psi_1 + \beta_{n+1} - \beta_n + \delta(n+1, n); \qquad 1 \leqslant n \leqslant N-1,\ \beta_1 = 0 \\ &\approx \psi_1 + \beta_{n+1} - \beta_n \qquad (9\text{-}19)\end{aligned}$$

In the above, the phase shifts necessary to steer the array towards direction u_0 and range r are included in the phase errors β_n to simplify notation. Further, ψ_1 is shorthand for the error-free correlation phase $\psi(d)$ (d = interelement spacing), and noise is ignored for the time being. The phase error at the reference element, β_1, has been arbitrarily set to zero. The approximation sign in the last portion of Eq. 9-19 indicates that the relationships hold to within the levels dictated by noise.

There are N unknown quantities in Eq. 9-19: β_n, $2 \leqslant n \leqslant N$ and ψ_1. However, there are only $(N-1)$ relationships to solve for these N unknowns. One of the unknowns can be eliminated if the unit-lag correlation phase ψ_1 is known or can be estimated. For most target scenarios, the unit-lag autocorrelation phase is approximately equal to kdu_0, where d is the interelement spacing and u_0 is the center of the field of view (the direction towards which the transmitter is pointed). Therefore, an estimate of ψ_1, denoted $\hat{\psi}_1$, is kdu_0. It is shown later than an incorrect estimate of ψ_1 only leads to a beampointing error.

Subtracting $\hat{\psi}_1 = kdu_0$ from all the $(N-1)$ relationships in Eq. 9-19 results in

$$\hat{\phi}(2, 1) - \hat{\psi}_1 = \beta_2 + (\psi_1 - \hat{\psi}_1) + \delta(2, 1)$$

$$\hat{\phi}(n, n-1) - \hat{\psi}_1 = \beta_n - \beta_{n-1} + (\psi_1 - \hat{\psi}_1) + \delta(n+1, n), \qquad 3 \leqslant n \leqslant N \qquad (9\text{-}20a)$$

For the present, assume that the unit-lag correlation phase estimate $\hat{\psi}_1$ is equal to ψ_1. As shown later, a small pointing error arises when $\hat{\psi}_1 \neq \psi_1$. Using $\hat{\psi}_1 = \psi_1$ and ignoring the noise terms $\delta(n+1, n)$ in Eq. 9-20a leads to

$$\begin{aligned} \beta_2 &\approx \hat{\phi}(2, 1) - \hat{\psi}_1 \\ \beta_n - \beta_{n-1} &\approx \hat{\phi}(n, n-1) - \hat{\psi}_1, \qquad 3 \leqslant n \leqslant N \end{aligned} \tag{9-20b}$$

where the approximation sign indicates that the equalities hold to within the limits dictated by noise. The effect of noise on the algorithm is discussed later. The $(N-1)$ relationships in Eq. 9-20b can be expressed in matrix notation as

$$\mathbf{Ax} \approx \mathbf{b} \tag{9-21a}$$

or

$$\begin{pmatrix} 1 & 0 & 0 & 0 & \cdots & 0 & 0 & 0 \\ -1 & 1 & 0 & 0 & \cdots & 0 & 0 & 0 \\ 0 & -1 & 1 & 0 & \cdots & 0 & 0 & 0 \\ \vdots & \vdots & \vdots & \vdots & \cdots & \vdots & \vdots & \vdots \\ 0 & 0 & 0 & 0 & \cdots & 0 & -1 & 1 \end{pmatrix} \begin{pmatrix} \beta_2 \\ \beta_3 \\ \beta_4 \\ \vdots \\ \beta_n \end{pmatrix} \approx \begin{pmatrix} \hat{\phi}(2, 1) - \hat{\psi}_1 \\ \hat{\phi}(3, 2) - \hat{\psi}_1 \\ \hat{\phi}(4, 3) - \hat{\psi}_1 \\ \vdots \\ \hat{\phi}(N, N-1) - \hat{\psi}_1 \end{pmatrix} \tag{9-21b}$$

where $\mathbf{A}$ is a $(N-1) \times (N-1)$ matrix, and $\mathbf{x}$ and $\mathbf{b}$ are $(N-1)$-dimensional column vectors.

The $(N-1)$ linear equations defined by Eq. 9-21 correspond exactly to those in Eq. 9-20b. Further, the matrix $\mathbf{A}$ does not depend on the data and the vector $\mathbf{b}$ is known because the $\hat{\phi}(n+1, n)$ are measured quantities and $\hat{\psi}_1 = ku_0 d$. Therefore the phase errors β_n can be obtained by solving the matrix equation 9-21, that is

$$\hat{\mathbf{x}} = \mathbf{A}^{-1}\mathbf{b} \tag{9-22}$$

where the $(N-1)$ vector $\hat{\mathbf{x}}$ consists of the phase corrections $\hat{\beta}_n$:

$$\hat{\mathbf{x}} = (\hat{\beta}_2\, \hat{\beta}_3 \ldots \hat{\beta}_N)^T \tag{9-23}$$

It is shown below that the phase corrections obtained from Eq. 9-23 lead to a small pointing error and can be expressed as

$$\hat{\beta}_n \approx \beta_n + (n-1)(\psi_1 - \hat{\psi}_1) \tag{9-24}$$

where the approximation sign indicates that the relationships hold to within the limits allowed by noise. The slope $(\psi_1 - \hat{\psi}_1)$ in the phase corrections shifts the image plane axis by an amount $\Delta u = (\psi_1 - \hat{\psi}_1)/kd$. The phase corrections in

Eq. 9-24 are similar to those obtained using the unit-lag SCA described in Section 9-1 (see Eq. 9-11). The only difference between the two sets of phase corrections is that the SCA phase corrections have a linear phase tilt of $n\psi_1$ whereas those in Eq. 9-24 have a phase tilt $(\psi_1 - \hat{\psi}_1)n$.

To prove Eq. 9-24, observe from Eq. 9-22 that $(\mathbf{A}\hat{\mathbf{x}} - \mathbf{b})$ is equal to the null vector. Therefore $(\mathbf{A}\hat{\mathbf{x}} - \mathbf{b})^T(\mathbf{A}\hat{\mathbf{x}} - \mathbf{b}) = 0$. Using 9-21 for $\mathbf{A}$ and $\mathbf{b}$ and setting $(\mathbf{A}\hat{\mathbf{x}} - \mathbf{b})^T(\mathbf{A}\hat{\mathbf{x}} - \mathbf{b})$ to zero leads to the following relationship between the phase corrections $\hat{\beta}_n$ and the measured correlation phases:

$$(\hat{\beta}_2 - \hat{\phi}(2, 1) + \hat{\psi}_1)^2 + \sum_{n=3}^{N} (\hat{\beta}_n - \hat{\beta}_{n-1} - \hat{\phi}(n, n-1) + \hat{\psi}_1)^2 = 0 \qquad (9\text{-}25a)$$

Use of Eq. 9-19 in Eq. 9-25a leads to

$$(\hat{\beta}_2 - \beta_2 + \hat{\psi}_1 - \psi_1)^2 + \sum_{n=3}^{N} (\hat{\beta}_n - \beta_n - \hat{\beta}_{n-1} + \beta_{n-1} + \hat{\psi}_1 - \psi_1)^2 = 0 \qquad (9.25b)$$

It is well known from polynomial theory that any set of N quantities $a_0, \cdots, a_{N-1}$, can be written in terms of a polynomial of order $(N-1)$, that is, $a_n = \sum_{k=0}^{N-1} c_k n^k$, where the c_k are the coefficients of the polynomial. Expanding the $(\hat{\beta}_n - \beta_n)$, $2 \leqslant n \leqslant N$, in this fashion leads to

$$\hat{\beta}_n - \beta_n = \sum_{k=0}^{N-2} c_k (n-1)^k; \qquad 2 \leqslant n \leqslant N \qquad (9\text{-}26)$$

The left side of Eq. 9-25 will be equal to zero if and only if each of the $(N-1)$ terms in the left side is zero because Eq. 9-25 is the sum of squares. Equating each of the $(N-1)$ terms in Eq. 9-25 to zero and substituting Eq. 9-26 for the $\hat{\beta}_n - \beta_n$ yields

$$\hat{\psi}_1 - \psi_1 = \beta_2 - \hat{\beta}_2 = -(c_0 + c_1 + \cdots + c_{N-2}) \qquad (9\text{-}27a)$$

$$\begin{aligned} \hat{\psi}_1 - \psi_1 &= \beta_n - \hat{\beta}_n - \hat{\beta}_{n-1} + \hat{\beta}_{n-1}, \qquad 3 \leqslant n \leqslant N \\ &= \sum_{k=0}^{N-2} c_k[(n-2)^k - (n-1)^k], \qquad 3 \leqslant n \leqslant N \\ &= -c_1 + c_2[(n-2)^2 - (n-1)^2] + \cdots, \qquad 3 \leqslant n \leqslant N \qquad (9\text{-}27b) \end{aligned}$$

Equation 9-27 provides $(N-1)$ relationships for the quantity $(\hat{\psi}_1 - \psi_1)$ in terms of the polynomial quantities c_k. A little thought will show that all the $(N-1)$ relationships will be simultaneously satisfied if and only if (a) $c_0 = 0$ and (b) $c_k = 0$, $k > 1$. Also note that $c_1 = \psi_1 - \hat{\psi}_1$. Using these values for the c_k in Eq. 9-26 yields $\hat{\beta}_n = \beta_n + (n-1)(\psi_1 - \hat{\psi}_1)$ as claimed in Eq. 9-24. That is, the phase corrections (Eq. 9-24) are equal to the phase errors to within a linear phase tilt. Of course, all the phase corrections are relative to the phase error at the reference element β_1, which was assumed to be zero.

9-3-3-2 Noise Analysis of the Unit-Lag SCA

Noise has not been considered so far; it is introduced now. In the presence of noise, the measured correlation phases $\hat{\phi}(n+1, n)$ are given by Eq. 9-20a. The phase errors are estimated by replacing the vector $\mathbf{b}$ in Eq. 9-21b by its estimate $\hat{\mathbf{b}}$. The difference $\hat{\mathbf{b}} - \mathbf{b}$, denoted $\delta\mathbf{b}$, is the noisy component of $\hat{\mathbf{b}}$ and is given by

$$\delta\mathbf{b} = \hat{\mathbf{b}} - \mathbf{b} = (\delta(2,1)\ \delta(3,2) \ldots \delta(N, N-1))^T \tag{9-28}$$

where $\delta(n, n-1)$ are the noise terms in the measured correlation phases $\hat{\phi}(n, n-1)$.

The phase error estimate vector $\hat{\mathbf{x}}$ is obtained from Eq. 9-22 by using $\hat{\mathbf{b}}$ instead of $\mathbf{b}$. That is, $\hat{\mathbf{x}} = \mathbf{A}^{-1}\hat{\mathbf{b}} = \mathbf{A}^{-1}(\mathbf{b} + \delta\mathbf{b})$. Because $\mathbf{A}^{-1}\mathbf{b}$ is the noise-free value of the estimated phase error vector, the error vector introduced due to noise, denoted $\mathbf{e} = \mathbf{A}^{-1}\delta\mathbf{b}$,

$$\mathbf{e} = \mathbf{A}^{-1}(\hat{\mathbf{b}} - \mathbf{b}) = \mathbf{A}^{-1}\delta\mathbf{b} \tag{9-29a}$$

$$= (\delta\beta_2\ \delta\beta_3 \ldots \delta\beta_N)^T \tag{9-29b}$$

where $\delta\mathbf{b}$ is defined by Eq. 9-28. The nth component of the $(N-1)$-dimensional vector $\mathbf{e}$ is the perturbation in the phase error estimate $\hat{\beta}_{n+1}$ introduced by noise, and is equal to $\delta\beta_{n+1} = \hat{\beta}_{n+1} - \beta_{n+1} - n(\psi_1 - \hat{\psi}_1)$.

The mean squared errors in the phase corrections can be quantified by computing the noise covariance matrix $E\{\mathbf{e}\,\mathbf{e}^T\}$, that is

$$E\{\mathbf{e}\,\mathbf{e}^T\} = \mathbf{A}^{-1}E\{\delta\mathbf{b}\,\delta\mathbf{b}^T\}(\mathbf{A}^T)^{-1} \tag{9-30}$$

In the above, the diagonal elements of $E\{\mathbf{e}\,\mathbf{e}^T\}$ correspond to the mean squared errors in the phase corrections $(= E\{\delta\beta_n^2\})$; the off-diagonal elements are the correlations $E\{\delta\beta_m\,\delta\beta_n\}$.

As mentioned in Section 9-1, the phase corrections obtained using the SCA (or equivalently Eq. 9-22) suffer from a serious limitation: the variance of the residual phase errors increases with element number n, that is, $E\{(\hat{\beta}_n - \beta_n)2\} \propto n$. As can be seen from Fig. 9-2, the limitation leads to a large loss in mainbeam gain when (a) the number of elements in the array is large and/or (b) the phase error in the unit-lag correlation phase estimates $\hat{\psi}_1$ is large. The limitation can be eased by using more correlation lags for estimating the phase corrections because the noise in the correlation lags $R(n+l, n)$, $l > 1$, are not completely correlated with the noise in $R(n+1, n)$.

This is demonstrated in Fig. 9-7, which plots the mainbeam gain as a function of the RMS phase error in the estimate of the unit lag correlation phase $\hat{\psi}(n+1, n)$ for various lags. The plot is for a periodic array with twenty elements.

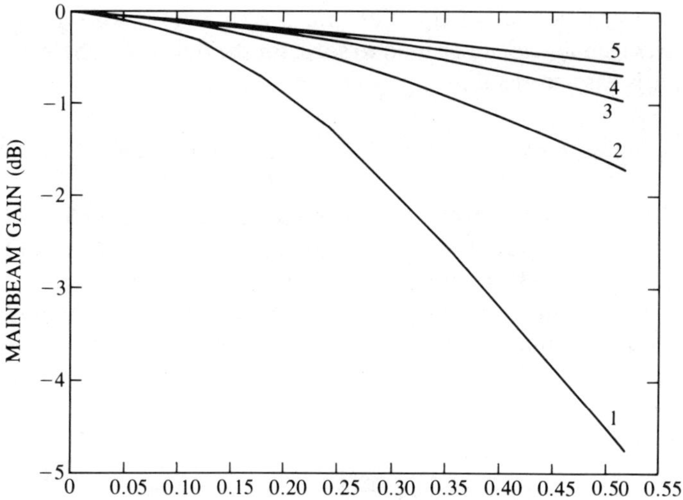

Figure 9-7 Normalized mainbeam gain (dB) versus rms phase error in the unit-lag correlation coefficient for the unit-lag and multiple-lag SCAs. Compare with the curve labeled $N = 20$ in Fig. 9-2.

The lowermost curve in Fig. 9-7 is for the unit-lag SCA; it corresponds to the curve labeled $N = 20$ in Fig. 9-2. As can be seen, the mainbeam gain decreases by about 5 dB when the rms phase error in $\hat{\psi}(n+1, n)$ is 0.5 radians. The remaining four curves are for lags 2, 3, 4, and 5, respectively. Notice that the mainbeam gain drops by less than 1 dB when more than two lags are used to estimate the phase errors β_n.

The curves in Fig. 9-7 were computed by assuming that the 20-element periodic array was used on both transmit and receive. Because the transmitting array was uniformly weighted and the clutter was assumed to be statistically homogenous, the ideal correlation function of the clutter echoes $R(x)$ was triangular, with a peak at the origin $x = 0$ and a null-to-null width of $x = (N-1)\lambda$, where $N = 20$. The correlation coefficients for all lags are thus real and positive. The variance of the phase error estimates $\hat{\beta}_n$ and their cross-correlations were calculated using Eq. 9-30 for the unit-lag SCA, and similar expressions were derived for the multiple-lag SCA. These statistics were then used for computing the loss in mainbeam gain. The curves will be different for different transmitting apertures; however, the general behavior will be similar. That is, the variance of the residual phase error can be decreased by using more lags. The multiple lag SCA is described next.

9-3-3-3 Multiple-Lag SCA

A method for incorporating the correlation phases for larger lags, $\hat{\phi}(n+l, n)$, $l > 1$, into Eq. 9-21 is described. It is shown that incorporating these lags leads to an overdetermined set of equations for the $(N-1)$ unknown phase errors; least-squares techniques can be used to solve for the unknown phase errors. From Eq. 9-19, it is easy to see that

$$\hat{\phi}(n+l, n) \approx \psi_l + \beta_{n+l} - \beta_n + \delta(n+l, n) \tag{9-31}$$

where ψ_l is the argument of the error-free correlation function for lag l and the $\delta(n+l, n)$ represent noise. The ψ_l, $l > 1$, are unknown quantities and have to be eliminated from Eq. 9-31 in order to derive relationships between the phase errors and the measured quantities $\hat{\phi}(n+l, n)$. An easy method for eliminating the ψ_l from Eq. 9-31 is to subtract two correlation phase estimates having the same lag. For example $\hat{\psi}(n+1+l, n+1) - \hat{\psi}(n+l, n)$ is independent of ψ_l, that is,

$$\begin{aligned}\gamma(n, l) &= \hat{\phi}(n+l+1, n+1) - \hat{\phi}(n+l, n)\\ &= \beta_{n+l+1} - \beta_{n+1} - \beta_{n+l} + \beta_n + \delta(n+l+1, n+1) - \delta(n+l, n)\\ &= \beta_{n+l+1} - \beta_{n+1} - \beta_{n+1} + \beta_n + \varepsilon(n, l)\\ &\approx \beta_{n+l+1} - \beta_{n+1} - \beta_{n+l} + \beta_n \end{aligned} \tag{9-32}$$

where the noise term $\varepsilon(n, l) = \delta(n+l+1, n+1) - \delta(n+l, n)$ has been ignored in the last line of Eq. 9-32.

Equation 9-32 relates measurable quantities $\gamma(n, l)$ with the unknown phase errors β_n. There are a total of $(N-l)$ correlation phases that can be measured for lag l, hence $(N-l-1)$ relationships of the type of Eq. 9-32 can be generated for lag l. These can be combined with the $(N-1)$ relationships between the unit-lag correlation phases and the phase errors in Eq. 9-20; this will increase the number of equations available to solve for the phase errors. The increased number of equations introduces an additional domain of averaging and hence improves performance with respect to noise. However, a little care must be exercised before the relationships in Eq. 9-32 are combined with those in Eq. 9-20 for the following reason.

It is shown below that the $(N-l-1)$ relationships between the lag l correlation phases and the phase errors in Eq. 9-32 are linear combinations of the unit-lag relationships shown in Eq. 9-20 except for noise and modulo 2π ambiguities. The modulo 2π ambiguities will have to be eliminated if both Eqs. 9-20 and 9-32 are to provide the same set of values for the phase error estimates. The following discussion develops a means for eliminating the modulo 2π ambiguities. Then, a method for estimating the phase errors using the correlation phases from lags 1 through L is developed.

First, observe from Eq. 9-32 that $\gamma(n, l) = \beta_{n+l+1} - \beta_{n+1} - \beta_{n+l} + \beta_n$ to within the limits allowed by noise. Further, observe from Eq. 9-19 that the difference between the unit-lag correlation phases $\hat{\phi}(n+l+1, n+l)$ and $\hat{\phi}(n+1, n)$ is also equal to the same quantity, that is, $\beta_{n+l+1} - \beta_{n+1} - \beta_{n+l} + \beta_n$. Hence, $\gamma(n, l)$ should be equal to $\hat{\phi}(n+l+1, n+l) - \hat{\phi}(n+1, n)$. This equivalence holds only if modulo 2π ambiguities are not present in the phase measurement process. With 2π ambiguities included, the equivalence between the $\gamma(n, l)$ and the measured unit-lag correlation phases can be expressed as

$$\begin{aligned}\gamma(n, l) &= \phi(n + l + 1, n + 1) - \phi(n + l, n) \\ &= \hat{\phi}(n + l + 1, n + l) - \hat{\phi}(n + 1, n) + 2\pi m \\ &\quad + \delta(n + l + 1, n + l) - \delta(n + 1, n) - \varepsilon(n, l)\end{aligned} \tag{9-33a}$$

where m is an integer and the $\varepsilon(\cdot, \cdot)$ and $\delta(\cdot, \cdot)$ represent noise. The first line of Eq. 9-33a is the definition of the $\gamma(n, l)$ (see Eq. 9-32) and the lower lines are an equivalent representation of $\gamma(n, l)$ in terms of the measured unit-lag correlation phases. Modulo 2π compatibility between the two quantities can be restored by subtracting $2\pi m$ from the $\gamma(n, l)$. The integer m can be determined when the absolute value of the noise term $\delta(n + l + 1, n + l) - \delta(n + 1, n) - \varepsilon(n, l)$ is smaller than π. This is because both $\gamma(n, l)$ and the $\hat{\phi}(n, n - 1)$ in Eq. 9-33 are measured quantities. The unwrapped value of $\gamma(n, l)$, denoted $\gamma(n, l)|_u$, can be expressed as

$$\gamma(n, l)|_u = \gamma(n, l) - 2\pi\hat{m} \tag{9.33b}$$

where

$$\hat{m} = \left.\frac{\gamma(n, l) - [\hat{\phi}(n + l + 1, n + l) - \hat{\phi}(n + 1, n)}{2\pi}\right|_{\text{round}} \tag{9.33c}$$

The subscript *round* in Eq. 9-33c denotes rounding to the nearest integer and $\hat{m}$ is an estimate of the integer m in Eq. 9-33a. Following the operation (9-33b), the $\gamma(n, l)|_u$ will be modulo 2π compatible with the unit-lag correlation phases. A method for using both the unit-lag phases and the $\gamma(n, l)$ in order to estimate the phase errors β_n is now developed.

Assume that the correlation phases from lags 1 through L are used for estimating the phase errors. The unit-lag correlation phases provide $(N - 1)$ relationships between the measured quantities $\hat{\phi}(n, n - 1)$ and the $(N - 1)$ unknown phase errors β_n, $2 \leqslant n \leqslant N$ (see Eq. 9-20). The lag l correlation phases provide $(N - l - 1)$ relationships between the measured quantities $\gamma(n, l)$, $1 \leqslant n \leqslant N - l - 1$, and the phase errors (see Eq. 9-32). Hence a total of $M_0 = L(N - 1) - L(L + 1)/2 + 1$ equations will be available to solve for the $(N - 1)$ unknown phase errors. These equations can be written in matrix notation in a manner similar to Eq. 9-21. The matrix **A** will have dimensionality $(M_0 \times N - 1)$ when lags 1 through L are used. For example, the matrix **A** and vector **b** for $L = 2$ are given by

$$\mathbf{A} = \begin{pmatrix} 1 & 0 & 0 & 0 & \cdots & 0 & 0 & 0 & 0 \\ -1 & 1 & 0 & 0 & \cdots & 0 & 0 & 0 & 0 \\ 0 & -1 & 1 & 0 & \cdots & 0 & 0 & 0 & 0 \\ \vdots & \vdots & \vdots & \vdots & \cdots & \vdots & \vdots & \vdots & \vdots \\ 0 & 0 & 0 & 0 & \cdots & 0 & 0 & -1 & 1 \\ -1 & 1 & 1 & 0 & \cdots & 0 & 0 & 0 & 0 \\ 1 & -1 & -1 & 1 & \cdots & 0 & 0 & 0 & 0 \\ \vdots & \vdots & \vdots & \vdots & \cdots & \vdots & \vdots & \vdots & \vdots \\ 0 & 0 & 0 & 0 & \cdots & 1 & -1 & -1 & 1 \end{pmatrix};$$

$$\mathbf{b} = \begin{pmatrix} \hat{\phi}(2, 1) - \hat{\psi}_1 \\ \hat{\phi}(3, 2) - \hat{\psi}_1 \\ \hat{\phi}(4, 3) - \hat{\psi}_1 \\ \vdots \\ \hat{\phi}(N, N-1) - \hat{\psi}_1 \\ \gamma(1, 2)|_u \\ \gamma(2, 2)|_u \\ \vdots \\ \gamma(N-3, 2)|_u \end{pmatrix} \tag{9-34}$$

where the $\gamma(n, l)|_u$ are the unwrapped values of the $\gamma(n, l)$ as defined by Eq. 9-33b. For larger lags, the matrix $\mathbf{A}$ can be constructed by observing which of the β_n appear in the various $\gamma(n, l)$. The phase corrections can be determined using

$$\hat{\mathbf{x}} = \mathbf{A}^{+}\mathbf{b} = [(\mathbf{A}^T\mathbf{A})^{-1}\mathbf{A}^T]\mathbf{b} \tag{9-35}$$

where $\mathbf{A}^{+}$ is the pseudo-inverse of $\mathbf{A}$ and $\mathbf{b}$ is a column vector of length M_0 containing (a) the $(N-1)$ quantities $\hat{\phi}(n+1, n) - \hat{\psi}(1)$, $2 \leqslant n \leqslant N$, and (b) the $(M_0 - N + 1)$ quantities $\gamma(n, l)$ after removal of the 2π ambiguities.

The matrix $\mathbf{A}^{+}$ can be precalculated and stored because $\mathbf{A}$ is independent of the data. Thus only $M_0(N-1)$ real multiplications are required to solve for the phase errors. This approach for determining the phase corrections is computationally more efficient than the iterative approach of Muller.

9-3-3-4 Noise Analysis of the MLSCA

The effect of noise on the MLSCA phase corrections can be quantified in a manner similar to the noise analysis of the unit-lag SCA, that is, the derivation of Eq. 9-30. It can be shown that the residual phase error covariance matrix for the multiple-lag SCA has the form

$$E\{\mathbf{e}\,\mathbf{e}^T\} = \mathbf{A}^+ E\{\delta\mathbf{b}\,\delta\mathbf{b}^T\}\mathbf{A}^{+T} \tag{9.36a}$$

where

$$\delta\mathbf{b} = (\delta(2,1)\,\delta(3,2)\ldots\delta(N,N-1)\,\varepsilon(1,2)\,\varepsilon(2,2)\ldots\varepsilon(N-3,2))^T \tag{9-36b}$$

$\varepsilon(n, l)$ is the noise present in the $\gamma(n, l)$ (see Eq. 9-32), and $L = 2$ has been assumed. The residual phase error variances and cross-correlations in Eq. 9-36 were used to compute the curves depicted in Fig. 9-7.

Least-squares theory provides the following interesting interpretation of the multiple-lag SCA. It can be used to show that the phase corrections obtained using Eq. 9-35 minimize the norm of the vector $\mathbf{Ax} - \mathbf{b}$. Therefore the solution vector $\hat{\mathbf{x}}$ is obtained by minimizing the following quantity, denoted F:

$$F = (\mathbf{Ax} - \mathbf{b})^T(\mathbf{Ax} - \mathbf{b}) = \sum_{n=2}^{N} (\hat{\beta}_n - \hat{\beta}_{n-1} - \hat{\phi}(n, n-1) + \hat{\psi}_1)^2$$
$$+ \sum_{l=2}^{L} \sum_{n=1}^{N-l-1} (\gamma(n, l)|_u - \hat{\beta}_{n+l+1} + \hat{\beta}_{n+1} + \hat{\beta}_{n+l} - \hat{\beta}_n)^2 \tag{9.37}$$

where the $\phi(n + 1, n)$ are defined by Eq. 9-19 and the $\gamma(n, l)|_u$ are defined by Eq. 9-33b. That is, the MLSCA estimates the phase corrections by minimizing the difference between the measured correlation phases $\hat{\phi}(n, n - 1)$ and $\gamma(n, l)|_u$ and the phase errors associated with these measured correlation phases.

In Eq. 9-37, terms containing the measured unit lag correlation phases and the multiple-lag correlation phases are given equal (unity) weighting in the minimization process. However, the noise in the measured correlation phases will in general increase with lag, hence it is appropriate to use smaller weights for larger lags. A weighted scheme for phase error compensation using the MLSCA is described next.

9-3-3-5 Weighted Multiple-Lag SCA

The variance of the noise terms in the correlation phases $\hat{\phi}(m, n)$ are of the form $E\{\delta^2(m, n)\} = 0.5[(R(0)/|R(m, n)|)^2 - 1]/K$, where K is the number of range bins [38]. From least-squares theory, weighting the various terms in Eq. 9-37 by the inverse of their variances provides better estimates of the phase errors than using the same weight for each quantity irrespective of its variance. Thus the terms in Eq. 9-37 containing the unit-lag correlation phases have to be weighted by $|R(2, 1)|^2$; terms containing the $\gamma(n, l)$ have to be weighted by $E\{\varepsilon^2(n, l)\} = 0.5|R(l + 1, 1)|^2$.⋆ Modifying Eq. 9-37 to accommodate this

⋆ From Eq. 9-32, observe that the $\gamma(n, l)$ contain two different lag l correlation phases, hence their variance is $2/|R(l + 1, 1)|^2$. The weighting coefficient for terms containing $\gamma(n, l)$ is the reciprocal of their variance.

weighting scheme yields

$$F = \sum_{n=2}^{N} (\hat{\beta}_n - \hat{\beta}_{n-1} - \hat{\phi}(n, n-1) + \hat{\psi}_1)^2 + \sum_{l=2}^{L} w \left\{ \sum_{n=1}^{N-l-1} (\gamma(n, l)|_u - \hat{\beta}_{n+l+1} + \hat{\beta}_{n+1} + \hat{\beta}_{n+l} - \hat{\beta}_n)^2 \right\} \tag{9-38a}$$

where

$$w_l = \frac{|R(l+1, 1)|^2}{|R(2, 1)|^2} \frac{(R^2(0) - |R(2, 1)|^2)}{(R^2(0) - |R(l+1, 1)|^2)} \tag{9.38b}$$

and $|R(l + 1, 1)|$ is the theoretical value of the magnitude of the correlation coefficient for lag l. The weights can be calculated using the estimated magnitudes of the correlation coefficients when theoretical values are not available. For a uniformly weighted transmit array, the theoretical value of $w_l = [(N - l)/(N - 1)]^2$ because the correlation function is triangular. Use of the theoretical value of w_l in Eq. 9-38 provides savings in computation because the pseudo-inverse of the matrix associated with the minimization to be performed in Eq. 9-38 can be precalculated and stored. This matrix can be constructed in a manner similar to the derivation of Eq. 9-34. For example, the matrix **A** and the vector **b** associated with the weighted least-squares minimization in Eq. 9-38 ($L = 2$) are given by

$$\mathbf{A} = \begin{pmatrix} 1 & 0 & 0 & 0 & \cdots & 0 & 0 & 0 & 0 \\ -1 & 1 & 0 & 0 & \cdots & 0 & 0 & 0 & 0 \\ 0 & -1 & 1 & 0 & \cdots & 0 & 0 & 0 & 0 \\ \vdots & \vdots & \vdots & \vdots & \vdots & \vdots & \vdots & \vdots & \vdots \\ 0 & 0 & 0 & 0 & \cdots & 0 & 0 & -1 & 1 \\ -\sqrt{w_2} & \sqrt{w_2} & \sqrt{w_2} & 0 & \cdots & 0 & 0 & 0 & 0 \\ \sqrt{w_2} & -\sqrt{w_2} & -\sqrt{w_2} & \sqrt{w_2} & \cdots & 0 & 0 & 0 & 0 \\ \vdots & \vdots & \vdots & \vdots & \vdots & \vdots & \vdots & \vdots & \vdots \\ 0 & 0 & 0 & 0 & \cdots & \sqrt{w_2} & -\sqrt{w_2} & -\sqrt{w_2} & \sqrt{w_2} \end{pmatrix} \tag{9-39a}$$

$$\mathbf{b} = \begin{pmatrix} \hat{\phi}(2,1) - \hat{\psi}_1 \\ \hat{\phi}(3,2) - \hat{\psi}_1 \\ \hat{\phi}(4,3) - \hat{\psi}_1 \\ \vdots \\ \hat{\phi}(N, N-1) - \hat{\psi}_1 \\ \sqrt{w_2}\,\gamma(1,2)|_u \\ \sqrt{w_2}\,\gamma(2,2)|_u \\ \vdots \\ \sqrt{w_2}\,\gamma(N-3,2)|_u \end{pmatrix} \tag{9-39b}$$

The phase error estimates for the weighted scheme are $\mathbf{A}^+\mathbf{b}$, where $\mathbf{A}^+$ is the pseudo-inverse of $\mathbf{A}$ in Eq. 9-39a. The noise properties of the algorithm can be determined using Eq. 9-36.

This completes the discussion on the multiple-lag SCA. The computational complexity of the algorithm is evaluated at the end of Section 9-6.

9-4 MULTIPLE-LAG SIMULATION EXPERIMENTS

Figure 9-8 is a computer simulation experiment in which transmitter diversity performed the decohering function. Two equal-strength point targets are located at $u = -0.3$. and 0.5. The solid curve shows the one-dimensional image of the two targets obtained using an 11-element periodic receiving array. The curve (b) shows the useless image that resulted when random phase errors were added at each receiver. The error distribution was uniform over the interval $\pm\pi$.

Images were then computed for five different locations of the transmitter. The transmitter locations were confined to the extent of the receive array. Cross-products obtained from the five sets of receiver measurements were averaged to provide the estimates of the correlation coefficients, which were then applied to the algorithm. The result is shown in curve c, which is a nearly identical copy of the original solid curve displaced 0.68 units of u to the right.

Figure 9.9 shows another experiment. In this experiment a continuous target from $u_1 = -0.55$ to $u_2 = -0.15$ having a random rough surface was simulated by one-hundred-unit amplitude and random phase scatterers randomly distributed from u_1 to u_2. A combination of range-bin diversity and transmitter diversity was employed. Two range bins were located 1300λ and 1340λ from the receiving array. The scatterer distributions in the two range bins were independent of each other. The receiving array consisted of 24 elements spaced one half-wavelength apart. The FOV was illuminated from three locations, the transmitters being placed at the center and the two ends of the array. Zero-mean

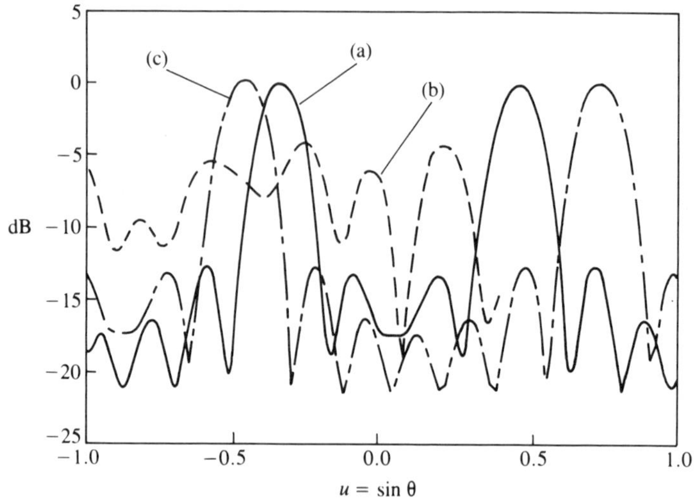

Figure 9-8 Computer simulation experiment to test the multiple-lag correlation algorithm. (a) Image with no phase errors. (b) Image with phase errors. (c) Image with phase errors corrected by the multiple-lag correlation algorithm. (From ref. [100])

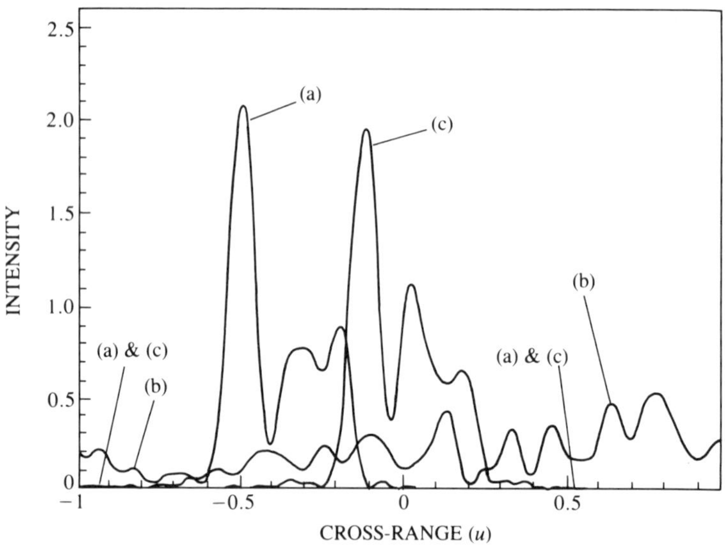

Figure 9-9 Demonstration of the multiple-lag calibrator for far-field targets. 100 coherent targets, located in the far field of the array between $u = -0.55$ and $u = -0.15$, are imaged using a 24-element filled periodic array. Curve (a) is the error-free image of the target scenario. Curve (b) is the image obtained after adding random phase errors at each receiver. Curve (c) is the image obtained after correcting for these errors using the first five lags. Sixth-order transmitter-location diversity was used for decorrelating the coherent targets. Observe that the images (a) and (c) are almost identical except for a shift $\delta u = 1/3$ "u" units in the image plane.

Gaussian noise having a variance of 0.01 was introduced at each receiver. The combination of three transmitters and two range bins provided sixfold diversity for decorrelating the target scenario.

Curve (a) in Figure 9-9 is the error-free image of the first range bin. Independent phase errors, uniformly distributed between $\pm\pi$, were then introduced at each receiver. Curve (b) is the image of the first range bin without correcting for the phase errors. This image does not resemble the target scenario whatsoever. Curve (c) is the image of the first range bin after correcting for the phase errors. This image is almost identical to the error-free image except for a shift $\Delta u = 1/3$ in the image plane.

9-5 COMPARISONS WITH OTHER ALGORITHMS

The experiments reported in this section fall into two groups. In the first three tests (Figs. 9-10 to 9-12) the targets satisfied the dominant scatter condition; in the remainder the low-variance beamformer condition was not met. The upper image of Fig. 9-10 was processed using the dominant scatterer algorithm; the lower with the MLSCA. The aircraft was a large jet at a distance of about 20 km from the Valley Forge laboratory. Because the dominant scatterer condition was satisfied, a satisfactory adaptively beamformed image was obtained. The bottom figure shows the image obtained using the same data set when the MLSCA algorithm was applied. It is evident that the two procedures were equivalent.

Figure 9-11 compares images of a Boeing 727 formed by four algorithms, the DSA, MSA, SCA, and MLSCA. A reference scatterer with very low normalized amplitude variance ($\sigma^2 = 0.05$) was on the target. All methods work well. The mainlobe gains of the adaptively formed arrays differ by no more than 0.5 dB.

Figure 9-12 compares images of the Limerick nuclear power plant obtained from the same four algorithms. Again the results are nearly alike. The array gains differ by less than 0.5 dB.

The situation changes when no low-variance range bin can be found. Figure 9-13 shows the use of bin 47, for which $\sigma^2 = 0.17$ is the reference bin. This high-variance bin resulted in failure of DSA as shown in part (a). SCA and MLSCA were both able to form images recognizable as airplanes.

Four images from another data set obtained from the same airplane are shown in Fig. 9-14. Again the SCA and MLSCA images are superior. Mainlobe gain is 3.9 dB greater than DSA and 3.3 dB greater than MSA.

Figure 9-15 shows DSA and SCA 1-D images of a corner reflector in which echoes from farmland were used for self-calibration. DSA fails completely; the DSA image bears no relation to that of a corner reflector, while the SCA image shows a prominent target. MLSCA performance is comparable to SCA. Both exhibit 4.5 dB greater mainlobe gain.

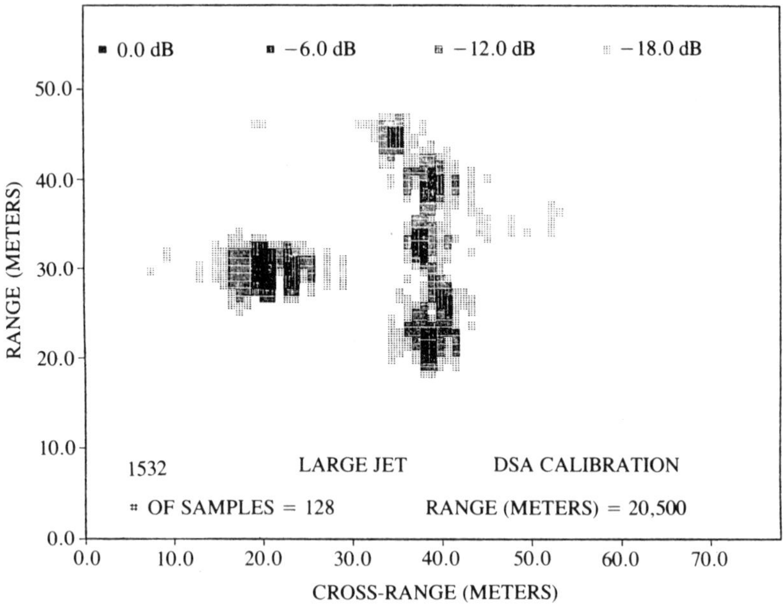

(a)

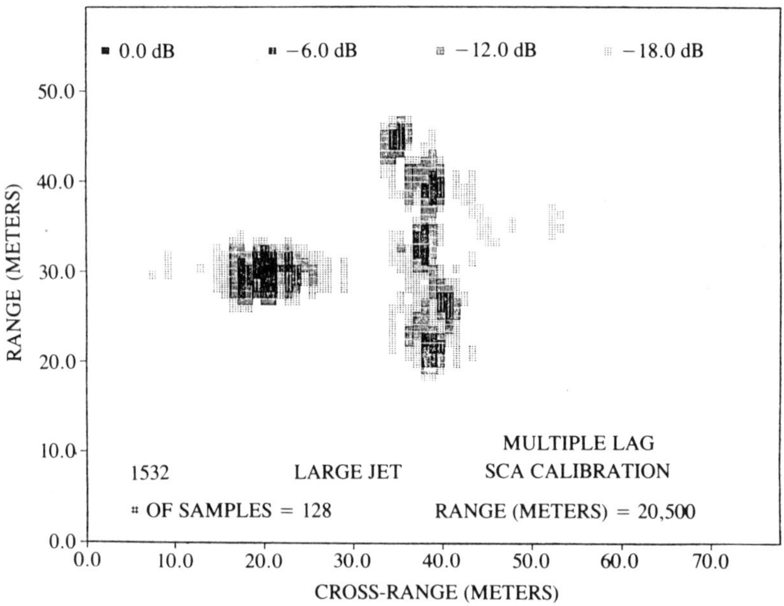

(b)

Figure 9-10 Radar camera ISAR images of a large jet aircraft at a distance of 20 km from Valley Forge Research Center. The range cell is 3 m. (a) DSA; (b) multiple-lag SCA. Both procedures worked well and yielded comparable images. (From ref. [109])

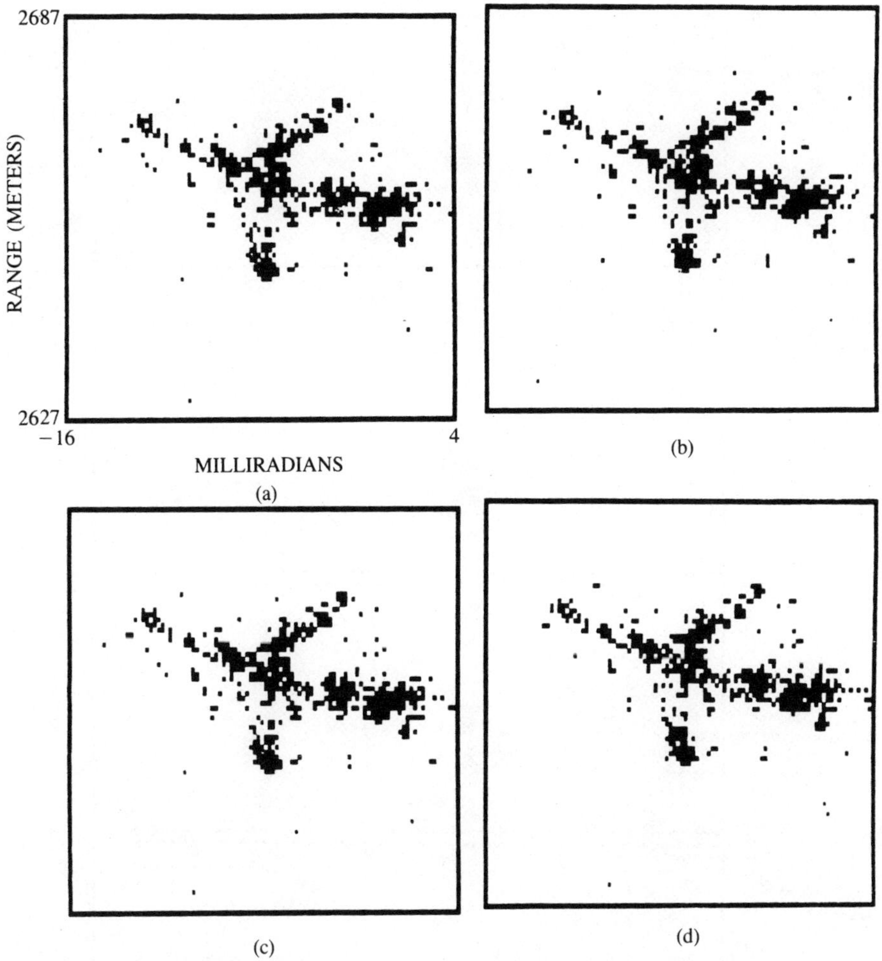

Figure 9-11 Range–azimuth image of a Boeing 727 aircraft obtained using the fourth quadrant (elements from 385 through 512). (a) DSA with bin 40 having amplitude variance 0.05 as the reference range bin. (b) MSA with bins 29, 40, and 42. (c) SCA with lag 1 and bins 37 through 40. (d) MMB with lag (1, 6, 28) and bins 37 through 40. (From ref. [47])

The observations from this and the preceding chapter lead to the following conclusions:

- If at least one range bin has a low-variance beamformer ($\sigma^2 < 0.12$), any algorithm is satisfactory. In general the DSA is the simplest to implement.
- If no low-variance bin can be found but one or more bins exhibit single targets with lower ratios of target to clutter strengths, then

  DSA will fail.

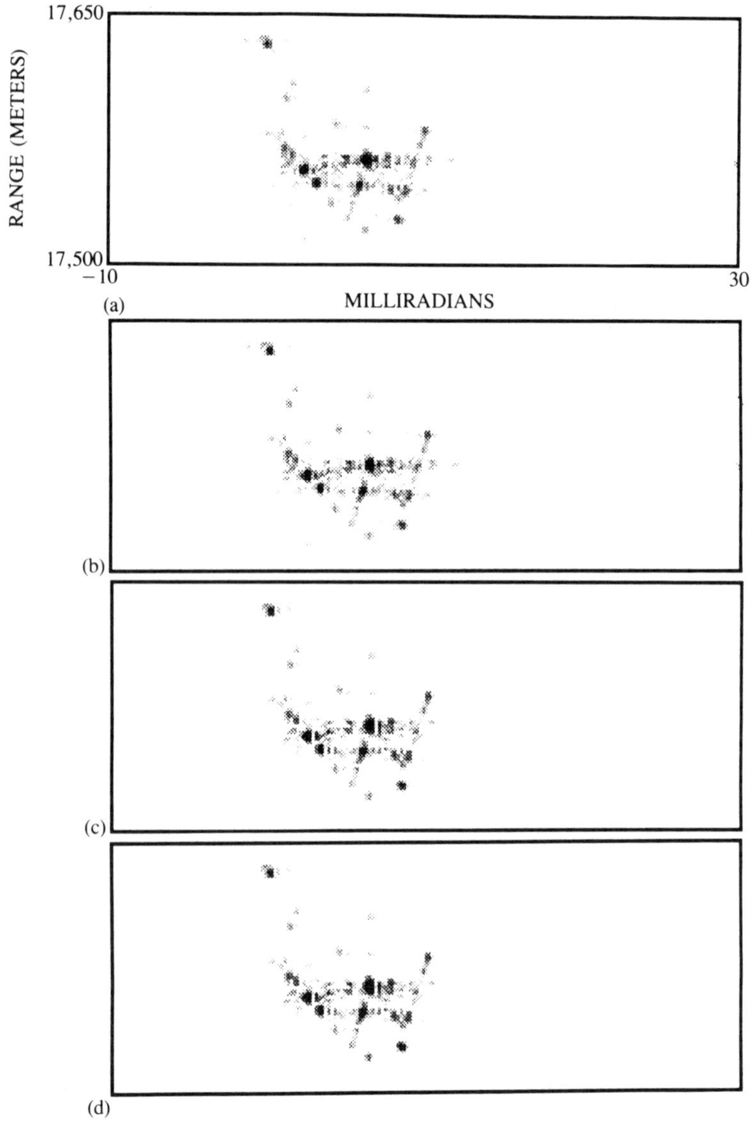

Figure 9-12 Range-azimuth images of Limerick nuclear power plant. (a) DSA with bin 67; (b) MSA with bins 29, 32, and 67. (c) SCA with bins 31 through 34; (d) MMB with bins 31 through 34, lag (1, 8, 92). (From ref. [47])

MSA will often succeed, particularly with the use of overlapping subarrays.
SCA and MLSCA will always succeed.

- If no bins exhibit single targets in clutter:
  DSA and MSA will fail.
  SCA and MLSCA will often succeed.

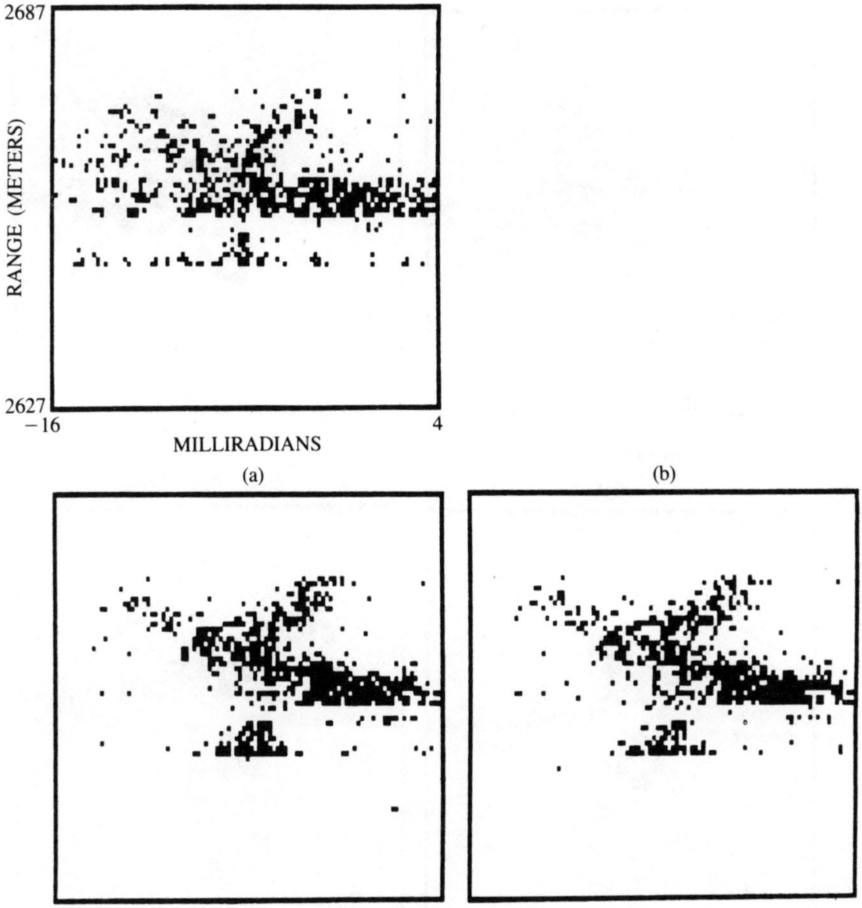

(c) (d)

Figure 9-13 Range–azimuth image of a Boeing 727 aircraft. (a) DSA with bin 47 having amplitude variance 0.17 as the reference range bin. (b) MSA is not available with bins 44 through 47. (c) SCA with lag 1 and bins 44 through 47. MMB with lag (1, 6, 28) and bins 44 through 47. (From ref. [47])

9-6 ALGORITHM BASED ON AN ENERGY CONSERVATION PRINCIPLE

9-6-1 Tsao's Concept

In a 1983 paper Tsao recognized that the angular width of an image formed by a diffraction-limited system was about one receive beamwidth larger than the size of the angular sector illuminated by the microwave transmitter; conversely, energy would fall outside the sector only if the aperture was distorted or the medium was turbulent or some other aberrative source was distorting the received signal data. Using Parceval's theorem, he argued that

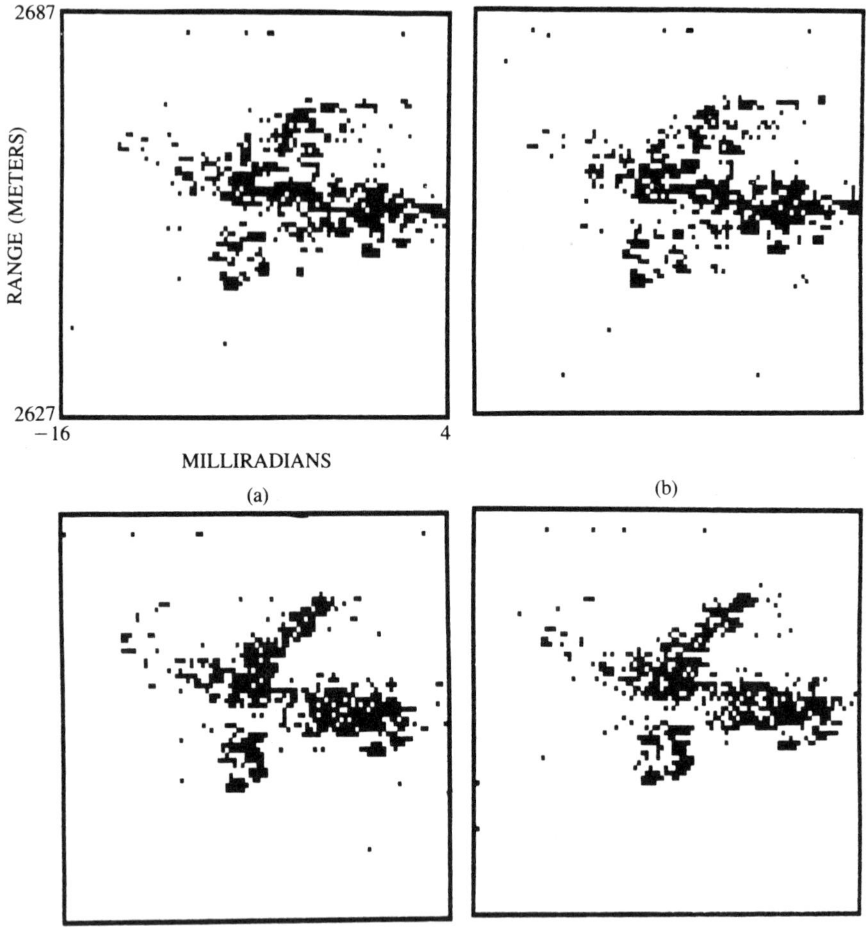

(a)

(b)

(c)

(d)

Figure 9-14 Range–azimuth image of a Boeing 727 aircraft obtained using the second quadrant (elements from 129 through 256). (a) DSA with bin 40 having amplitude variance 0.15 as the reference range bin. (b) MSA with bins 38, 39, and 40. (c) SCA with lag 1 and bins 39 through 42. (d) MMB with lag (1, 6, 28) and bins 39 through 42. (From ref. [47])

the total image energy, being equal to the total signal energy in the aperture, was independent of the distortion and therefore independent of the distribution of the image energy in the image plane. By comparing the total aperture plane energy to that portion of the image energy that fell in the illuminated sector, he obtained an objective function suitable for feedback-controlling the weight vector of the phased array [119].

Figure 9-16 illustrates Tsao's idea. Energy from a small transmitter illuminates a sector $\Delta\theta$ or, its equivalent, Δu, where $u = \sin\theta$ and θ is the angle from the array normal. Echoes from several range bins are imaged by the system. In the

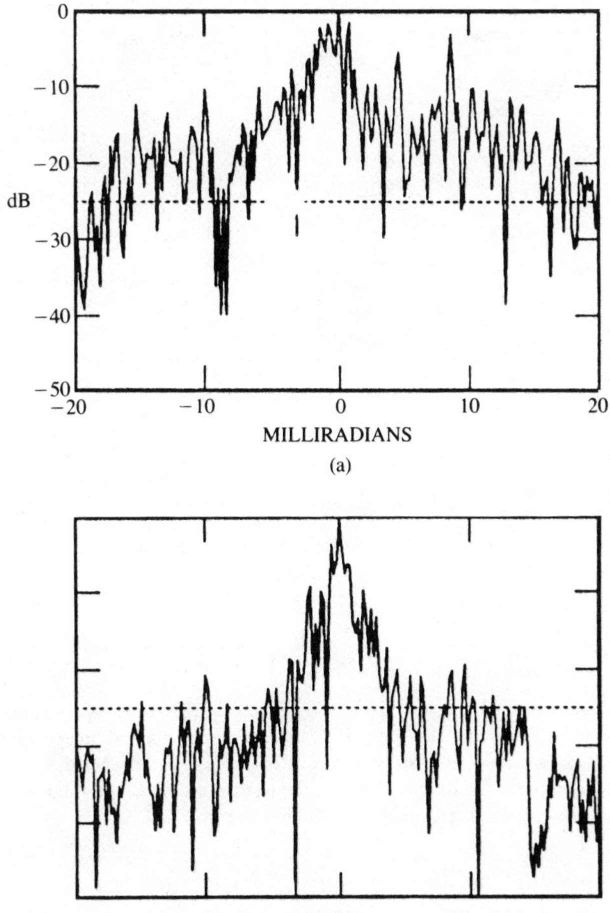

Figure 9-15 Cross-range images of corner reflector when array is calibrated on echoes from farmland. (a) DSA; (b) SCA. (From ref. [47])

absence of all sources of distortion, for example, medium turbulence, element-position error within the aperture, and channel-to-channel tuning errors in the receivers, all the energy falls within the illumination sector. When phase aberrations occur from any of these sources, however, the image energy fails to cohere perfectly in azimuth and as a consequence some portion of the energy falls outside the illuminated sector.

The ECA procedure measures the image energy within the illuminated sector. The sector is defined by the angular region $u_1 \leqslant u \leqslant u_2$, $|u| \leqslant 1$, and $1 \leqslant k \leqslant K$ range bins. Call the energy in this sector $E_{\Delta u}$. It equals $\sum_{k=1}^{K} \int_{u_1}^{u_2} |\hat{s}_k(u)|^2 \, du$. Let E_A be the total energy in the aperture from the same set of range bins. In Tsao's algorithm these energies are compared and the difference is reduced to zero by

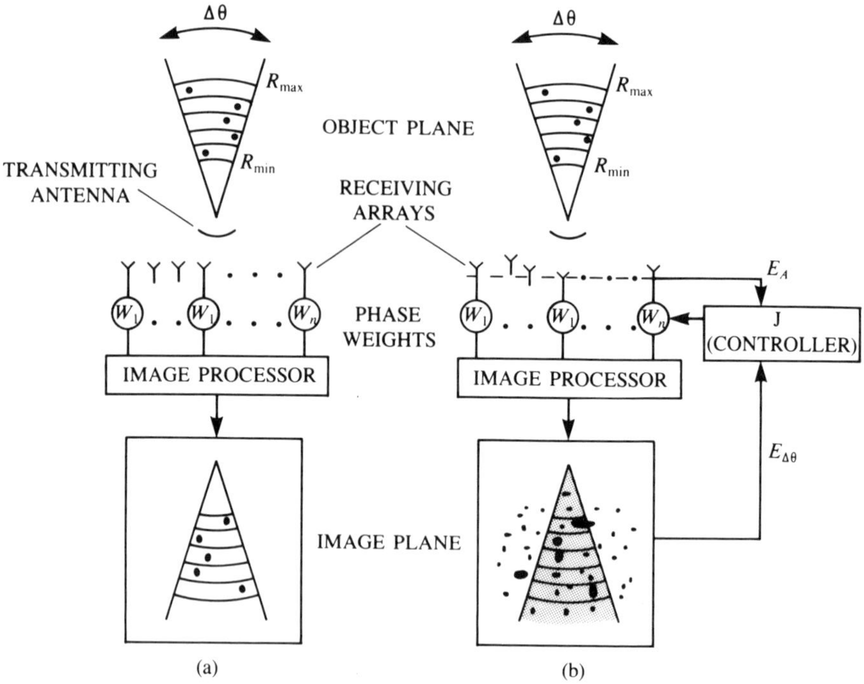

Figure 9-16 Illustration of Tsao's energy conservation principle. A transmitter illuminates scatterers in sector $\Delta\theta$. (a) All the energy in the diffraction-limited image of an unaberrated receiving aperture falls within $\Delta\theta$. (b) The image from an aberrated aperture extends beyond $\Delta\theta$. Tsao's algorithm compares the image energy in the shaded sector to the total signal energy in the aperture and adjusts the weights to make the difference zero.

feedback-controlling the array weights. Two factors can cause the image to extend outside the interval $[u_1, u_2]$. The first is the lowpass filtering imposed by the aperture, which smears the energy the order of one beamwidth. This factor only marginally extends the image outside $[u_1, u_2]$. The second factor is the phase aberration. When the phase errors are distributed over $\pm\pi$ the image will spread over all u.

A procedure similar to Tsao's is in use at the Very Large Array (VLA) of the national Radio Astronomy Observatory in Socorro, New Mexico [116]. Tsao's algorithm is also similar to certain iterative bandlimited extrapolation methods used to reconstruct signals from partial data, e.g., Papoulis [75], Gerchberg and Saxton [31] and Feinup [29], [30].

9-6-2 Noise-Free Performance

This section shows the noise-free performance of Tsao's algorithm when calibrated by small numbers of discrete scatterers [76]. Comparison with the unit-lag and multiple-lag SCA, with noise and with distributed clutter, is given in Section 9-6-4.

A 30-element linear receiving array with 2λ interelement spacing is assumed. The backscattered radiation from ten range bins is calculated and is used to drive the system, that is, to obtain the weight factor. Each range bin contains two scatterers that are uniformly distributed between ± 0.0835 in u. Their locations are random in this interval. Their amplitudes are chosen from 0 to 1 and their phases are uniform in the interval $\pm\pi$. Errors are introduced in the placement of the positions of the antennas, resulting in highly distorted beam patterns. Following sufficient iterations of the algorithm, the radiation patterns are restored.

Three cases are shown in the figures. Figures 9-17a and b compare the undistorted radiation pattern, the pattern after the array was distorted, and the correction pattern when the RMS element position error is $\lambda/8$ in both dimensions. A displacement error of $\lambda/8$ corresponds to a phase error variance σ_ϕ^2 of 0.617 square radians. The expected value of the array gain relative to the design value is $\exp(-\sigma_\phi^2)$, which evaluates to a loss of 2.6 dB. The solid curve of Fig. 9-17a shows the radiation pattern after the distortion is introduced. The gain of the main lobe has dropped by the expected amount. More serious is the side radiation pattern, which has changed from a sinc function to a random pattern whose peak value in the angular interval shown has risen to about 8 dB below the main lobe. Figure 9-17b makes the same comparison after Tsao's algorithm has been applied. The curves are almost identical in the main lobe and differ insignificantly in the side radiation pattern. Figures 9-17c and d show the ideal scanned images and the images following distortion and compensation for a range bin having two closely spaced scatterers of comparable strength. Figures 9-17e and f show the same for two separated scatterers with a 20-dB intensity difference. Figures 9-18 and 9-19 show the performance when the RMS element position error increases to $\lambda/4$ and $\lambda/2$, respectively.

Table 9-1 summarizes the experimental comparisons between the properties of the error-free antenna and the distorted antennas for the three noise-free cases. The performance is exceedingly satisfactory even for an rms position error as large as 0.5 wavelengths. It is interesting to note that the performances of the $\lambda/4$ and $\lambda/2$ errors are similar; in some categories one array is better, in the other categories the other array is superior. It is believed that this is because the distributions of the phase errors of both cases are very broad.

Figure 9-20 shows the learning curves for the three cases. The ordinate is a function of the image energy outside the illumination sector. The abscissa is the number of iterations of the algorithm. The system rapidly begins to work, accomplishes most of the task in five or so iterations, during which time the main beam is reconstructed, and then slows to work on the sidelobes.

Figure 9-21 plots the residual phase errors across the array for the three cases. The phase error variances are listed in Table 9-1.

9-6-3 A Computationally Efficient ECA Procedure

The procedure was developed by Subbaram in [111]. Let $e_k(x_n)$, $1 \leqslant n \leqslant N$, represent the error-free echoes received at the N array elements from a source

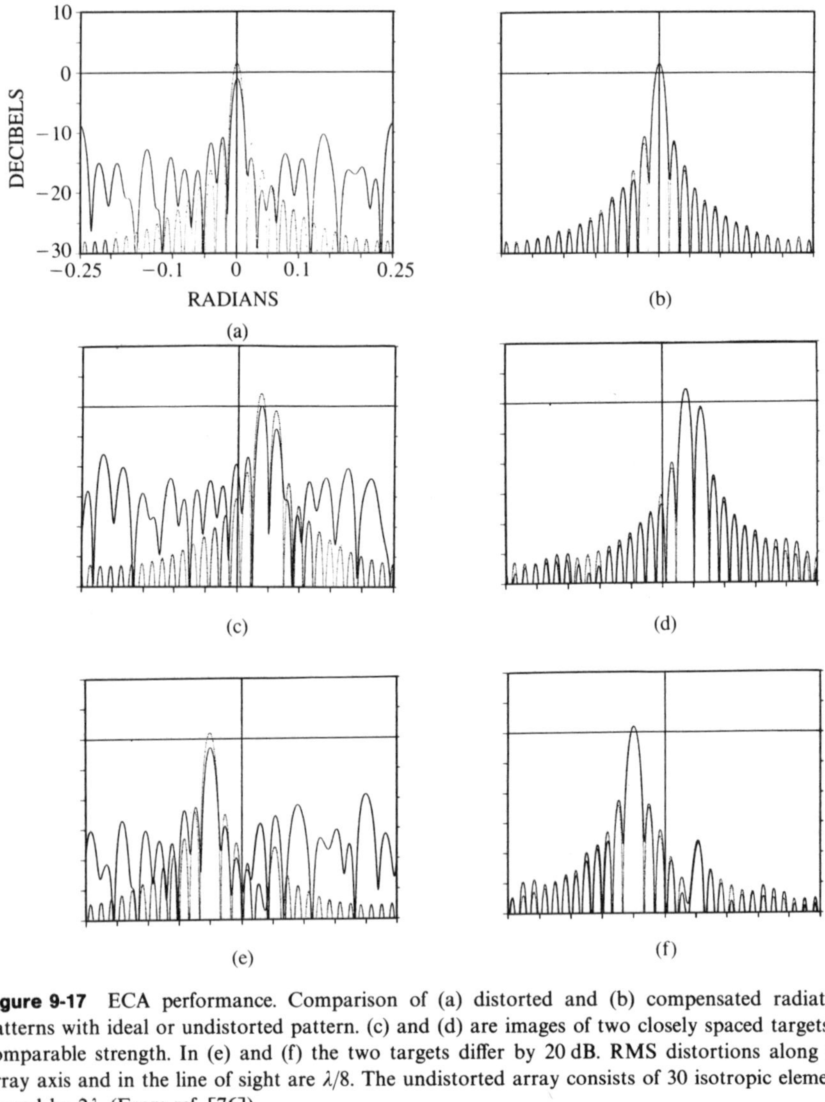

Figure 9-17 ECA performance. Comparison of (a) distorted and (b) compensated radiation patterns with ideal or undistorted pattern. (c) and (d) are images of two closely spaced targets of comparable strength. In (e) and (f) the two targets differ by 20 dB. RMS distortions along the array axis and in the line of sight are $\lambda/8$. The undistorted array consists of 30 isotropic elements spaced by 2λ. (From ref. [76])

distribution $s_k(u)$ in the kth range bin. The measured data $\hat{e}_k(x_n) = e_k(x_n)\exp(j\beta_n)$, where β_n is the phase error at the nth receiver. Let $\hat{\beta}_n$ represent the as yet unknown estimate of the phase errors. The estimated image, with the phase corrections included, can be expressed as

$$\hat{s}_k(u) = \sum_{n=1}^{N} \hat{e}_k(x_n)\, e^{-j(kux_n + \hat{\beta}_n)} \tag{9-40}$$

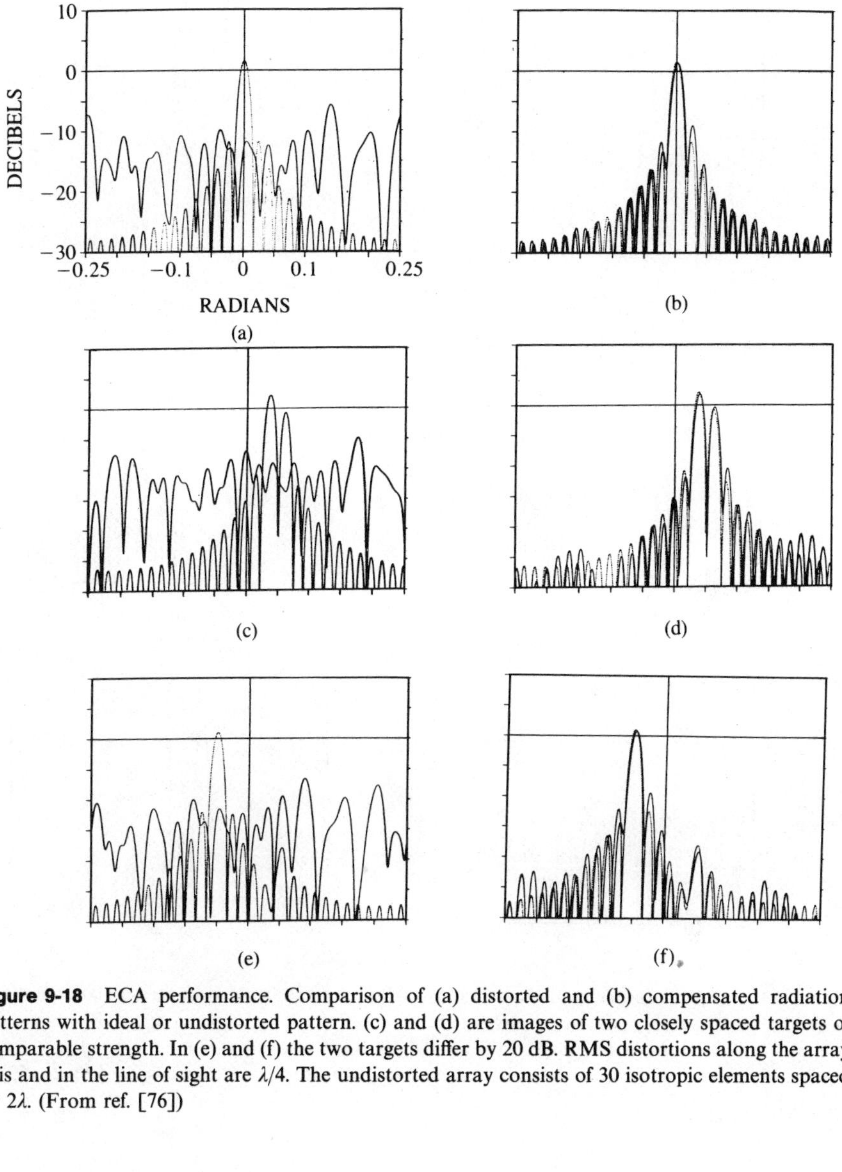

Figure 9-18 ECA performance. Comparison of (a) distorted and (b) compensated radiation patterns with ideal or undistorted pattern. (c) and (d) are images of two closely spaced targets of comparable strength. In (e) and (f) the two targets differ by 20 dB. RMS distortions along the array axis and in the line of sight are $\lambda/4$. The undistorted array consists of 30 isotropic elements spaced by 2λ. (From ref. [76])

The ECA requires data from several range bins to reliably estimate the phase errors. This was also found earlier to be a characteristic of the unit-lag and multiple-lag SCAs. Assume that the ECA attempts to estimate the phase errors by maximizing the energy in only a single range bin within the illuminated sector. The energy, by Parceval's theorem, is $\sum_{n=1}^{N} |\hat{e}_1(x_n)|^2/N$. It is possible that the system can incorrectly deduce that the phase errors are $\hat{\beta}_n = \text{Phase}\ \{\hat{e}_k(x_n)\} - ku_i x_n$, in which case all the energy in this given range bin can become concentrated in a single direction u_i. This solution is only correct when $s_k(u) = \delta(u - u_i)$.

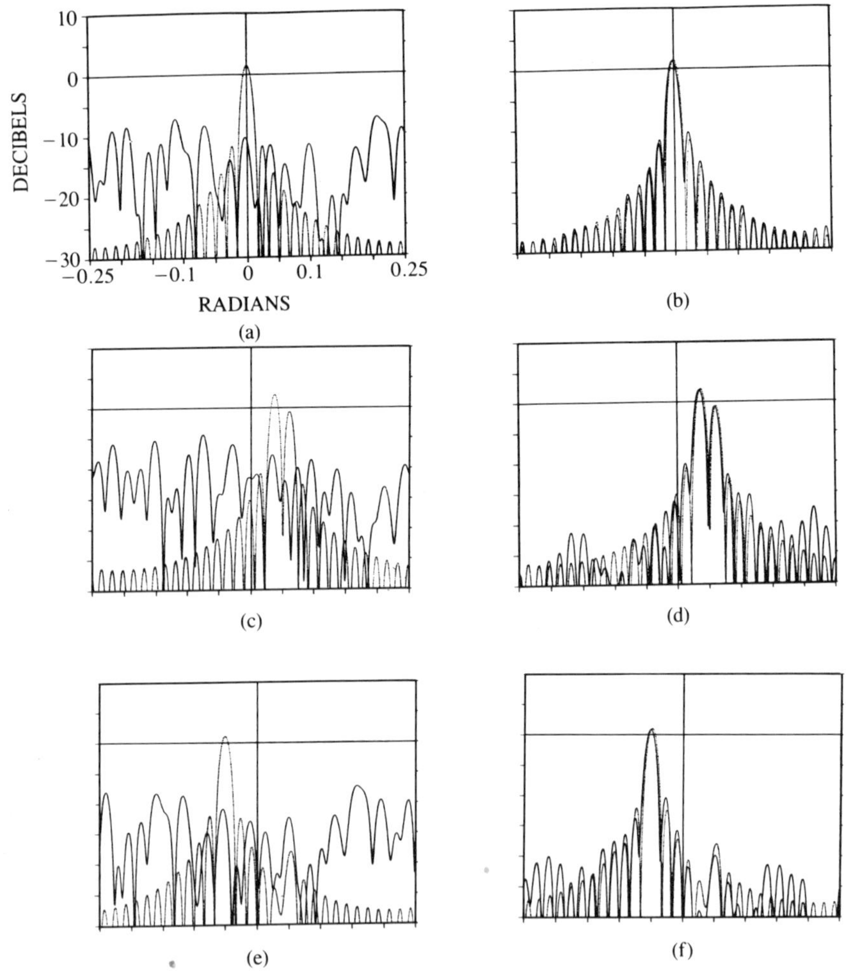

Figure 9-19 ECA performance. Comparison of (a) distorted and (b) compensated radiation patterns with ideal or undistorted pattern. (c) and (d) are images of two closely spaced targets of comparable strength. In (e) and (f) the two targets differ by 20 dB. RMS distortions along the array axis and in the line of sight are $\lambda/2$. The undistorted array consists of 30 isotropic elements spaced by 2λ. (From ref. [76])

However, this same set of phase corrections cannot simultaneously concentrate the energies in all range bins into a single direction. Thus use of several range bins is necessary to ensure that the algorithm provides a useful set of $\hat{\beta}_n$.

The manner in which the ECA estimates the phase corrections using data from multiple range bins is now described. Let $\hat{e}_k(x_n)$ represent the echo at the nth array element from the kth range bin, whose source distribution is $s_k(u)$. Let K represent the total number of range bins. The ECA determines the phase corrections $\hat{\beta}_n$ by maximizing the energy within the illuminated sector $u_1 \leqslant u \leqslant u_2$

TABLE 9-1 Comparison Between an Error-free Antenna and Three Distorted Antennas after Application of Tsao's Self-calibration Algorithm[a]

	RMS Position Error		
	0.125λ	0.25λ	0.5λ
Loss in mainlobe gain (dB)	0.102	0.23	0.18
Increase in PSL of the radiation pattern (dB)	1.20	2.80	1.35
Increase in ASL (dB)	0.122	0.145	0.174
Phase error variance after self-calibration (sq. radians)	0.026	0.085	0.079
Percentage of image energy falling outside the target region (average of 10 range bins)	0.130	0.391	1.196
Fractional beampointing error (average of 5 range bins)	0.034	0.135	0.153

[a] The errors are uncorrelated from element to element and the variances are equal in x and y

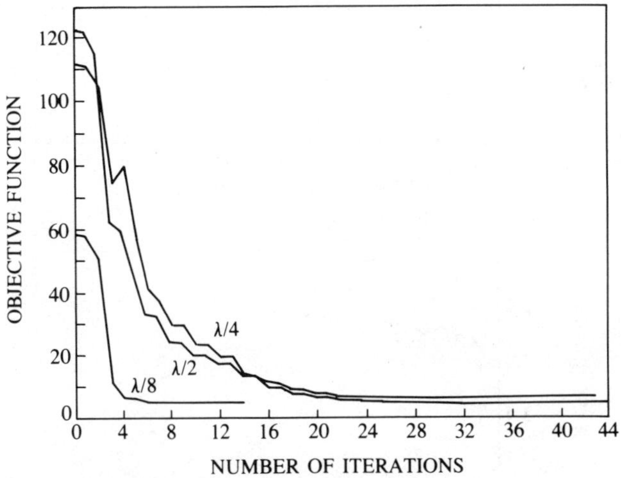

Figure 9-20 ECA learning curves for rms array errors of $\lambda/8$, $\lambda/4$, and $\lambda/2$. The ordinate is the fraction of image energy outside the illumination sector. The abscissa is the number of iterations. (From ref. [76])

over all the K range bins, that is, by maximizing the following objective function with respect to the phase corrections $\hat{\beta}_n$:

$$F = \sum_{k=1}^{K} \int_{u_i}^{u_2} |\hat{s}_k(u)|^2 \, du \tag{9-41}$$

where $\hat{s}_k(u)$ is the estimate of the image in the kth range bin and is given by Eq. 9-40. Substituting Eq. 9-40 into Eq. 9-41 and performing the integration over u yields

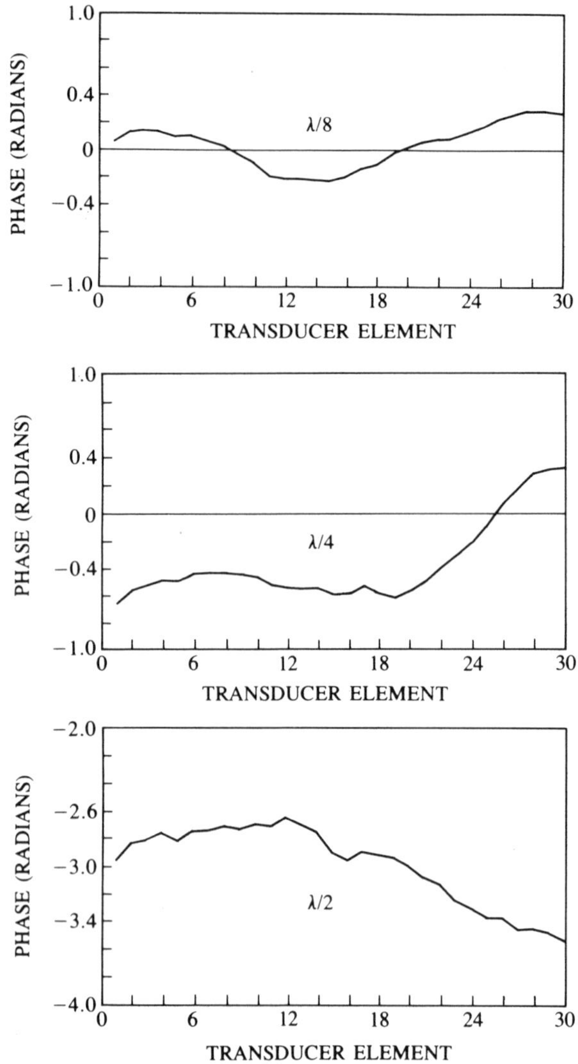

Figure 9-21 Residual phase errors across the receiving array for rms element-position errors of $\lambda/8$, $\lambda/4$, and $\lambda/2$. (From ref. [76])

$$F = (u_2 - u_1) \sum_{k=1}^{K} \sum_{m=1}^{N} \sum_{n=1}^{N} \hat{e}_k(x_m) e_k^*(x_n) \times e^{j(\hat{\beta}_n - \hat{\beta}_m)} e^{jk(u_1 + u_2)(x_n - x_m)/2} \operatorname{sinc}\left\{\frac{k(u_2 - u_1)(x_n - x_m)}{2}\right\} \tag{9-42}$$

where sinc(x) equals $\sin(x)/x$.

The only unknown quantities in Eq. 9-42 are the phase errors $\hat{\beta}_n$. The expression for F can be simplified by combining all the known quantities in Eq. 9-42 into a single term. To this end, let

$$\rho(m, n) = (u_2 - u_1)\, e^{jk(u_1 + u_2)(x_n - x_m)/2} \operatorname{sinc}\left\{\frac{k(u_2 - u_1)(x_n - x_m)}{2}\right\} \sum_{k=1}^{K} \hat{e}_k(x_m)\, \hat{e}_k^*(x_n) \tag{9-43}$$

Substitution of Eq. 9-43 into Eq. 9-42 yields the following expression for the objective function F:

$$F = \sum_{m=1}^{N} \sum_{n=1}^{N} \rho(m, n)\, e^{j(\hat{\beta}_n - \hat{\beta}_m)} \tag{9-44}$$

A computationally efficient iterative algorithm for maximizing F is now described. Before proceeding with the maximization, it is instructive to examine the behavior of F as a function of $\hat{\beta}_i$, when all the other phase corrections $\hat{\beta}_n$, $n \neq i$, are held constant. Using Eq. 9-44, it can be shown that

$$\begin{aligned} F(\hat{\beta}_i) &= A_0 + e^{j\hat{\beta}_i} \sum_{\substack{m=1\\ m\neq i}}^{N} \rho(m, i)\, e^{-j\hat{\beta}_m} + e^{-j\hat{\beta}_i} \sum_{\substack{n=1\\ n\neq i}}^{N} \rho(i, n)\, e^{j\hat{\beta}_n} \\ &= A_0 + 2A_i \cos(\hat{\beta}_i + \phi_i) \end{aligned} \tag{9-45}$$

where the fact that $\rho(m, n) = \rho^*(n, m)$ was used (see Eq. 9-43), A_0 is a quantity independent of $\hat{\beta}_i$, defined by

$$A_0 = \sum_{n=1}^{N} \rho(n, n) + \sum_{\substack{m=1\\ m\neq i}}^{N} \sum_{\substack{n=1\\ m\neq i, m\neq n}}^{N} \rho(m, n)\, e^{j(\hat{\beta}_n - \hat{\beta}_m)} \tag{9-46a}$$

and

$$A_i = \left| \sum_{\substack{m=1\\ m\neq i}}^{N} \rho(m, i)\, e^{-j\hat{\beta}_m} \right| \tag{9.46b}$$

$$\phi_i = \text{Phase}\left\{ \sum_{\substack{m=1\\ m\neq i}}^{N} \rho(m, i)\, e^{-j\hat{\beta}_m} \right\} \tag{9.46c}$$

From Eq. 9-45, observe that $F(\hat{\beta}_i)$ reaches its optimal value when $\hat{\beta}_i = -\phi_i$, that is,

$$\hat{\beta}_i\big|_{\text{opt}} = -\phi_i \tag{9-47}$$

The maximization procedure can now be described in terms of the following steps:

1. Compute the $\rho(m, n)$, $1 \leqslant m, n \leqslant N$, using Eq. 9-43. This requires approximately $2N^2K$ real multiplications and N^2K real additions.
2. Choose $\hat{\beta}_1 = 0$ and use the unit-lag SCA phase corrections as the starting values for B_n, $n > 1$. These phase corrections can easily be computed because the unit-lag correlations $\rho(m + 1, m)$ have already been computed in step (1).
3. Use the starting values of the $\hat{\beta}_i$ in order to compute the A_i and ϕ_i, $i > 1$, using Eqs. 9-46b and 9-46c, respectively. This requires N^2 complex multiplications and additions (N for each value of i). Also compute F using Eq. 9-44 (N^2 real multiplications and additions).
4. Maximize F sequentially with respect to the β_i, $2 \leqslant i \leqslant N$, while holding the rest of the β_n, $n \neq i$, constant. Use Eq. 9-47 to compute the optimal $\hat{\beta}_i$. Also update F, and the A_n and ϕ_n, $n \neq i$, using

$$\begin{aligned} F_{\text{new}} &= F_{\text{old}} - 2A_i \cos(\hat{\beta}_i|_{\text{old}} + \phi_i) + 2A_i \\ A_n\, e^{j\phi_n}|_{\text{new}} &= A_n\, e^{j\phi_n}|_{\text{old}} - \rho(i, n)(e^{-j\hat{\beta}_i}|_{\text{old}} - e^{-j\hat{\beta}_i}|_{\text{new}}) \end{aligned} \tag{9-48}$$

   As can be seen from Eq. 9-48, each update requires two complex multiplications and additions. A total of $(N - 1)(N - 2)$ updates need to be performed for one iteration; this requires $2(N - 1)(N - 2) \approx 2N^2$ complex multiplications and additions.
5. Compare the current value of F with its value in the previous iteration. Repeat step (4) if F has increased significantly. Stop the iterations if the two values of F are comparable.

9-6-4 Comparison of ECA Performance with Unit- and Multiple-lag SCA

The following computer simulation experiments compare the performances of the unit-lag and multiple-lag SCA with the ECA for varying signal-to-noise ratios and for various types of target scenarios. Performance is evaluated in terms of the sidelobe levels obtained after phase error compensation. A 32-element filled and periodic array is used in the simulations. Uniform weighting is used on transmit while a -40 dB Dolph–Chebehev taper is used to synthesize a low sidelobe receiving pattern. The results are taken from [111].

It is shown that while the unit-lag SCA is capable of accurately estimating the phase errors at high SNRs, the other two algorithms may have to be used to estimate the phase errors at low SNRs (<0 dB). The following simulations demonstrate that both the MLSCA and the ECA provide accurate estimates

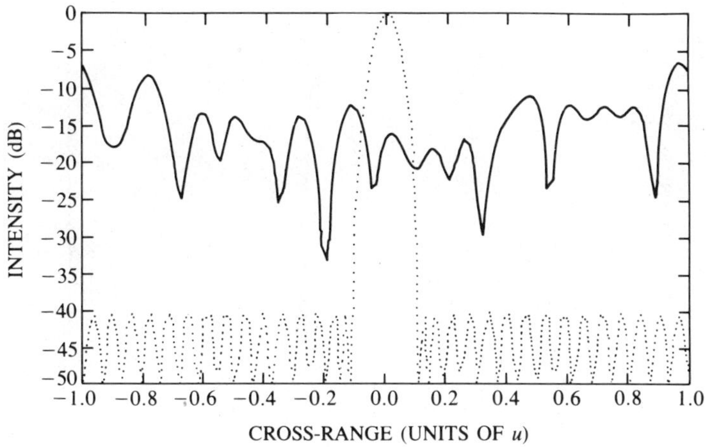

Figure 9-22 Array pattern with and without phase error. The dotted line represents the pattern of a 32-element phased array with a −40 dB sidelobe taper. The solid line is the pattern when random phase errors, uniformly distributed between $\pm\pi$, are introduced at each element. Notice the absence of a main beam.

of the phase errors when the SNR is such that the unit-lag SCA phase corrections are inaccurate.

In the following simulation, phase errors randomly distributed between $\pm\pi$ were introduced at each receiver. Figure 9-22 shows the low design sidelobe pattern of a 32-element array (dotted line) and the pattern obtained after introducing the random phase errors (solid line). Note that this pattern has no main beam because of the large phase errors.

Data from 300 clutter range bins were used for estimating the phase errors. The clutter angular intensity spectrum was assumed to be uniform in the region $|u| \leqslant 1$. Because the transmitting array was uniformly weighted, the spatial correlation of the clutter echoes is triangular, with a peak at the origin $x = 0$ and a null-to-null width of $x = (N - 1)\lambda$. The correlation coefficients for all lags are thus real and positive. The clutter-to-noise ratio per element per range bin was −3 dB.

Figure 9-23a compares the performance of the unit-lag SCA (dotted line) with the SCA that uses 10 correlation lags. Both reconstruct the main lobe and have similar sidelobe properties; the multilag process is considerably superior in the sidelobe region close to the main lobe.

Figure 9-23b compares the multiple-lag SCA (10 lags) pattern (solid line) with the ECA (dotted line). Note that in this case there is not much difference between the two patterns. This is because the transmitter and receiver beamwidths were the same and as a consequence the correlation function was real and positive for all lags.

Table 9-2 summarizes the results of this computer simulation experiment in terms of the rms residual phase error, mainbeam gain and peak sidelobe level.

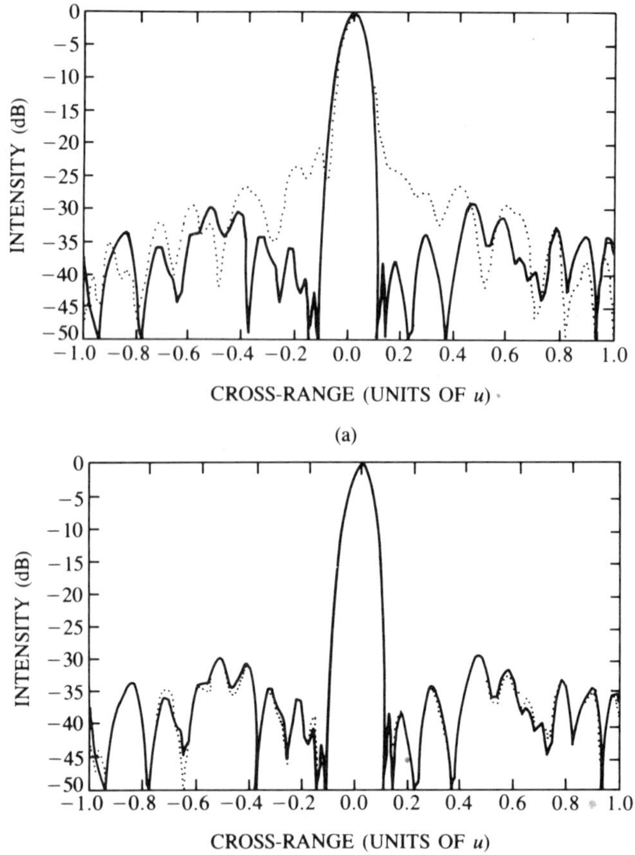

(b)

Figure 9-23 Array patterns after compensating for phase errors introduced in Fig. 9-22. (a) Dotted line is the pattern after phase error compensation using the unit-lag SCA. Solid line is the pattern for the SCA that uses 10 correlation lags. (b) Solid line is the same as above (SCA with 10 lags). Dotted line is the pattern after phase error compensation using the energy conservation algorithm.

TABLE 9-2 Parameters of experiment of Figs. 9-22 and 9-23

Algorithm	*RMS Residual Phase Error (deg)*	*Mainbeam Gain (dB)*	*Peak Sidelobe (dB)*
SCA (lag = 1)	10.9	−0.21	−20.4
SCA (lag = 2)	8.2	−0.12	−22.8
SCA (lag = 5)	5.2	−0.04	−29.7
SCA (lag = 10)	4.6	−0.03	−29.1
SCA (lag = 15)	5.0	−0.03	−28.7
ECA	4.4	−0.02	−29.5

The same 300 sets of clutter echoes were for all cases. The unit- and dual-lag phase corrections lead to a large PSL; the SCA with lags larger than 5 and the ECA perform comparably.

This experiment showed that the SCA for high lags performs comparably with the ECA under low clutter-to-noise conditions. Because the computational complexity of the SCA is smaller than the ECA, the SCA is the preferred method in this case. The following experiment shows that the ECA works under conditions where both the unit- and multiple-lag SCA algorithms fail.

The previous simulation was conducted using a transmit/receive array, that is, the sizes of the transmit and receive arrays were equal. As shown in Chapter 8 (see the discussion following Eq. 8-20 and Fig. 8-3), the correlation distance of clutter echoes for a transmit aperture of size L_t is approximately L_t, that is, the correlation between any two elements separated by a distance smaller than L_t will be large. However, when the transmitting aperture is small, such as in bistatic systems, only those correlation lags smaller than the length of the transmit aperture can be used for phase error compensation. As shown by the following experiment, the SCA performs poorly in these circumstances, whereas the ECA provides accurate estimates of the phase errors.

A 32-element receiving array with a -40 dB Dolph-Chebchev taper was also used in this simulation. Random phase errors having an rms value of 5° were introduced at each of the 32 receivers. The main effect of these small errors upon the pattern is in the sidelobe region. Figure 9-24a shows the resulting pattern. The PSL has increased to -25 dB due to the phase errors.

Data from 100 range bins were used for phase error compensation using the unit-lag SCA and the ECA. The number of point targets in each range bin was a uniformly distributed random integer between 1 and 10. The location of the point targets was uniformly distributed in the region $|u| \leqslant 0.5$, the extent of the illuminated patch. A one-wavelength-long transmitting antenna can be used for illuminating this region. Only the unit-lag and dual-lag correlation coefficients have lag distances less than or equal to λ for a filled array. Thus only these lags can be used by the SCA for phase error compensation because the correlation distance for a target scene confined to $|u| \leqslant 0.5$ is λ.

Figure 9-24b compares the performance of the unit-lag SCA (dotted line) with the ECA (solid line). The PSL of the pattern obtained using the SCA is -22 dB, which is larger than the -25 dB PSL without phase error compensation. The ECA pattern is close to the -40 dB design Dolph–Chebchev pattern. The various parameters of this simulation are summarized in Table 9-3. The ECA is the preferred algorithm in this case.

9-7 OBSERVATIONS

The SCAs use the spatially stationary nature of the correlation between the samples of the radiation field measured at aperture locations x_1 and x_2, $R(x_1, x_2) = E\{e(x_1)e^*(x_2)\}$, to estimate the phase errors. That is, the SCAs

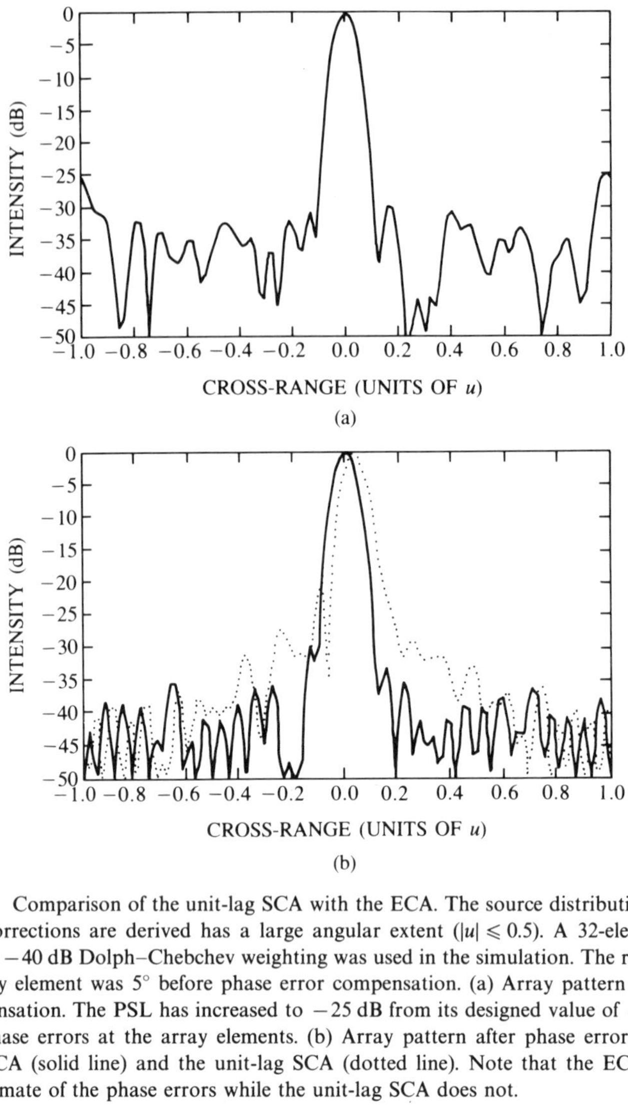

Figure 9-24 Comparison of the unit-lag SCA with the ECA. The source distribution from which the phase corrections are derived has a large angular extent ($|u| \leqslant 0.5$). A 32-element receiving array with a -40 dB Dolph–Chebchev weighting was used in the simulation. The rms phase error at each array element was 5° before phase error compensation. (a) Array pattern without phase error compensation. The PSL has increased to -25 dB from its designed value of -40 dB due to the small phase errors at the array elements. (b) Array pattern after phase error compensation using the ECA (solid line) and the unit-lag SCA (dotted line). Note that the ECA provides an accurate estimate of the phase errors while the unit-lag SCA does not.

TABLE 9-3 Parameters of the experiment of Fig. 9-24

Algorithm	*RMS Residual Phase Error (deg)*	*Mainbeam Gain (dB)*	*Peak Sidelobe (dB)*
No compensation	5.2	−0.02	−25
SCA (lag = 1)	11.0	−0.16	−22
ECA	2.1	−0.01	−33

require $R(x, x_2)$ to be a function of $(x_1 - x_2)$ alone and not both x_1 and x_2 in the absence of phase errors (all $\beta_n = 0$). An additional restriction is imposed when the SCAs are used to calibrate *aperiodic* arrays: the phase of $R(x_1 - x_2)$ is required to be a linear function of $(x_1 - x_2)$.

It was demonstrated in [5] that the phase of $R(x_1 - x_2)$ is linear with x when the maximum lag for which the correlation coefficient is calculated, denoted x_{max}, is smaller than the mainlobe width of $R(x)$. As mentioned earlier, this mainlobe width is approximately the length of the transmitting aperture L_t. Therefore the SCAs can successfully calibrate phased arrays when $x_{max} < L_t$.

The unit-lag SCA requires fewer computations than both the MLSCA and the ECA. It should therefore be used whenever it provides adequate performance, that is, when (a) $x_{max} < L_t$ and (b) the SNR in the unit-lag correlation estimates is large. The MLSCA improves performance with respect to noise and should be used if the unit-lag SCA does not provide adequate performance.

Both the unit-lag and multiple-lag SCA will fail when L_t is smaller than the maximum spacing between adjacent elements. The ECA should be used in these circumstances. It requires more computations than both the unit-lag and multiple-lag SCAs.

Table 9-4 summarizes the various aspects of phase error compensation using

TABLE 9-4 Evaluation of various phase error compensation algorithms. N = number of elements; K = number of range bins or transmitter locations; L = number of lags for multiple-lag SCA; I = number of ECA iterations; L_t = length of transmitting aperture; x_{max} = maximum lag distance for which the correlation coefficient is calculated.

	Unit-lag SCA	*MLSCA*	*ECA*
High SNR, $L_t > x_{max}$	Good	Good	Good
Low SNR, $L_t > x_{max}$	Fair	Good	Good
$L_t < x_{max}$	Bad	Worse	Good
Multiples	$4KN$	$4KNL + N^2L$	$2N^2K + 2N^2I$

the unit-lag SCA, the multiple-lag SCA, and the ECA. SNR refers to the signal-to-noise ratio in the estimates of the correlation coefficients. It is inversely proportional to the square of the abscissa of Fig. 9-7 for the range of phase errors shown.

The computational complexities of the various algorithms were derived in the following manner. A total of $4K$ real multiplications and additions are required to estimate each correlation coefficient used by the SCAs (see Eq. 9-2); hence approximately $4NKL$ real multiplications are required to estimate the correlation coefficients for the unit-lag and multiple-lag SCAs. $(N - 1)$ more additions are required to estimate the phase errors using the unit-lag SCA (see

Eq. 9-5); approximately N^2L multiplications and additions are required to estimate the phase errors using the MLSCA using Eq. 9-35. Thus the unit-lag and multiple-lag SCAs require approximately $4NK$ and $4NKL + N^2L$ multiplications, respectively. The ECA first computes the $\rho(m, n)$ in Eq. 9-43 using $2N^2K$ real multiplications. In addition, $2N^2$ multiplications per iteration are required to update the quantities A_n and ϕ_n using Eq. 9-48b. Hence the ECA requires $2N^2K + 2N^2I$ multiplications, where I is the number of iterations performed by the ECA.

10

DATA REDUCTION

The product of three parameters determines the data handling requirement of a large phased array system. As shown in Fig. 4-16, the first is the number of antenna elements N, which equals the number of receiver channels in the system. The second is the bandwidth B of the radiated signal, which is the rate at which the received echoes must be sampled.⋆ The third is the number of bits per sample b into which the samples are quantized.

We begin by examining the product NBb, which is the rate at which bits must be delivered to and processed in the signal processor. That this product can grow unreasonably large in a large imaging system unless special steps are taken to control it is very easy to demonstrate. For example, a phased array system with a filled aperture designed for a $1\,\text{m}^2$ resolution cell at a distance of 10 km has the following characteristics:

Interelement spacing $d = \lambda/2$
Array length $L = dN = \lambda N/2$
Beamwidth $\Delta\theta \cong \lambda/L = 2/N$
Cross-range resolution $\Delta S = R\,\Delta\theta = 2R/N$

⋆ The Nyquist sampling rate is twice the modulation bandwidth, or $2B$ data samples per second, which appears to be twice the rate indicated above. However, each sample of the RF (or IF) echo is a complex amplitude consisting of an amplitude and a phase. Alternately, if sampling is done after coherent quadrature demodulation, two channels of video data are obtained. Either way, two data samples are taken at each sampling instant, thereby satisfying the Nyquist condition at a sample rate of B.

Therefore $$N = \frac{2R}{\Delta S} = 2 \times 10^4$$

Range resolution $\Delta R = c/2B$

Therefore $$B = c/2R = 150 \text{ MHz}$$

Bits per sample per each I and Q channel $= 6$, therefore $b = 12$.

Consequently $NBb = 3.6 \times 10^{13}$ bps in this illustration. Such a number is unrealistically large. For example, it is $3\frac{1}{2}$ orders of magnitude larger than the radio frequency used in the imaging experiments shown throughout this book. Practicality demands its reduction.

Experience with experimental phased-array antenna systems at Valley Forge Research Center (VFRC) has shown that aperture thinning can reduce N by about two orders of magnitude (or about four orders of magnitude when the array is two-dimensional), and in each case tested the result has been highly consistent with theory. Theory indicates that the limiting reduction is not the thinning factor, however, but the number of elements that remain after thinning, for this number dictates the sidelobe properties. Thus giant arrays can be designed with even larger savings [94].

This chapter discusses the remaining two means of data reduction. Bit compression of a sample, which is accomplished by data quantization, is shown to offer about one order of magnitude in data reduction. Bit compression also offers an opportunity for reducing the complexity of the signal processor through the elimination of all complex multiplications. This chapter also treats the third factor, B, and shows how reduction of the spectral occupancy of the signal within the overall passband of the system can reduce the data rate requirement.

10-1 DATA QUANTIZATION

A high-resolution imaging system necessarily requires a very wide data bandwidth for data acquisition and image processing. The bit rate required by a microwave imaging system for data acquisition was derived for filled periodic arrays at the end of Chapter 4 (4-10). This bit rate is $Rcb/(\Delta R)^2$, where R is the range to the illuminated patch, c is the velocity of light, b is the number of bits used for representing a single complex data sample, and ΔR is the resolution in both the range and cross-range dimensions. A filled periodic array imaging a target scene 10 km away in order to resolve targets 1 m apart in range and cross-range should acquire data at a rate of $3 \times 10^{12} b$ bits per second. This is a very large data rate. It can be significantly reduced by thinning the receiving array. It can be further reduced by using a small number of bits for representing each data sample.

Conventional high-resolution microwave imaging systems use 6 to 8 bits for representing each quadrature component of the received data, that is, they represent each complex data sample using 12 to 16 bits. This chapter shows that using fewer bits per data sample does not significantly degrade image quality. In addition, the reduction in the number of bits per sample can be exploited to simplify the architecture of the signal processor used for image formation.

Data quantization can be performed in several ways. One technique is through the use of hard limiting at the array front end, which strips all amplitude from the received target echoes while retaining the phase information undistorted. As shown in [51], hard limiting results in the original image plus distortion contributions consisting of convolutions of the original image. It was later shown that, under certain SNR conditions, signals at the output of the array elements can be hard limited without significant image degradation [94]. Hard limiting a temporal waveform is known to result in intermodulation products between different spectral components of the waveform. In a similar fashion, hard limiting at the array element results in intermodulation components due to pairs of plane wave components arriving from different directions. These beat products exist at all harmonic sum and difference angles of the arrival directions, and give rise to image artifacts. However, careful attention to SNR relations constrains buildup of the intermodulation components to less than the output noise level [94].

An important paper by Oppenheim and Lim discussed the dominant influence of phase in imagery creation [72]. No solid theory was presented, but several intellectual arguments and numerous experimental examples were offered advancing the primacy of phase.* Experimental work on phase-only holography performed between the late 1960s and early 1980s encouraged advancement of the idea [27], [61], [67], [80]. Reference [54] evaluated the effect of phase quantization on such holograms.

In contrast to the success with hard limiting, experiment easily demonstrates that images formed from amplitude-only information are useless. Amplitude plus sign, on the other hand, or equivalently amplitude plus one bit of phase, provides a useful and often satisfactory image [123], [124]. Multibit phase quantization is analyzed in a clear and concise fashion in [32], for which [23] offers a compact derivation. In [70], a discussion and experimental simulation results are offered on phase-only image reconstruction oriented towards SAR, a closely related subject. Experimental SAR data were used to test various data quantization techniques in [52].

The first part of this chapter presents a theory on the effect of hard limiting and several other types of data distortion on image quality. The theory was developed in [101]. The correlation between the undistorted and distorted images is used as a measure of image quality in order to evaluate the different methods of data distortion. Experimental microwave data and a subjective

* An extensive bibliography of 42 papers is found in [72].

viewing experiment are then used for justifying the use of the correlation metric as a measure of image quality.*

Next, an optimal method for quantizing the data given the number of bits used for representing each data sample is presented. This method was developed in [59]. Finally, a method that exploits the small number of bits used for representing the data samples in order to reduce computations and the gate count and power requirements of the signal processor is presented. This material is developed using results from [59], [101], [110] and [126].

10-2 THEORY

10-2-1 Image Quality Measure

Let $e(x)$ represent the electric field at the aperture and let $e(x)w(x) = f(x) = f$ denote the aperture excitation due to the electric field. $w(x)$ is the aperture weighting function and f is used as an abbreviation for $f(x)$. The proper image, denoted I_f, is the magnitude of the Fourier transform of $f(x)$, that is, $I_f = |\mathscr{F} f|$ using the operator notation of Section 2-4.

Data quantization results in data distortion, that is, the distorted data $g = g(x)$ are equal to Df, where D is a distortion operator. The distorted image, denoted I_g, is the magnitude of the Fourier transform of g, that is, $I_g = |\mathscr{F} g|$.

Clearly, I_g is a degraded version of the proper image I_f. The measure of degradation in image quality introduced in [101] and used in this chapter is the normalized zero-lag correlation between I_f and I_g:

$$\rho = \frac{\overline{I_f I_g}}{I_{f\,\mathrm{rms}} I_{g\,\mathrm{rms}}} = \frac{\overline{|\mathscr{F} f||\mathscr{F} g|}}{I_{f\,\mathrm{rms}} I_{g\,\mathrm{rms}}} \tag{10-1}$$

where the overbar means that an average is taken over all picture elements in the images and the rms notation denotes root mean squared, that is, $I_{f\,\mathrm{rms}} = \sqrt{\overline{I_f^2}}$. Note that the normalized mean squared difference between I_f and I_g is equal to $2(1 - \rho)$.

$\mathscr{F} f$ and $\mathscr{F} g$ are complex numbers; hence $I_f I_g = |\mathscr{F} f||\mathscr{F} g| \geqslant |(\mathscr{F} f)(\mathscr{F} g)^*|$, and

$$\rho \geqslant \frac{\overline{|(\mathscr{F} f)||(\mathscr{F} g)^*|}}{I_{f\,\mathrm{rms}} I_{g\,\mathrm{rms}}} \tag{10-2}$$

* The data sets used consisted of ground-based microwave images of terrain or aircraft. These images are dominated by strong isolated reflectors, unlike airborne SAR data, which are highly speckled, having a local variance approximately equal to the mean. The correlation metric has not been tested for SAR images and consequently its validity for this use is unresolved.

Applying Parceval's theorem to the numerator and denominator yields

$$\rho \geqslant \frac{\overline{|fg^*|}}{f_{\text{rms}}\, g_{\text{rms}}} = \rho_{lb} \tag{10-3}$$

where ρ_{lb} is the lower bound of ρ and the overbar of a function f or g in Eq. 10-3 means averaging in the aperture, that is,

$$\overline{fg^*} = \frac{1}{L}\int_{-L/2}^{L/2} f(x)g^*(x)\,dx \tag{10-4a}$$

for a continuous aperture of length L and

$$\overline{fg^*} = \frac{1}{N}\sum_{n=1}^{N} f_n g_n^* \tag{10-4b}$$

for a discrete aperture with N elements.

Thus the lower bound on the similarity between the undistorted and distorted images is the magnitude of the normalized zero-lag correlation between the undistorted and distorted data.

10-2-2 Test of Quality Measure

To be a useful quality measure, Eq. 10-3 must be nearly independent of the specific source distribution responsible for the radiation field at the aperture. Otherwise, if a given distortion produces markedly different correlation coefficients when applied to input data derived from markedly different source distributions, there would be little value to this measure as an indicator of the effect of distortion. Equation 10-3 was tested in [101] for its sensitivity to the nature of the scene by applying four different target models or source distributions to two extreme distortion-inducing cases. The results indicate remarkably low sensitivity to the target model, and are presented below.

Let the amplitude and phase of the undistorted data $f = f(x)$ be denoted by $A = A(x)$ and $\phi = \phi(x)$, respectively, that is, let

$$f = A\exp(j\phi) \tag{10-5}$$

Initially, the sensitivity of ρ to source distribution will be evaluated for two extreme distortion cases, viz. hard limiting (phase-only) and amplitude-only distortion.

10-2-2-1 Distortion Models

(*1*) *Hard limiting*. For this type of distortion, $g = e^{j\phi}$ where ϕ, of course, is $\phi(x)$. Substituting this result into Eq. 10-3 yields the correlation for phase-only distortion, denoted ρ_1, as

$$\rho_1 \geqslant \frac{\overline{|(Ae^{j\phi})(e^{-j\phi})|}}{A_{\rm rms}} = \frac{\bar{A}}{A_{\rm rms}} \tag{10-6}$$

Thus the image correlation for hard limiting is lower-bounded by the simple ratio of the average of the aperture excitation to its rms value.

(2) *Amplitude only.* Here $g = A$ and the image correlation, denoted ρ_2, is

$$\rho_2 \geqslant \frac{\overline{|A^2 e^{j\phi}|}}{\overline{A^2}} \tag{10-7}$$

where only amplitude information has been preserved. When the amplitude and phase across the aperture are statistically independent, ρ_2 simplifies to

$$\rho_2 \geqslant \frac{\overline{|A^2 e^{j\phi}|}}{\overline{A^2}} = \overline{|e^{j\phi}|} \tag{10-8}$$

The values of ρ_1 and ρ_2 will now be computed for four target models. It is shown that they are almost insensitive to the type of target model.

10-2-2-2 Target Models

(*i*) *Single target of amplitude A at $u = u_0$.* The spatial description of the target is $A\delta(u - u_0)$, and the aperture excitation is $f(x) = A\exp(jkxu_0)$. Thus $\bar{A} = A$, as does $A_{\rm rms}$. Hence, from Eq. 10-6,

$$\rho_1 = 1 \tag{10-9a}$$

On the other hand, ρ_2 is very small in general. Inserting Eq. 10-7 or 10-8 in Eq. 10-4a yields

$$\rho_2 \geqslant \frac{1}{L}\left|\int_{-L/2}^{L/2} e^{jku_0x}\,dx\right| = \left|\frac{\sin(\pi L u_0/\lambda)}{(\pi L u_0/\lambda)}\right| \tag{10-9b}$$

Note that ρ_2 rapidly asymptotes to zero as the target angle off-broadside becomes more than a few beamwidths. Thus hard-limiting preserves the image of a single target, whereas amplitude-only destroys it.

(*ii*) *Two targets of equal strength.* Let the complex target amplitudes be $A\exp(j\phi_1)$ and $A\exp(j\phi_2)$ and let their locations be u_1 and u_2. Assume that the target phases are independent random variables uniformly distributed between $-\pi$ and π. The aperture excitation is

$$f(x) = A\exp[j(ku_1x + \phi_1)] + A\exp[j(ku_2x + \phi_2)]$$

Using the fact that the target phases are independent $U(-\pi, \pi)$ random variables, [101] shows that $\bar{A} = (4A)/\pi$ and $A_{\text{rms}} = \sqrt{2}A$. Thus

$$\rho_1 \geqslant \frac{\bar{A}}{A_{\text{rms}}} = \frac{2\sqrt{2}}{\pi} \simeq 0.900 \tag{10-10a}$$

The lower bound for ρ_2 evaluates to

$$\rho_2 \geqslant \frac{4}{\pi^2}\left|\operatorname{sinc}\frac{\pi L(u_1 + u_2)}{2\lambda} - \tfrac{1}{6}\operatorname{sinc}\frac{\pi L(3u_1 - u_2)}{2\lambda} - \tfrac{1}{6}\operatorname{sinc}\frac{\pi L(3u_2 - u_1)}{2\lambda}\right| \tag{10-10b}$$

which has the same characteristic as Eq. 10-9b, that is, it asymptotes rapidly to zero as the targets move off broadside or as the separation between them increases. Thus hard-limiting again preserves the image, whereas amplitude-only destroys it.

(*iii*) *Multiple targets.* Now let there be many targets characterized by real amplitudes A_i, phases ϕ_i, and locations u_i. The aperture excitation is $f(x) = \sum_i A_i \exp[j(ku_i x + \phi_i)]$. Assuming that none of the targets dominates the sum of all of the rest, the quadrature components of the aperture excitation are Gaussian and independent (central limit theorem), the pdf of the aperture amplitude is Rayleigh, and the phase is $U(-\pi, \pi)$. Consequently, $\bar{A} = (\sqrt{\pi/2})A_{\text{rms}}$, and

$$\rho_1 \geqslant \frac{\bar{A}}{A_{\text{rms}}} = \frac{\sqrt{\pi}}{2} \tag{10-11a}$$

and

$$\rho_2 = 0 \tag{10-11b}$$

(*iv*) *Many targets plus one large target.* The aperture excitation in this case is

$$f(x) = A_0\, e^{jku_0 x} + \sum_{i=1}^{I} A_i\, e^{j(ku_i x + \phi_i)}$$

where A_0 is the amplitude of the largest target, u_0 is its location and the A_i, ϕ_i, and u_i are the amplitudes, phases, and locations of the smaller targets. Let $\overline{a^2} = \sum_{i=1}^{I} A_i^2$. When A_0^2 is at least 10 dB larger than $\overline{a^2}$, the lower bounds of ρ_1 and ρ_2 evaluate to [101]

$$\rho_1 \geqslant \frac{A_0}{\left(A_0^2 + \dfrac{\overline{a^2}}{2}\right)^{0.5}} \tag{10-12a}$$

and

$$\rho_2 \geqslant \frac{A_0^2}{A_0^2 + \dfrac{a^2}{2}} \left| \operatorname{sinc}\left(\frac{\pi L u_0}{\lambda}\right) \prod_{i=1}^{I} J_0(A_i/A_0) \right| \tag{10-12b}$$

where $J_0(x)$ is the zeroth-order Bessel function of the first kind and $|J_0(\cdot)| \leqslant 1$. Note that ρ_1 is close to unity. The expression for ρ_2 asymptotes toward zero even more rapidly than the ρ_2 for the single target case given by Eq. 10-9b.

Table 10-1 summarizes the image correlation for the four target models. Two facts are evident. First the correlation coefficients ρ_1 and ρ_2 are nearly independent of the scene, which is essential for an estimator of the degradation of image quality of a microwave scene. Second, hard limiting hardly alters the image, which quantitatively affirms the known fact that the amplitude of the input data carries little of the image information.

TABLE 10-1 Lower bound on image correlation when input data are distorted. Hard limiting (ρ_1) and amplitude only (ρ_2) [101]

Target models	*Hard limiting* $\rho_1 \geqslant$	*Amplitude only* $\rho_2 \geqslant$
Single target	1.0	≈ 0
Two equal strength targets	0.90	≈ 0
Multiple targets	0.89	0
Many targets plus one large target (A_0 is 10 dB greater than A_{rms})	0.98	≈ 0

From ref. [101]

10-3 OTHER DISTORTION MODELS

Figure 10-1 illustrates the coding or transformation of f into g by the various distortions discussed in this paper. Point 0 represents the undistorted signal $f = A\exp(j\phi)$. Point 1 represents the hard limited transformation $g = \exp(j\phi)$. Point 2 represents amplitude only coding $g = A$. The remaining distortions are discussed below.

Figure 10-2 shows two-dimensional microwave images of a scene in which the input data were subjected to the various transformations. The ordinate is the distance from the array in meters and the abscissa is azimuth in milliradians. The scales are adjusted so that a square target would appear as a square in the image. The scene is the housing development of single homes in Phoenixville, PA, located 4.5 km away from the imaging system that was shown earlier in Fig. 2-12. The microwave array, operating at X-band ($\lambda = 3$ cm) is 83 m long, giving a cross-range beam dimension of 1.8 m. The size of the range cell is 3 m.

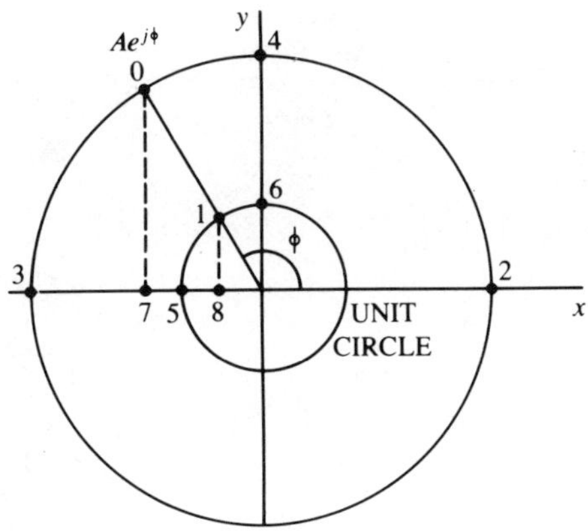

Figure 10-1 The nonlinear transformation of the received data $f = A\,e^{j\phi}$ (point 0) to g (points 1 through 8). (From ref. [101])

Figure 10-2a corresponds to point 0 (no distortion), 10-2b to point 1 (hard limiting), 10-2c to point 2 (amplitude only). As expected, Fig. 10-2b looks nearly identical to 10-2a while 10-2c is very different. The measured image correlation (ρ) (10-1) is indicated in the legend of the figure, along with the lower-bound theoretical estimate based on the multiple-target model (ρ_{lb}), and the measured correlation with the average value of the image removed (ρ').★ The effect of other types of distortion on image quality is examined below.

(3) *Amplitude with sign.* Unlike amplitude-only distortion, amplitude with the addition of a sign bit provides images that are not badly distorted versions of the original image [123], [124]. For this case, $g = \pm A$, where the positive sign is chosen if $-\pi/2 < |\phi| \leqslant \pi/2$ and the negative sign otherwise. Point 3 in Fig. 10-1 illustrates the coding for this case. A complex sample $A\exp(j\phi)$ is shown in the third quadrant. Whereas amplitude-only transforms the signal to point 2 (the intersection of the positive real axis with the circle of radius A), amplitude plus sign locates g at point 3. The transformation can be written as $g = A\,\mathrm{sgn}[\cos\phi]$.

The image correlation differs from Eq. 10-8 only in the numerator. It is either $|\overline{\exp j\phi}|$ or $|\overline{\exp j(\phi - \pi)}|$ according to whether $\cos\phi$ is positive or negative. The result is a phase error distribution of $\psi = \phi$ or $\psi = \phi - \pi$ that is considerably compressed relative to amplitude-only distortion. When ϕ is uniform in the interval $[-\pi, \pi]$, the phase error ψ is uniform in the interval between $[-\pi/2, \pi/2]$. For this distribution, $|\overline{\exp j\psi}|$ is no longer zero. Instead, the correlation increases to

$$\rho_2 \geqslant \left| \frac{1}{\pi} \int_{-\pi/2}^{\pi/2} e^{j\psi}\, d\psi \right| = \frac{2}{\pi} = 0.637 \tag{10-13}$$

★ $\rho' = [\overline{(I_f - \overline{I_f})][(I_g - \overline{I_g})}]/\sigma_{I_f}\sigma_{I_g}]$

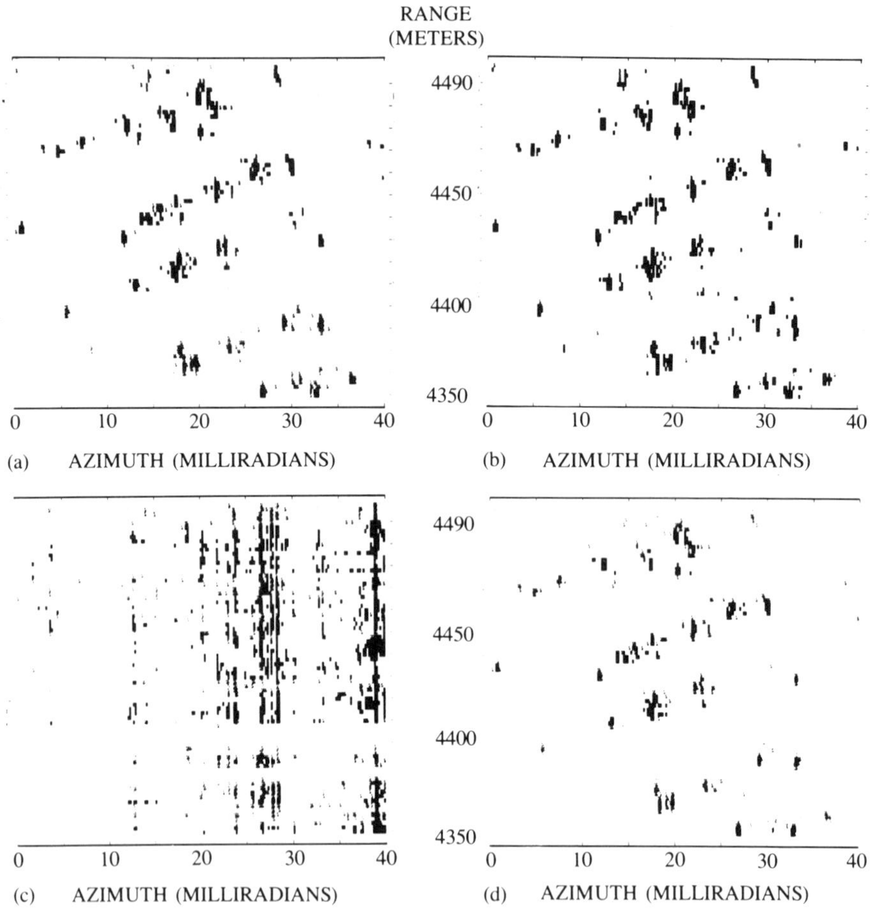

Figure 10-2 Two-dimensional high-resolution microwave images of a housing development of single homes in Phoenixville, PA. Range resolution = 3 m, cross-range resolution = 1.8 m, $\lambda = 3$ cm. (a) Undistorted input data. (b) Input data hard limited ($\rho_1 = 0.889$; $\rho_{lb} = 0.886$; $\rho'_1 = 0.826$). (c) Amplitude only ($\rho_2 = 0.555$; $\rho_{lb} = 0$; $\rho'_2 = 0.178$). (d) Amplitude with one bit of phase ($\rho_3 = 0.933$;

(continues on facing page)

Thus the addition of only one bit of phase increases the correlation from zero to greater than 60%.

The experimental image for amplitude plus sign is shown in Fig. 10-2d; it bears a remarkable similarity to the original image of Fig. 10-2a.

(4) *Amplitude with two bits of phase.* Point 4 of Fig. 10-1 illustrates the value of g for this case. The phase error pdf compresses further to $U(-\pi/4, \pi/4)$ and the correlation rises to $2\sqrt{2}/\pi = 0.900$. The image of Fig. 10-2e is nearly indistinguishable from Fig. 10-2a.

(5) *Hard limiting with one bit of phase.* This case is represented by point 5

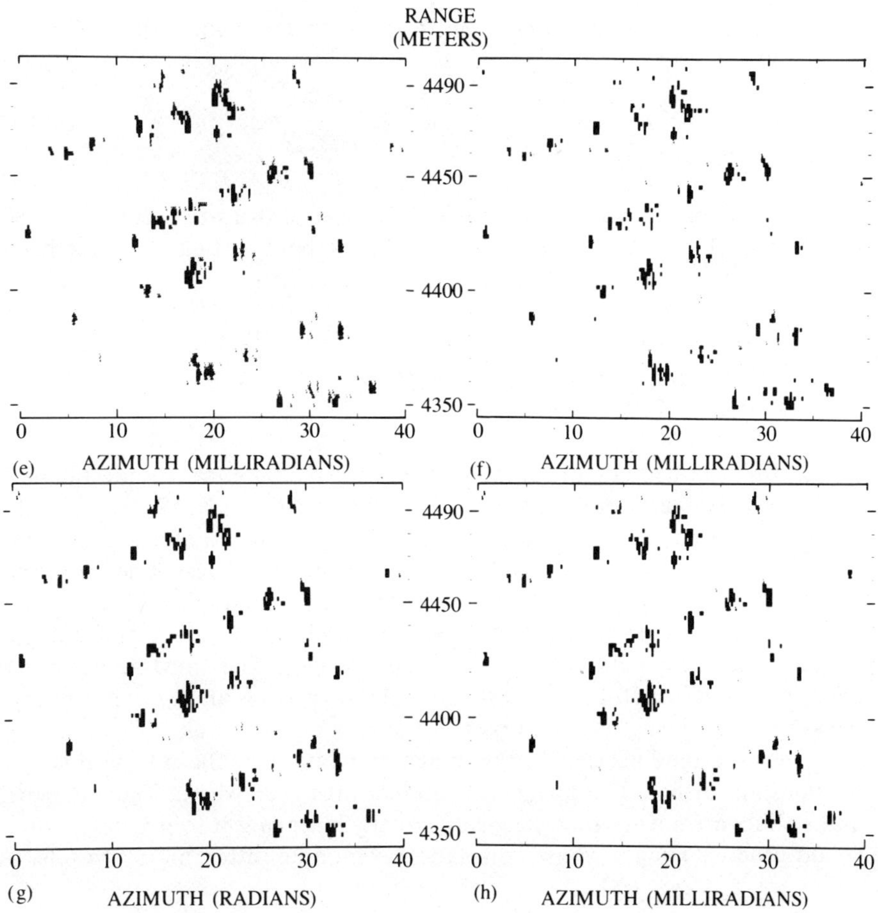

(legend continued from facing page)
$\rho_{lb} = 0.637$; $\rho'_3 = 0.900$). (e) Amplitude with two bits of phase ($\rho_4 = 0.980$; $\rho_{lb} = 0.900$; $\rho'_4 = 0.971$). (f) Hard limited with one bit of phase ($\rho_5 = 0.823$; $\rho_{lb} = 0.564$; $\rho'_5 = 0.723$). (g) Hard limited with two bits of phase ($\rho_6 = 0.872$; $\rho_{lb} = 0.798$; $\rho'_6 = 0.801$). (h) Single quadrature component ($\rho_7 = 0.856$; $\rho_{lb} = 0.707$; $\rho'_7 = 0.776$). (From ref. [101])

of Fig. 10-1. The correlation is

$$\rho_5 \geqslant \frac{\bar{A}}{A_{\text{rms}}} \, |\overline{e^{j\psi}}| = \frac{\sqrt{\pi}}{2} \frac{2}{\pi} = 0.564 \tag{10-14}$$

because $\bar{A} = (\sqrt{\pi}/2)A_{\text{rms}}$ for a Rayleigh distribution and because the phase error has a $U(-\pi/2, \pi/2)$ pdf. The image is shown in Fig. 10-2f.

(6) *Hard limiting with two bits of phase*. This is represented by point 6 of Fig. 10-1. The lower bound on the correlation rises to $\sqrt{2/\pi} = 0.798$. The image is shown in Fig. 10-2g.

(7) *A single quadrature component.* In this case $g = A\cos\phi$ (point 7 in Fig. 10-1) and

$$\rho_7 \geqslant \frac{\overline{|(A\,e^{j\phi})(A\cos\phi)|}}{A_{\mathrm{rms}}(A_{\mathrm{rms}}/\sqrt{2})} = \frac{\overline{A^2}/2}{\overline{A^2}/\sqrt{2}} = 0.707 \qquad (10\text{-}15)$$

The image is shown in Fig. 10-2h. This type of distortion is discussed in detail in [101]. It can be further modified by hardlimiting the data before quadrature detection as shown by point 8 of Fig. 10-1.

10-4 EXPERIMENTAL VALIDATION OF THE CORRELATION METRIC

Figure 10-3 (dashed curve) compares the lower-bound correlation estimate to the measured image correlation coefficient for the several data distortion models discussed. The estimator is below the experimental value in all cases. The data are for three different microwave scenes. The first is the housing developments of Fig. 10-2. The houses and their relative locations are typical of a suburban development of single homes. The second scene is from an older residential part of the same town with houses of various designs. The third scene is very high-resolution of a 50-ft radar dish. Each datum point is an N-sample average, the sample size being indicated by the numbers on the graph.

A heuristic modification to the image correlation coefficient improves the relationship further. The images are nonnegative everywhere. Therefore their mean values are not zero. Consequently, totally independent source distributions would produce images whose correlations would be rather high. Consider an

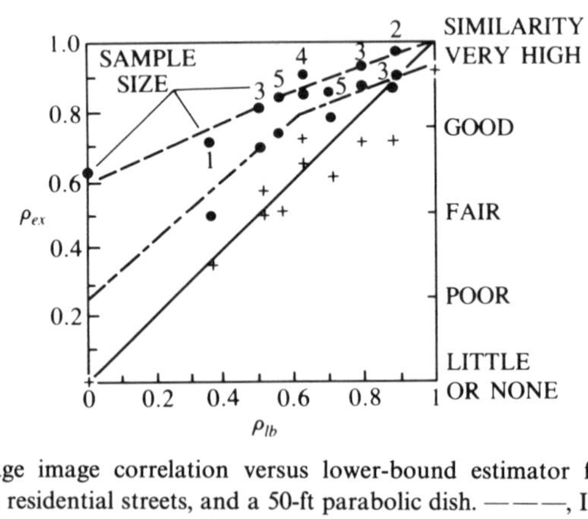

Figure 10-3 Average image correlation versus lower-bound estimator for suburban housing development, urban residential streets, and a 50-ft parabolic dish. ———, Image correlation coefficient $\rho = \overline{I_f I_g}/I_{f\,\mathrm{rms}} I_{g\,\mathrm{rms}}$. — — — —, Zero-mean correlation coefficient $\rho' = \overline{(I_f - \overline{I_f})(I_g - \overline{I_g})}/\sigma_{I_f}\sigma_{I_g}$. +++, Subjective viewing experiment, 63 subjects, suburban housing only. Solid line, unity slope line. (From ref. [101])

image consisting of equal numbers of black and white pixels. Randomly redistributing the pixel intensities to form a new image destroys all similarity between the two images. Yet, assigning a value of unity for the white pixels and a value zero to the black ones leads to an image correlation coefficient of 0.25.

It seems reasonable to remove the mean values of each image prior to correlation, in which case the image correlation in the above example becomes zero. The correlation in the aperture plane of the complex electric fields derived from two independent scenes is also zero. One would expect, therefore, the aperture plane correlation coefficient (10-3) to correspond more closely to the mean-free image correlation coefficient than the one calculated from Eq. 10-1.

The third (primed) value in the legend of Fig. 10-2 is the measured mean-free image correlation coefficient. It is plotted against the aperture-plane coefficient (10-3) in the dash-dot curve in Fig. 10-3 for the three scenes described above. It is clear that the theoretical coefficient is a better representation of the mean-free image correlation coefficient.

The correlation measure was also tested subjectively by 63 viewers who compared Fig. 10-2a to 12 images derived from the distorted data. The viewers decided whether the members of each pair were very similar, whether similarity was good, fair, poor, or whether there was little or no similarity. The judgments were weighted 1.0, 0.75, 0.5, 0.25, and 0.0, respectively. The average response for each pair is plotted as plus signs in Fig. 10-3. The fit to the unity slope line is satisfactory for most of the tests, leading to the conclusion that ρ_{Ib} properly characterizes the degraded image.

10-5 REDUCTION OF COMPUTATIONAL REQUIREMENT

The image obtained using the quantized data is the Fourier transform of the data, that is, $G(u) = |\mathscr{F} g|$, where $\mathscr{F}$ is the Fourier transform operator. More specifically, the image plane can be divided into M different points at $u = u_m$, $1 \leqslant m \leqslant M$, where the M points are separated by approximately one beamwidth of the receiving array. Then, $G(u_m)$ can be written as

$$G(u_m) = \sum_{n=1}^{N} g_n \exp(-jku_m x_n) \tag{10-16}$$

for a linear array with N elements located at x_n. The g_n in Eq. 10-16 are the distorted data, and are Fourier transformed to compute all the M image pixels $G(u_m)$. Clearly, a total of $4N$ real multiplications are required to compute each image pixel $G(u_m)$. The computational burden can be significantly reduced if the exponential kernel $\exp(-jku_m x_n)$ is also quantized. Let η_n represent the phase error introduced by quantizing the kernel. Then the (m, n)th term in Eq. 10-16 becomes

$$g_n \exp[-j(ku_m x_n - \eta_n)] = (g_n \exp(j\eta_n)) \exp(-jku_m x_n)$$

The first factor on the right can be treated as a new distorted signal. The signal distortion is a function of the scene being imaged while the kernel distortion is deterministic *a priori*. These errors being independent, their averages factor, changing the phase factor of the lower-bound correlation estimate to

$$|\overline{e^{j(\phi+\eta)}}| = \overline{e^{j\phi}}\;\overline{e^{j\eta}}$$

Table 10-2 lists the combined effect of signal distortion and phase quantization of the kernel for a variety of cases.

TABLE 10-2 Joint effect of signal distortion and phase distortion in the image processor. ρ_{lb} is the lower bound estimate of the image-correlation coefficient [101]

Signal distortion			
Amplitude	*Phase*	*Phase quantization in processor*	ρ_{lb}
None	None	1 bit	$2/\pi = 0.637$
None	None	2 bits	$2\sqrt{2}/\pi = 0.900$
None	1 bit	1 bit	$4/\pi^2 = 0.406$
None	1 bit	2 bits	$4\sqrt{2}/\pi^2 = 0.573$
None	2 bits	1 bit	$4\sqrt{2}/\pi^2 = 0.573$
None	2 bits	2 bits	$8/\pi^2 = 0.811$
Hard limit	None	1 bit	$1/\sqrt{\pi} = 0.564$
Hard limit	None	2 bits	$\sqrt{2/\pi} = 0.798$
Hard limit	1 bit	1 bit	$2/\pi\sqrt{\pi} = 0.359$
Hard limit	1 bit	2 bits	$2\sqrt{2}/\pi\sqrt{\pi} = 0.508$
Hard limit	2 bits	1 bit	$2\sqrt{2}/\pi\sqrt{\pi} = 0.508$
Hard limit	2 bits	2 bits	$4/\pi\sqrt{\pi} = 0.718$

From ref. [101]

Compressing the input signal and the exponential kernel of the image processor each into a 1-bit or 2-bit phasor results in an extraordinarily simple computation process. All complex multiplications are replaced by binary 1-bit additions using counters as shown below. Consider the 2-bit × 2-bit case with hard limiting on the signal. The input signal at the nth element $A_n e^{j\phi_n}$ is replaced by $g_n = 1, j, -1$, or $-j$, according to its quadrant. The unit phasor $\exp(-jku_m x_n)$ is similarly replaced by $p_{m,n} = \pm 1, \pm j$ depending on its quadrant. The complex multiplications required to compute $G(u_m)$ are the product of two unit phasors, each of which takes on one of the four values ± 1 and $\pm j$. Hence their product is also equal to one of the four values ± 1 and $\pm j$. Computation of this product does not require any multiplication; instead it requires simple binary logic circuits such as AND and OR gates. The output of these logic circuits increments one of four counters, each of which stores the total number of products that are equal to ± 1 and $\pm j$, respectively. Let $X_{-1,m}$, $X_{1,m}$, $X_{-j,m}$, and $X_{j,m}$ be the number of counts in the four counters after the N input data samples are operated upon. The mth image pixel strength is the magnitude of the sum product samples $g_n p_{m,n}$ which is simply

$$G(u_m) = |(X_{1,m} - X_{-1,m}) + j(X_{j,m} - X_{-j,m})| \tag{10-17}$$

Figure 10-4 shows the image of the housing development processed according to Eq. 10-17. Note its similarity with the original image of Fig. 10-2a.

Quantizing the kernel in the image processor adds to the distortion in the image. The processing load can be reduced by replacing the complex multiplications in Eq. 10-16 with real additions even without quantizing the Fourier kernel at the expense of an increase in memory requirements. This can be seen by noting that the complex multiplications in Eq. 10-16 multiply the quantized data by unit phasors $\exp(-jku_m x_n)$ that are independent of the data samples. Hence all possible values of the product $g_n \exp(-jku_m x_n)$ can be stored before the image is computed. If the complex data g_n are quantized to b bits, the product can take on one of 2^b values for each combination (m, n). Hence a total of $MN2^b$ complex storage locations are required for storing all possible values of the product in Eq. 10-16. For 100 elements ($N = 100$) and 100 beams ($M = 100$), a total of 20×2^b kilobytes of memory are required when the input data are compressed to b bits each. This evaluates to 80 kilobytes for $b = 2$, a number easily met using current state-of-the-art memory addressing devices and RAM. However, note that the memory requirement increases exponentially

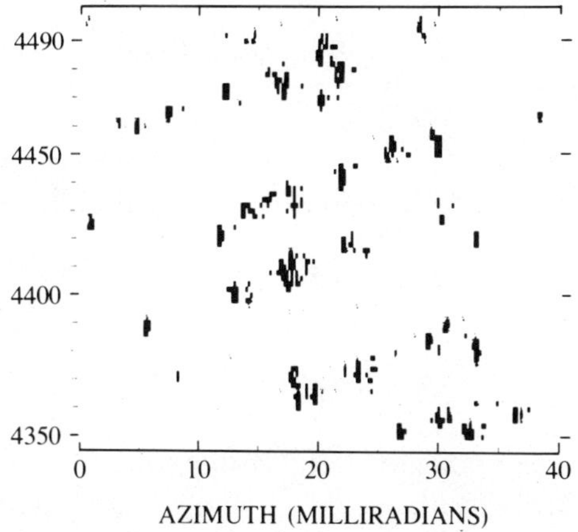

Figure 10-4 Same scene as Fig. 10-2. The image is formed using a single quadrature component and quantizing the exponential kernel of the image processor to two bits (see Eq. 10-17). (From ref. [101])

with b; hence the method may not be practical when b is large. Signal processing architectures that replace all complex multiplications by additions using the methods described above are developed in [110] and [126].

10-6 OPTIMUM DATA QUANTIZATION

Several methods for quantizing the received data were discussed in Section 10-3. This section develops a rule for dividing up the total number of bits between amplitude and phase so as to maximize the image quality measure ρ in Eq. 10-3. A target scene consisting of many targets, none of which dominates the sum of the rest, is assumed. Such an assumption is realistic for microwave imaging because a majority of the range bins in the field of view consist of this type of a target scenario. Under this assumption, the amplitude and phase of the received echoes are statistically independent.

Earlier sections in this chapter reached the following conclusions. Hard limiting preserves much of image quality, whereas amplitude-only information destroys it. Hard limiting with 1 bit of phase sometimes suffices. Hard limiting with 2 bits of phase often results in acceptable imagery. This section extends this list of conclusions as follows. A minimum of 4 bits of information per complex data sample guarantees a relatively good image provided that 3 bits are used for phase and one for amplitude. Eight bits are required to ensure a high-fidelity image. When the number of bits is no more than 8, the number of bits assigned to phase should be two larger than the number of bits assigned to amplitude. The number of phase and amplitude bits should be equal for finer quantization.

When the amplitude and phase of the undistorted signal are statistically independent, the lower bound on the image correlation, ρ_{lb}, defined by Eq. 10-3, is

$$\rho_{lb} = \frac{\overline{|fg^*|}}{f_{\text{rms}}\, g_{\text{rms}}} = \frac{\overline{AA'}\,\overline{|\exp[j(\phi - \phi')]|}}{A_{\text{rms}} A'_{\text{rms}}} = \rho_a \rho_\phi \tag{10-18}$$

where

$$f = A \exp(j\phi), \qquad g = A' \exp(j\phi') \tag{10-19}$$

$$\rho_a = \frac{\overline{AA'}}{A_{\text{rms}} A'_{\text{rms}}} \tag{10-20a}$$

$$\rho_\phi = \overline{|\exp[j(\phi - \phi')]|} \tag{10-20b}$$

where f and g are the undistorted and distorted data, respectively. The next two subsections describe methods for optimally quantizing the amplitude and phase given N_a amplitude levels and N_ϕ phase levels, respectively. Then, a method for choosing N_a and N_ϕ, given a total of $(N_a + N_\phi)$ levels is developed. The development and results are taken from [59].

10-6-1 Optimum Amplitude Quantizer

Let N_a levels be available for quantizing the amplitude. Further, let the amplitude axis be split up into N_a regions $[a_0, a_1), [a_1, a_2), \ldots, [a_{N_a-1}, a_{N_a})$, where $[a_0, a_1)$ means the region $a_0 \leqslant A < a_1$. The amplitude can range between zero and infinity; hence $a_0 = 0$ and $a_{N_a} = \infty$. Finally, let the amplitude of the distorted data g be denoted a'_i when the amplitude A falls in the region $[a_{i-1}, a_i)$. The objective is to determine the a_i and the a'_i that maximize the amplitude correlation coefficient in Eq. 10-20a.

This can be accomplished in two steps. The first step defines the N_a regions $[a_{i-1}, a_i)$, $1 \leqslant i \leqslant N_a$. The next step determines the amplitude of the distorted data a'_i for each of the N_a regions.

The N_a regions are obtained by maximizing the entropy of the quantized output. The probability of the amplitude lying in the ith region, denoted p_i, is

$$p_i = P\{a_{i-1} \leqslant A < a_i\} = \int_{a_{i-1}}^{a_i} p(a)\, da, \qquad 1 \leqslant i \leqslant N_a \tag{10-21}$$

where $p(a)$ is the pdf of the undistorted amplitude A. Since A is Rayleigh, $p(a) = (a/\sigma^2)\exp(-a^2/2\sigma^2)$, where σ^2 is the variance of A. The entropy of the quantized data, denoted H_a, is

$$H_a = -\sum_{i=1}^{N_a} p_i \log_2(p_i) \tag{10-22}$$

It is easily shown that H_a is maximized when all the p_i are equal, that is, when $p_i = 1/N_a$, all i. Further, substitution of the Rayleigh pdf into Eq. 10-21 leads to $p_i = \exp(-a_{i-1}^2/2\sigma^2) - \exp(-a_i^2/2\sigma^2)$. Invoking $p_i = 1/N_a$, all i, into the above leads to the following values of a_i [59]:

$$a_i = \sigma(-2\ln(1 - i/N_a))^{0.5}, \qquad 1 \leqslant i \leqslant N_a - 1 \tag{10-23}$$

with $a_0 = 0$ and $a_{N_a} = \infty$.

The next step consists of determining the amplitude of the distorted data a'_i for each of the N_a intervals $[a_{i-1}, a_i)$. This is accomplished by maximizing the amplitude correlation ρ_a in Eq. 10-20a. It is shown in [59, Appendix A] that ρ_a is maximized when

$$a'_i = E\{A \mid a_{i-1} \leqslant A < a_i\} = \hat{a}_i; \qquad 1 \leqslant i \leqslant N_a \tag{10-24}$$

In other words, the optimal quantizer output a'_i for region i is just the expected value of the amplitude in that region.

The optimal amplitude quantizer when a_i and a'_i are chosen using Eqs. 10-23 and 10-24, respectively, yields an amplitude correlation

$$\rho_a = \frac{\overline{AA'}}{A_{\text{rms}}A'_{\text{rms}}} \approx 1 - \exp(-(1.056N_a^{0.95} + 2.173)) \tag{10-25}$$

where the numerator was derived using (a) $\overline{AA'} = \sum_1^{N_a} \Pr\{a_{i-1} \leqslant A < a_i\} E\{AA' \mid a_{i-1} \leqslant A < a_i\} = \sum_1^{N_a} p_i a'_i \hat{a}_i = \sum_1^{N_a} \hat{a}_i^2/N_a$, and (b) a Rayleigh distribution to calculate $\hat{a}_i$.

Thus the optimum amplitude quantizer is defined by Eqs. 10-23 and 10-24. The optimum value of ρ_a is given by Eq. 10-25.

It can be shown that the global maximum of ρ_a is not attained using this procedure [59]. This is because the intervals $[a_{i-1}, a_i)$, $1 \leqslant i \leqslant N_a$, were chosen not by maximizing ρ_a but by maximizing the entropy of the quantized data. However, as noted in [59], the procedure outlined above attains a value of ρ_a that is almost equal to its global maximum. The loss incurred in choosing a suboptimal ρ_a is traded off by the ease in choosing the N_a intervals $[a_{i-1}, a_i)$ (10-23) and the quantizer outputs a'_i (10-24) for each of these intervals.

10-6-2 Optimum Phase Quantizer

The optimal phase quantizer is derived in a similar manner to the optimum amplitude quantizer. Assume, as before, that the aperture excitation is due to a collection of scatterers, none of which dominates the rest. Then the aperture phase measurements are uniformly distributed between $-\pi$ and π.

Let N_ϕ levels be available for quantizing the phase. The phase quantizer that maximizes ρ_ϕ is obtained by splitting the phase axis into N_ϕ regions $[\phi_0, \phi_1), [\phi_1, \phi_2), \ldots, [\phi_{N_\phi - 1}, \phi_{N_\phi})$, of equal probability as follows:

$$\phi_i = -\pi + \frac{2\pi i}{N_\phi}; \qquad 0 \leqslant i \leqslant N_\phi \tag{10-26}$$

In other words, the optimum phase quantizer splits the phase axis into N_ϕ regions of equal angular extent (equal probability), thereby maximizing the entropy of the quantized phase. The quantizer output for each of those regions that maximizes ρ_ϕ, denoted ϕ'_i, is [59]

$$\phi'_i = E\{\phi \mid \phi_{i-1} \leqslant \phi < \phi_i\} = -\pi + \frac{\pi(2i-1)}{N_\phi}; \qquad 1 \leqslant i \leqslant N_\phi \tag{10-27}$$

That is, the output phase for the ith region is the midpoint of the ith region. The value of ρ_ϕ after optimal phase quantization is [59]

$$\rho_\phi = |\overline{e^{j(\phi - \phi')}}| = \frac{\sin(\pi/N_\phi)}{\pi/N_\phi} = \text{sinc}(\pi/N_\phi) \tag{10-28}$$

10-6-3 Optimum Quantizer

The last two sections presented quantization techniques that maximize ρ_a and ρ_ϕ given N_a amplitude levels and N_ϕ phase levels. A method of choosing N_a and N_ϕ, given a total of b bits for representing each complex data sample, so as to maximize ρ_{lb}, is now described. Because $\rho_{lb} = \rho_a \rho_\phi$ where ρ_a and ρ_ϕ are independent, it can be maximized by maximizing the product $\rho_a \rho_\phi$. Note that ρ_a is a function only of the number of amplitude levels N_a (see Eq. 10-25), and ρ_ϕ a function of only the number of phase levels N_ϕ (see Eq. 10-28). Hence, given the total number of levels available for representing each data sample ($N_a + N_\phi$), ρ can be maximized by examining the value of $\rho_a \rho_\phi$ for all combinations of (N_a, N_ϕ) such that $N_a + N_\phi$ = constant.

Table 10-3 shows the product of $\rho_a \rho_\phi$ for various numbers of amplitude and phase bits. Assume that three bits are available for representing each data sample. These bits can be divided in four ways between amplitude and phase, viz. three bits of amplitude, two bits of amplitude and one bit of phase, and so on. Among the four combinations, note that ρ reaches its maximum value of 0.8646 when one bit is assigned to amplitude and two bits are assigned to phase. This is the optimal quantizer given a total of three quantization bits per data sample. Similarly, when the number of quantization bits b is between four and eight, the optimal quantizer uses $(b/2 - 1)$ bits for amplitude and $(b/2 + 1)$ bits for phase (even b), or $(b - 1)/2$ bits for amplitude and $(b + 1)/2$ bits for phase (odd b).

Also note that $\rho_{max} = 0.9359$ with just four bits per data sample. Thus good imagery can be obtained using as few as four bits per data sample. Further note that $\rho_{max} = 0.9927$ when eight bits are used to represent each data sample, implying that no more than eight bits are required for very high-fidelity imaging.

TABLE 10-3 $\rho_a \rho_\phi$ **vs. number of amplitude and phase bits per sample [59]**

	H_Φ							
H_A	0	1	2	3	4	5	6	7
0	0.0000	0.5641	0.7978	0.8636	0.8805	0.8848	0.8858	0.8861
1	0.0000	0.6114	0.8646	0.9359	0.9542	0.9588	0.9600	0.9603
2	0.0000	0.6272	0.8870	0.9601	0.9789	0.9836	0.9848	0.9851
3	0.0000	0.6330	0.8952	0.9689	0.9879	0.9927	0.9939	0.9942
4	0.0000	0.6352	0.8983	0.9723	0.9913	0.9961	0.9973	0.9976
5	0.0000	0.6360	0.8995	0.9736	0.9926	0.9975	0.9987	0.9990

From ref. [59]

10-7 CONCLUSIONS REGARDING BIT COMPRESSION BY DATA QUANTIZATION

This chapter quantitatively confirms the assertions of previous writers that the phase of the signal received at the aperture contains the preponderance of the information useful for image formation. Further, an easily calculable image correlation metric has been established for judging the extent of the distortion in the image due to an arbitrary distortion of the signal at the aperture or of the imaging kernel in the signal processor. This is potentially useful in establishing tolerances in the data-gathering portion of a system as well as the signal-processing portion. As an example, given some known or measurable nonlinearity at the front end of a measuring system, it is possible to judge whether such nonlinearity is tolerable or whether it must be removed.

A method for optimally distributing the total number of bits available for representing each data sample between amplitude and phase has also been developed. This optimal quantizer determines the number of amplitude and phase bits by maximizing the lower bound of the correlation metric. The optimum quantizer is examined in greater detail in [60].

Clearly, data quantization eases the data rate requirements between the antenna elements and the signal processor. Most imaging systems depend upon quadrature coherent demodulation of the received signal. The quadrature components are each sampled and quantized to five or more bits, the large number of bits per sample being insurance against image degradation. Yet, as demonstrated by Fig. 10-2e, hard limiting plus one bit of phase produces an image having remarkably good quality. For those situations where the quality illustrated is adequate, about an order of magnitude of data reduction at the signal processor is aohieved. The reduction represents a huge saving in the required data rate between the elements and the signal processor.

Data quantization can also be used to ease the computational requirements of the signal processor, especially when the imaging kernel is also quantized. This is demonstrated by Fig. 10-4, where both the data and the imaging kernel were represented using two bits of phase each. The resulting signal processor (see Eq. 10-17) was remarkably simple: all complex additions and multiplications were eliminated and the processor consisted of counters only.

From Fig. 10-2, it is evident that the small loss in correlation between the proper and distorted images hardly effects the broad features of the image. The performance of functions akin to photo interpretation, classification of scenes, differentiation between categories of targets, and so on, are little disturbed. However, it is equally evident from the figures that isolated pixels appear to move, change in intensity, and sometimes appear where they were not present before. Such detailed changes in image content signal two types of limitations: false target detection and reduced target dynamic range. These subjects require further study. However, in view of the large potential reductions in data rate and computations, this theory offers material benefit in microwave imaging systems design.

10-8 REDUCTION IN THE SPECTRAL OCCUPANCY OF THE PASSBAND

A lesson applicable to the sampling-rate problem can be learned from the fact that N, the number of elements and receiving channels, can be reduced by two or more orders of magnitude by thinning the array. Thinning can be done periodically or aperiodically. Figure 10-5 illustrates the two cases. A periodically thinned array (upper) has equally spaced elements in which the interelement spacing $d > \lambda/2$. The radiation field exhibits grating lobes spaced by λ/d. The grating lobes can introduce multiple target images, which generally must be avoided. Figure 4-14b illustrated this undesirable phenomenon. Sometimes the extra images can be eliminated by designing the transmitter beam to be narrower than the spacing between grating lobes. This technique is illustrated in the upper right and was used in the imaging experiment of Fig. 5-13.

Aperiodic thinning destroys grating lobes but raises the sidelobe level. Random thinning is one form of aperiodic thinning. The radiation pattern is shown in the lower right; it is the Fourier transform of the randomly thinned aperture distribution. Its main lobe has width λ/L and its side radiation pattern

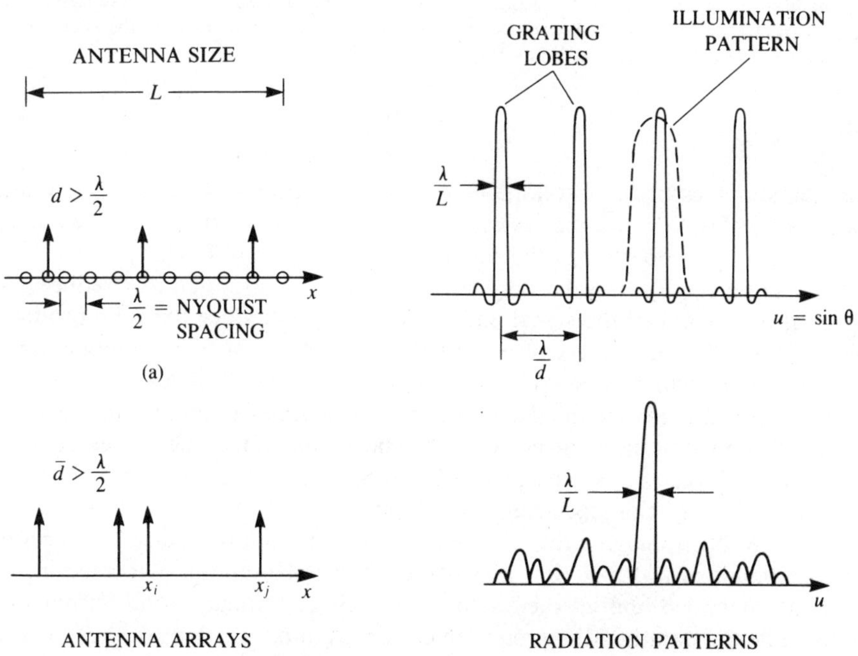

Figure 10-5 (a) Thinned periodic array and (b) random array and their radiation patterns. The grating lobes of the thinned periodic array disappear when the element locations are randomized. The energy of the grating lobes spreads uniformly in $u = \sin\theta$. The result is high sidelobes.

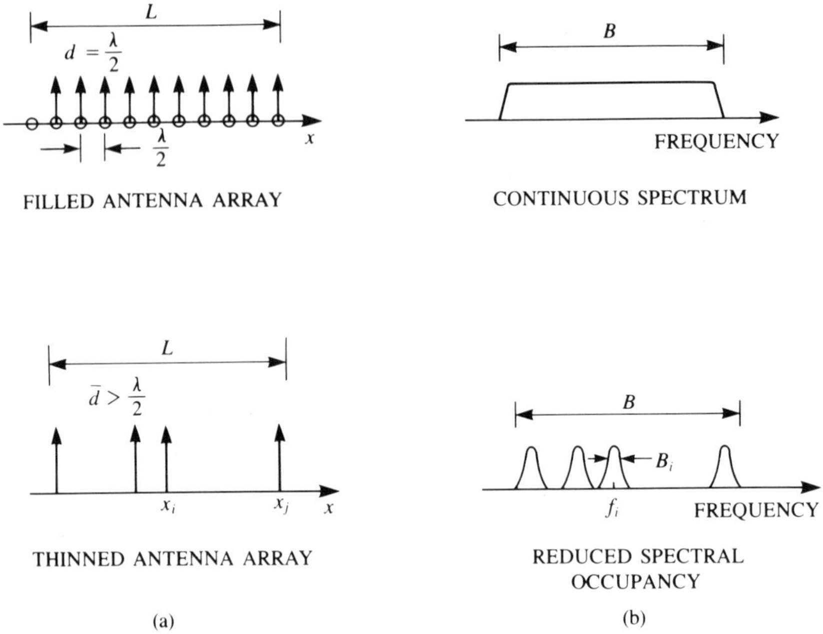

Figure 10-6 (a) Filled and (b) thinned arrays and their spectral analogs. The range resolutions of both spectra are the same ($\Delta R = R/2B$) but the data handling requirement of the system with the lower right-hand spectrum is reduced by the factor $\sum B_i/B$.

has the statistics of the envelope of narrowband Gaussian noise. Its average sidelobe level is N^{-1} and the peak sidelobe is typically 10 dB higher. The theory of the sidelobe properties of the random array is given in [94].

The same theory may be applied in the frequency domain to reduce the spectral occupancy of the signal and, by so doing, to reduce the NBb product. Figure 10-6 illustrates the concept. In the upper left is a filled antenna array of length L for which $N = L/d = 2L/\lambda$ and the beamwidth is $\Delta\theta = \lambda/L$. Below it is a thinned array of the same length with a smaller number of elements $N < 2L/\lambda$ and with the same beamwidth. The analog of the filled antenna array is the continuous signal spectrum of width B in the upper right. A pulse waveform having this spectrum has duration B^{-1} and a range resolution $\Delta R = c/2B$. Now imagine that the continuous spectrum is replaced by a set of narrow bands of width B_i (lower right). Let the B_i be randomly located in the spectral interval B and let their sum $\sum_i B_i < B$. The range resolution remains $c/2B$ while the data handling requirement drops in proportion to $\sum_i B_i/B$. The penalty is a deterioration of the temporal sidelobes of the signal waveform, which is the Fourier transform of the signal spectrum. Its properties can be deduced from the analogous properties of the random array.

The first step is to model the spectrum as a set of delta functions in the

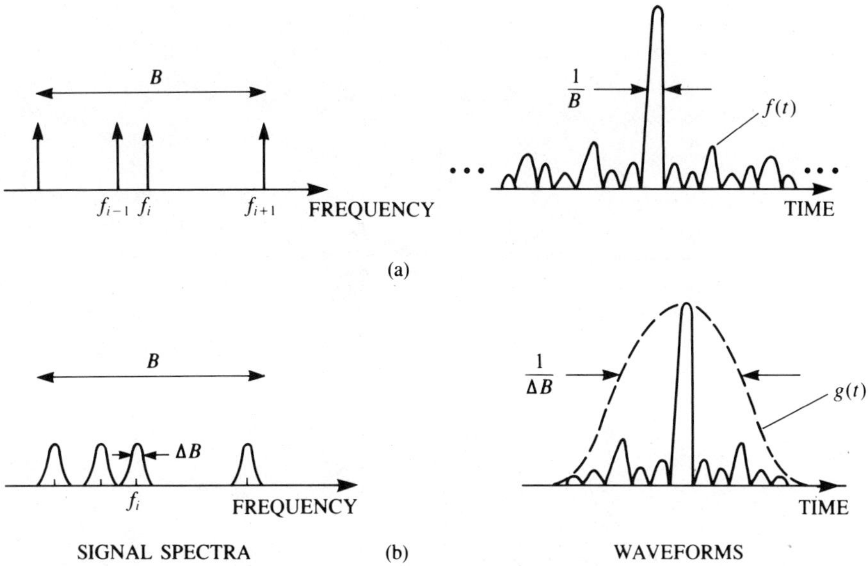

Figure 10-7 (a) A δ-function spectrum and its waveform. The pulse duration is B^{-1}. The sidelobes extend indefinitely and have the same statistics as the random array. (b) Each tone is broadened to ΔB. The pulse duration and the sidelobe statistics are unchanged within $(\Delta B)^{-1}$; outside, the sidelobes fall rapidly toward zero.

spectral interval B, that is,

$$F(f) = \sum_{i}^{M} \delta(f - f_i), \qquad \text{all } f_i \text{ in interval } B \tag{10-29}$$

which is analogous to the thinned antenna array with isotropic antenna elements defined by a weighting function or aperture distribution

$$w(x) = \sum_{j}^{N} \delta(x - x_j) \tag{10-30}$$

The second step is to realistically broaden the spectral components and recalculate the waveform properties. The first step is shown in Fig. 10-7a. The waveform is the sum of cw tones, each of infinite extent, and the sum is of infinite extent. It consists of a central pulse of duration B^{-1}, plus sidelobes that extend indefinitely in both directions. If the frequencies f_i are randomly selected, the statistics in the sidelobe region are the same as those of the random array: The average power level is M^{-1}, where M is the number of tones, and the peak is nominally 10 dB larger.

Broadening each tone to a width ΔB, as shown in the lower left, can be thought of as convolving the δ-function spectrum with a narrow passband $G(f)$

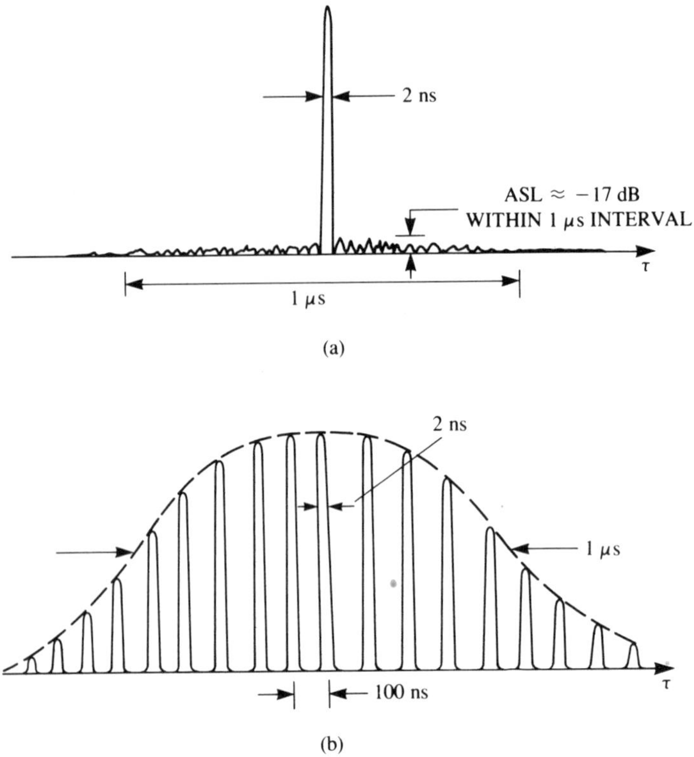

Figure 10-8 Signal waveforms for 10:1 spectral thinning. $B = 500$ MHz, $\Delta B = 1$ MHz, $M = 50$ tone bursts. (a) Random distribution of tone frequencies within B. (b) Periodic.

having spectral width ΔB. The waveform

$$\mathscr{F}\{F(f) * G(f)\} = \mathscr{F}\{F(f)\}\mathscr{F}\{G(f)\} = f(t)g(t) \tag{10-31}$$

where $\mathscr{F}$ is the Fourier transform, $f(t)$ is the upper waveform on the right, and $g(t)$ is the envelope shown dashed below. Within the time interval $1/\Delta B$ the properties of the wave are nominally those associated with the δ-function spectrum; outside, the wave rapidly drops toward zero.

As an example, consider the case in which 50 tones, each 1 μs in duration, are randomly distributed in frequency over a 500-MHz interval in the spectrum. The bandwidth of each tone is 1 MHz and the occupied portion of the 500-MHz spectrum is 50 MHz. The spectral thinning factor is 10, which is exactly the reduction in the data handling requirement. The waveform is shown in Fig. 10-8a. The pulse duration is 2 ns, corresponding to the 500-MHz spectral width, the range resolution is 30 cm, and the sidelobes persist for about 1 μs. The average sidelobe level is approximately −17 dB within the 1-μs interval and it falls rapidly toward zero outside.

10-9 PERIODIC SPECTRAL THINNING

Periodic thinning does not lead to a satisfactory waveform design. Consider again the example above. The 50 tones would be spaced 10 MHz apart. The grating lobes would be spaced by the reciprocal of 10 MHz, or 100 ns. Assuming that each "tone" is a burst of duration 1 μs, the waveform would have the same envelope as in the earlier example. The overall duration would be 1 μs. Within it would be approximately 10 short pulses each of 2 ns duration. This highly unsatisfactory waveform is sketched in Fig. 10-8b.

The number of grating lobes is the thinning factor. Hence the larger the thinning factor, the poorer the waveform.

A spectrum with periodic structure can be useful, however, provided that spectral thinning is not the objective. One method for achieving very wide bandwidth with practical equipment utilizes a periodic spectrum. It is described in [125]. The multiple tones fill the band, implying that the frequency spacing Δf is the reciprocal of the pulse duration δ. The tones are not radiated simultaneously, however. Figure 10-9 shows the waveform and Fig. 10-10 is its time–frequency diagram. One tone is generated and transmitted at the time of occurrence of each radar trigger pulse. The pulse duration is δ. The first pulse is radiated with center frequency f_0. Its echoes are coherently detected, sampled,

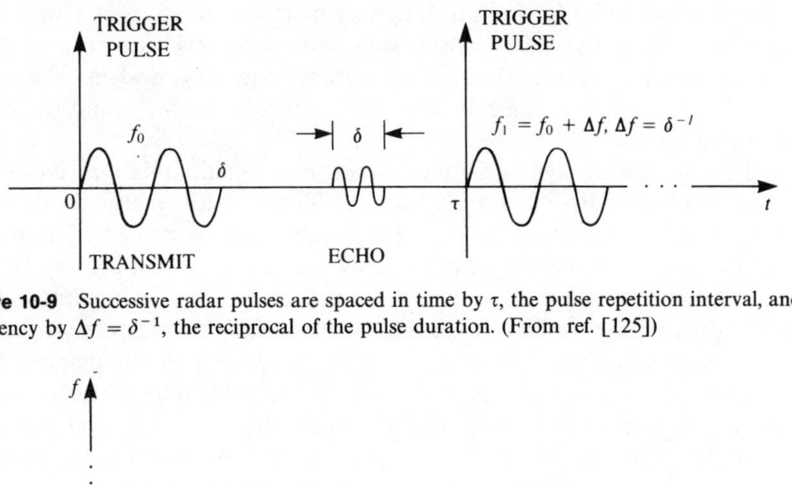

Figure 10-9 Successive radar pulses are spaced in time by τ, the pulse repetition interval, and in frequency by $\Delta f = \delta^{-1}$, the reciprocal of the pulse duration. (From ref. [125])

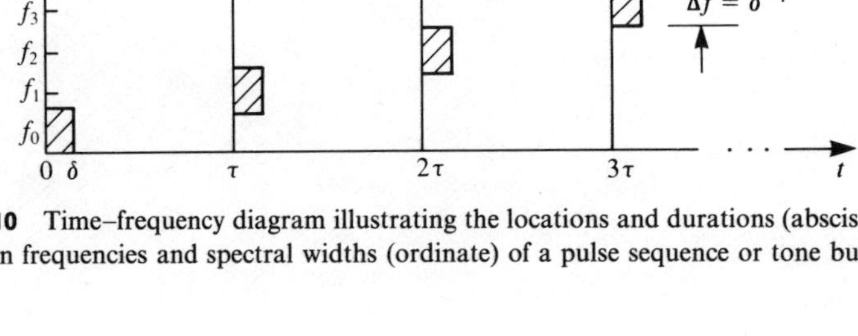

Figure 10-10 Time–frequency diagram illustrating the locations and durations (abscissa) and the transmission frequencies and spectral widths (ordinate) of a pulse sequence or tone burst.

digitized, and stored. The second pulse is transmitted after one pulse repetition interval τ; its frequency is $f_1 = f_0 + \Delta f = f_0 + \delta^{-1}$. The second echo trace is also coherently detected, sampled, digitized, and stored. The third echo is shifted in frequency to $f_2 = f_0 + 2\delta^{-1}$, and so forth. The last pulse in the sequence is at frequency $f_{M-1} = f_0 + (M - 1)\delta^{-1}$ and occurs at $t = (M - 1)\tau$. Provided that the targets are stationary, the M echo traces may be added coherently to

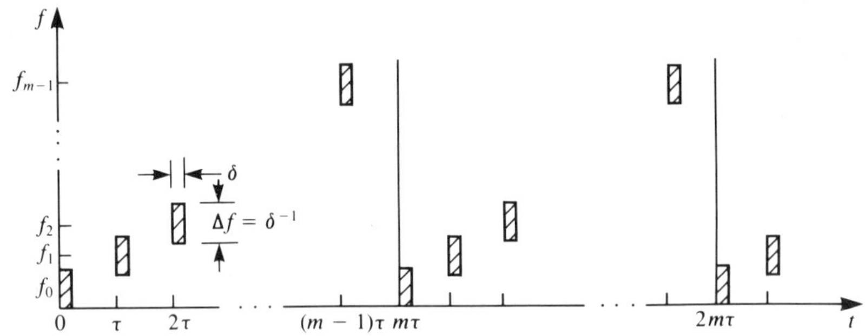

Figure 10-11 Frequency–time diagram of an M-step frequency waveform.

form a composite echo having resolution properties derived from a waveform of bandwidth $B = M\Delta f$. The transmission pattern repeats after each M-pulse sequence tone burst. Figure 10-11 shows the repetition pattern when the number of tones is M, the burst length is $M\tau$, the bandwidth is $M\delta^{-1}$, and the range resolution is $c\delta/2M$.

Modern frequency synthesis techniques have made this procedure for achieving wide bandwidth very attractive. There is a penalty to be paid, however. The effective interpulse period is lengthened from τ to $M\tau$. This means that when targets are moving, their echoes are effectively sampled at $1/M$ of the rate. Ambiguities in the Doppler spectrum occur at frequency intervals of $(M\tau)^{-1}$ rather than τ^{-1}. To compensate for this deficiency, the pulse repetition frequency can be increased by a factor of M. While this design change solves the Doppler ambiguity problem, it reduces the unambiguous maximum radar range by exactly the same factor and the radar becomes useful only for very short range. The use of this technique enforces a compromise between resolution, Doppler ambiguity, and maximum unambiguous range.

11

SUPERRESOLUTION

11-1 INTRODUCTION

The nominal beamwidth λ/L of an aperture of length L at wavelength λ is not an immutable fact of nature but is a fact of the signal processing of natural objects such as lenses. The operation of a lens or a telescope mirror or a microwave dish or an acoustic transducer is governed by diffraction theory, which predicts the nominal standard beamwidth given above, and precludes the recognition of two targets as distinct and separate if they are closer than this value. The method by which a lens operates upon a radiation field to produce an image is, of course, described by diffraction theory or approximately by its far-field equivalent, Fourier theory.

When an aperture is a sampled device such as a phased array, it is no longer fundamental that the image-forming signal processing correspond to diffraction theory. The question naturally arises: is there some other procedure that offers superior resolving power? The development of such algorithms and their performance is the subject of modern *spectrum estimation*, as it is called in time-series literature, or *superresolution*, an apt term in the spatial signal processing or imaging field. The pioneering work of Burg [15] sparked the field of high-resolution spectrum estimation or superresolution, which is the subject of this chapter.

As before, let $e(x)$ be the one-dimensional radiation field in the axis containing the array, caused by a one-dimensional complex source distribution $s(u)$ at some arbitrary far-field distance. Then $e = \mathscr{F}^{-1}s$, where $\mathscr{F}^{-1}$ represents the inverse Fourier transform operator. The conventional imaging procedure yields an estimate $\hat{s}(u)$ by forming the Fourier transform of the product $e(x)$

and the aperture weighting function $w(x)$. Thus

$$\hat{s}(u) = \int_0^L w(x)\, e(x)\, e^{-jkux}\, dx \tag{11-1}$$

The image $I(u) = |\hat{s}(u)|^2$ is an estimate of the desired power density spectrum of the source $S(u) = |s(u)|^2$. To complete this review of the earlier chapters, $\hat{s}(u)$ can also be represented in conventional form as

$$\hat{s}(u) = f * s \tag{11-2}$$

where $f = \mathscr{F}w$ is the radiation pattern of the aperture and the symbol "$*$" denotes convolution.

Let the aperture be a periodic array with N elements with an interelement spacing d. Then

$$I(u) = \left| \sum_{n=0}^{N-1} i_n\, e^{-jkndu} \right|^2 \tag{11-3}$$

where $i_n = w_n e_n$, $e_n = e(x = x_n)$, and similarly for w_n.

In matrix notation

$$I(u) = |\mathbf{i}^T \mathbf{b}(u)|^2 \tag{11-4a}$$

where $\mathbf{i}$ and $\mathbf{b}(u)$ are N-dimensional column vectors

$$\mathbf{i} = [i_0 \quad i_1 \quad \cdots \quad i_{N-1}]^T \tag{11-4b}$$

$$\mathbf{b}(u) = [1 \quad e^{-jkud} \quad \cdots \quad e^{-jku(N-1)d}]^T \tag{11-4c}$$

and the superscript T means transpose.

The image $I(u)$ in Eq. 11-4 has the same form as the spectrum of a moving average (MA) process. It can be rewritten in polynomial form and represented by $(N - 1)$ zeros or roots through the transform of the real angular variable u to the complex variable

$$z = e^{-jkud} \tag{11-5}$$

When expressed as a function of z the image $I(z)$ becomes

$$\begin{aligned} I(z) = |\hat{s}(z)|^2 &= \left| \sum_{n=0}^{N-1} i_n z^n \right|^2 \\ &= |i_{N-1}(z - z_1)(z - z_2)\cdots(z - z_{N-1})|^2 \end{aligned} \tag{11-6}$$

where z_n is the nth root of the polynomial $\hat{s}(z)$.

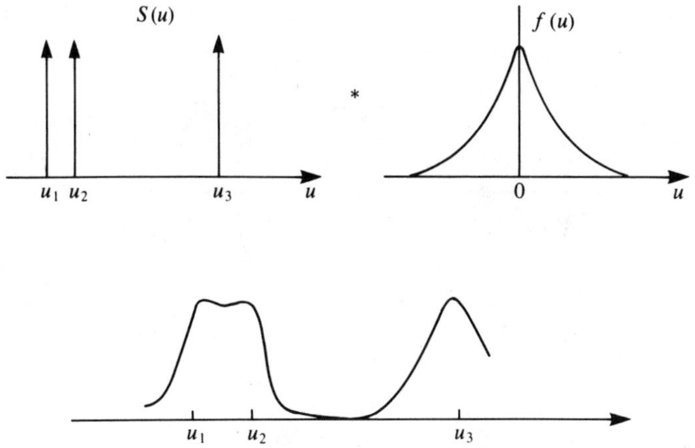

Figure 11-1 $\hat{s}(u) = \sum_{t=1}^{3} A_t e^{j\phi_t} f(u - u_t)$.

$\hat{s}(z)$, defined by Eq. 11-6, has no poles and has $(N - 1)$ zeros at $z = z_n$. The significance of $\hat{s}(z)$ being an all-zero function is elucidated in the following paragraphs.

The $(N - 1)$ zeros of $\hat{s}(z)$ locate the nulls and deep dips in the real image through the relationship (11-5). Of primary interest to the user, however, are not the nulls in the pattern but the peaks, for they disclose the directions of arrivals of the energy and are bearings to the targets. These directions of arrival are indicated by the absence of zeros. Because of the convolution (11-2), $\hat{s}(u)$ is not sharp or peaked in the target directions and the image crests can be no narrower than the beamwidth λ/L associated with $w(x)$. Hence targets closer in angle than λ/L tend to merge into a single peak as the illustration in Fig. 11-1 shows. Given the three targets shown there as δ-functions, the complex image is the sum of three radiation patterns displaced by the target locations and weighted by their complex strengths.

The peaks of $\hat{s}(z)$ at the target locations are approximately quadratic and hence quite broad. In contrast, the nulls of $\hat{s}(z)$, which occur when $-kud$ is close to the argument of the z_n are approximately linear and therefore much narrower than the peaks. Figure 11-2 shows the zeros of $\hat{s}(z)$, which would be on or about the unit circle. $\hat{s}(z)$ is the product of the vectors from all the zeros to the head of the unit phasor $z = e^{-jkud}$, which lies on the unit circle. The real variable u is the argument of z divided by $-kd$.

$\hat{s}(z)$ is small as the phasor rotates to the neighborhood of a zero and grows larger between the zeros. Its magnitude is sketched as the distance from the unit circle in Fig. 11-3. The square of the distance shown in Fig. 11-3 is the image given by Eq. 11-6. It is the angular spectrum wrapped around the unit circle. Lobes between zeros point in the target directions. The lobes are $\approx \lambda/L$ wide and are approximately quadratic near the crests. Hence the resolution is

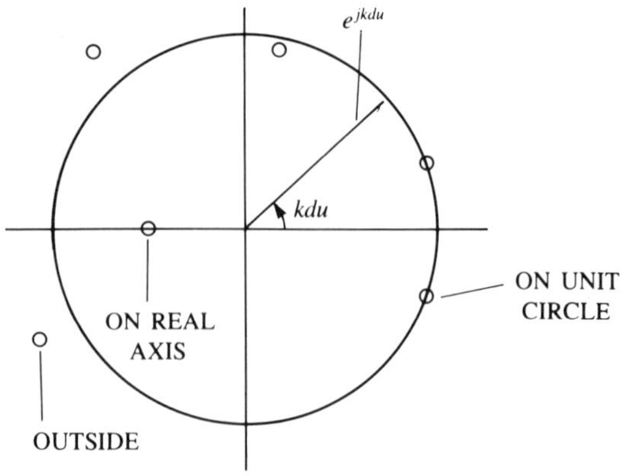

Figure 11-2 The unit circle and zeros of $\hat{s}(z)$. $z = e^{-jkud}$.

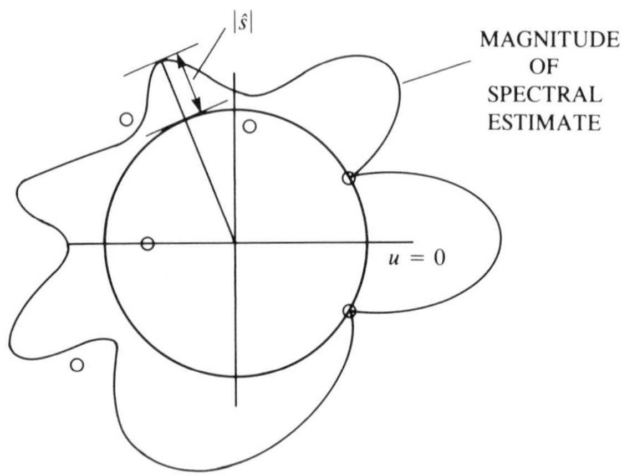

Figure 11-3 $\hat{s}$ is the product of vectors from $z = e^{-jkud}$ to all the zeros.

limited to $\approx \lambda/L$ and the accuracy of angle estimation is limited by the noisiness of the measurements.

The image, on the other hand, can be very crisp or sharp near the zeros, which lie between target directions. It would be desirable to interchange the properties between the lobe crests and the nulls, that is, we would like the sharpness or crispness of the pattern to be associated with the directions to the targets, not between them.

This requires the use of poles rather than zeros to represent $\hat{s}$ (or $I = |\hat{s}|^2$). It is evident that when the target or source distribution is represented by a set of T δ-functions (point targets),

$$s(u) = \sum_{t=1}^{T} A_t\, e^{j\psi_t}\delta(u - u_t) \tag{11-7}$$

of amplitudes A_t, phases ψ_t, at $u = u_t$, a better estimator of the angular spectrum would have the form

$$|\hat{s}|^2 = \frac{1}{|\mathbf{a}^T \mathbf{b}(u)|^2} \tag{11-8a}$$

where $\mathbf{b}(u)$ is a beamsteering vector defined by Eq. 11-4c and $\mathbf{a}$ is an N-dimensional weight vector

$$\mathbf{a}^T = [a_0 \quad a_1 \quad \cdots \quad a_{N-1}] \tag{11-8b}$$

The spectrum of Eq. 11-8a has the form of the reciprocal of Eq. 11-4a. Because the roots of the polynomial are in the denominator, $\hat{s}(u)$ is an all-pole model of $s(u)$. All-pole models are called *autoregressive* (AR). The AR weights are chosen so that the zeros of the denominator of Eq. 11-8a occur at the target locations. Therefore the angular spectrum $\hat{s}(u)$ will exhibit sharp peaks at the target locations. In contrast, target locations are represented by broad maxima in the image of Eq. 11-4a. The sharpness of the AR spectrum peaks makes it possible to resolve targets separated by amounts smaller than a beamwidth λ/L.

Section 11-2 describes a method for determining the all-pole weight vector $\mathbf{a}$ using least-squares methods.

Least-squares methods usually require the second-order statistics of the aperture data, that is, they require estimates of the data correlation $E\{e_m e_n^*\}$. The usual procedure for estimating the correlation of incoherent radiation fields is to sample e_m and e_n at several instants and average the product $e_m e_n^*$ over these statistically independent snapshots.

The luxury of averaging multiple snapshots is not available in coherent radar because the received echoes do not change from pulse to pulse. Incoherent transmitter location diversity, described in Chapter 6, can be used to estimate the correlation function. This is because the phase relationships between coherent targets change with transmitter location. Hence transmitting sequentially from different locations generates independent snapshots of the radiation field of the coherent targets.

Transmitter location diversity belongs to a class of techniques that can be used to decorrelate the radiation fields of coherent target scenarios. These techniques change the location of the array origin from snapshot to snapshot, either synthetically as in transmitter diversity, or by moving the array. The change in array–target geometry also changes the phase relationships between the coherent targets from snapshot to snapshot. Independent snapshots of the coherent target scenario can be generated in this fashion. The manner in which the target phase relationships vary with the location of the array origin is described in Section 11-3. The results of Section 11-3 are applied to periodic arrays in Section 11-4, where a technique called *spatial smoothing* [26], [62], [87] is used to synthesize several snapshots of the data correlation from the echoes due to a single transmitted pulse. It is shown that spatial smoothing is equivalent to the modified covariance method (or forward–backward linear prediction (FBLP) method) described in [71], [122].

Section 11-5 briefly describes the superresolution (SR) techniques called the *maximum likelihood method* (MLM) [18] and *MUSIC* [84].

Spatial smoothing techniques require a Fourier transform relationship between the aperture data and the angular spectrum. Microwave imaging with large apertures is usually performed in the near field. Section 11-6 shows that adaptive beamforming can be used to transform the measured near-field data to the far field in order to conform with the requirements of spatial smoothing.

Practical applications of SR in coherent microwave imaging are discussed in Section 11-7 and SR techniques are evaluated using experimental data in Section 11-8. Readers familiar with SR theory and interested primarily in the experimental applications can proceed directly to Section 11-7.

11-2 ESTIMATION OF THE AUTOREGRESSIVE WEIGHT VECTOR

A time series and its spectrum are Fourier transform pairs. So are the data measured at a periodic aperture and the corresponding far-field angular spectrum. Thus frequency estimation using time series data is equivalent to angular spectrum estimation using the data measured at a periodic aperture [11]. Most of the spectral analysis algorithms developed in time series literature are therefore directly applicable to angular spectrum estimation using periodic apertures.

Consider a target scenario consisting of a set of T point targets having amplitudes A_t, phases ψ_t at $u = u_t$, $1 \leqslant t \leqslant T$. Let an N-element periodic array with $\lambda/2$ interelement spacing be used to estimate the angular spectrum of the target scenario. The field at the nth element, denoted e_n, is

$$e_n = \sum_{t=1}^{T} A_t\, e^{j\psi_t}\, e^{jku_t x_n} = \sum_{t=1}^{T} A_t\, e^{j\psi_t}\, e^{j\pi n u_t}; \qquad 0 \leqslant n \leqslant N-1 \qquad (11\text{-}9)$$

where $x_n = n\lambda/2$ was used to arrive at the second equality.

In the absence of noise, e_n can be predicted exactly using a weighted sum of the echoes at the previous T elements [40]. In other words,

$$e_n = -\sum_{t=1}^{T} a_t e_{n-t} \qquad \text{or} \qquad \sum_{t=0}^{T} a_t e_{n-t} = 0 \qquad \text{with } a_0 = 1 \qquad (11\text{-}10)$$

The negative sign in the first part of Eq. 11-10 is used to conform with standard SR notation. The T unknown filter coefficients a_t can be determined by exploiting the model of Eq. 11-10, that is, by assuming that the nth sample can be exactly predicted using the previous T samples. Reference [62] describes a method for calculating the a_t using $2T$ noiseless data samples and the model of Eq. 11-10. These values of a_t make the denominator of the all-pole spectrum in Eq. 11-8a equal to zero at $u = u_t$, all t. Thus the target locations can be inferred by determining the peaks of $|\hat{s}|^2$ in Eq. 11-8a.

In practice, data are contaminated by additive noise and the model of Eq. 11-10 will only be approximate. The measured data at the nth element with additive noise introduced, denoted $\hat{e}_n$, is

$$\hat{e}_n = e_n + \text{noise} \tag{11-11}$$

where the noise samples are independent complex Gaussian random variables, each with zero mean and variance σ_n^2. The data samples $\hat{e}_n$ can be modeled using an autoregressive moving average (ARMA) process as follows [40]:

$$\sum_{t=0}^{T} a_t \hat{e}_{n-t} = \sum_{t=0}^{T} b_t \varepsilon_{n-t} \tag{11-12}$$

where the ε_n are independent samples of a zero-mean complex Gaussian random process having unit variance, the a_t are called the AR parameters with $a_0 = 1$, and the b_t are called the MA parameters. The model of Eq. 11-12 is called an *ARMA model of order* (T, T), where the first T corresponds to the number of AR parameters (equal to the model order of the AR part of the ARMA process) and the second T to the number of MA parameters (equal to the model order of the MA part of the process).

Because $\hat{e}_n$ is a random process, its angular spectrum is the expected value of the squared magnitude of its Fourier transform. Both sides of Eq. 11-12 represent convolutions in the aperture domain, that is, multiplications in the Fourier domain. Taking the Fourier transform of both sides of Eq. 11-12 results in

$$\hat{s}(u) = \frac{B(u)}{A(u)}\,\varepsilon(u) \tag{11-13}$$

where $\hat{s}(u)$, $A(u)$, $B(u)$, and $\varepsilon(u)$ represent the Fourier transforms of $\hat{e}_n$, a_n, b_n, and ε_n, respectively. Taking the expectations of the squared magnitude of both sides of Eq. 11-13 yields

$$I(u) = E\{|\hat{s}(u)|^2\} = \left|\frac{A(u)}{B(u)}\right|^2 \overline{|\varepsilon(u)|^2}$$

$$= \left|\frac{b_0 + b_1 e^{-j\pi u} + \cdots + b_T e^{-j\pi u T}}{1 + a_1 e^{-j\pi u} + \cdots + a_T e^{-j\pi u T}}\right|^2 \overline{|\varepsilon(u)|^2} \tag{11-14}$$

where the overbar on the random process $|\varepsilon(u)|^2$ is used to denote its average value. Thus the ARMA model leads to an angular spectrum that is the ratio of two polynomials, each of order T.

$I(u)$ represents the angular spectrum of the noisy data set and is completely specified by the ARMA parameters a_t and b_t. The least-squares estimate of the ARMA parameters can be estimated by solving a set of highly nonlinear equations: this process is computationally cumbersome [13]. In contrast, it is easy to compute the least-squares estimate of a purely autoregressive process. Thus it is highly desirable to model the data as a realization of an AR process. This is accomplished by using the equivalence between an ARMA process and an AR process of possibly infinite order, as guaranteed by the Wold decomposition theorem [13]. Therefore the ARMA process of order (T, T) in Eq. 11-12 is equivalent to an AR process of possibly infinite order. In practice, it is sufficient to truncate the infinite-order AR process to a finite model order $P \geqslant T$, especially when the SNR is large because the higher-order AR coefficients asymptote to zero. Approximating the ARMA process of Eq. 11-12 with an AR process of order $P \geqslant T$ leads to the following model for the data $\hat{e}_n$:

$$\sum_{p=0}^{P} a_p \hat{e}_{n-p} = b_0 \varepsilon_n \tag{11-15}$$

where ε_n is a zero-mean complex Gaussian random variable with unit variance and $|b_0|^2$ is called the *mean squared error*. The model of Eq. 11-15 results in the following angular spectrum for the data:

$$I(u) = \frac{|b_0|^2 \overline{|\varepsilon_n|^2}}{|\sum_{p=0}^{P} a_p e^{-j\pi u p}|^2} = \frac{|b_0|^2}{|\mathbf{a}^T \mathbf{b}(u)|^2} \tag{11-16}$$

where the last equality arises because $\overline{|\varepsilon_n|^2} = 1$, $\mathbf{a}$ is the $(P + 1)$-dimensional weight vector

$$\mathbf{a}^T = [a_0 \quad a_1 \quad \cdots \quad a_P] \tag{11-17}$$

$a_0 = 1$, $\mathbf{b}(u)$ is the $(P + 1)$-dimensional beamsteering vector

$$\mathbf{b}^T(u) = [1 \quad e^{-j\pi u} \quad e^{-j2\pi u} \quad \cdots \quad e^{-jP\pi u}] \tag{11-18}$$

and $|b_0|^2$ is the mean squared error.

Notice that the angular spectrum of Eq. 11-16 is of the same form as the all-pole spectrum of Eq. 11-8. It is completely specified by the P AR parameters and $|b_0|^2$. It is shown below that these unknown parameters can be easily computed by solving a set of linear equations in an efficient manner. The Yule–Walker equations [13] relate the unknown parameters and the correlation function of the data (either known or estimated) using the following matrix equation:

$$\mathbf{R}\mathbf{a} = |b_0|^2 \mathbf{i}_1 \tag{11-19a}$$

where $\mathbf{a}$ is the $(P + 1)$-dimensional vector in Eq. 11-17, $\mathbf{i}_1 = [1, 0, \ldots, 0]^T$ is also a $(P + 1)$-dimensional vector, and $\mathbf{R}$ is a $(P + 1) \times (P + 1)$ matrix with elements

$$R(i, j) = E\{e_i e_j^*\}; \qquad 0 \leqslant i, j \leqslant P \tag{11-19b}$$

Solving Eq. 11-19a using Gaussian elimination requires $O(P^3)$ operations ($O(P^3)$ is to be read as order P^3). However, the structure of the correlation matrix $\mathbf{R}$ can be exploited to generate more computationally efficient solutions. Note that $\mathbf{R}$ is a Hermitian ($R(i, j) = R^*(j, i)$) and positive semidefinite (determinant ($\mathbf{R}$) $= |\mathbf{R}| \geqslant 0$) matrix because it is a correlation matrix. Further, because the data $\hat{e}_n$ are realizations of a second-order stationary process,

$$R(m, m + k) = R(n, n + k), \qquad \text{all } m, n \qquad \text{or} \quad R(i, j) = R(i - j) \tag{11-20}$$

The second-order stationarity property of the data makes $\mathbf{R}$ a Toeplitz matrix (all elements along any diagonal are equal). It is therefore desirable to estimate the data correlations so that the resulting $\mathbf{R}$ is a positive semidefinite, Hermitian, and Toeplitz matrix. The Levinson algorithm [57] can be used to estimate the AR parameters when $\mathbf{R}$ is Toeplitz and Hermitian. It requires $O(P^2)$ operations, which is approximately $1/P$ times smaller than the number of operations required by Gaussian elimination. In addition to being more computationally efficient, the Levinson algorithm also provides an estimate of the model order P when it is not known *a priori*, as is usually the case.

In summary, data from a target scene consisting of T targets can be modeled as an AR process of order $P \geqslant T$. The AR parameters can be estimated by solving Eq. 11-19 or using the Levinson algorithm when the estimated data correlations have the property (11-20). What remains is the problem of estimating the data correlations $R(i, j)$. This problem is addressed in the remaining part of this section and the following two sections in the context of coherent microwave imaging.

Statistically independent snapshots of an incoherent target scenario can be obtained by illuminating the field of view at different times. Hence the data correlations $R(i, j)$ for incoherent targets can be easily estimated by averaging the statistically independent snapshots.

Unfortunately, the data correlations cannot be estimated by averaging independent snapshots for a coherent target scenario because the target echoes do not change from pulse to pulse. However, there are several techniques that can be utilized to make the coherent targets appear incoherent to the receiving array. They are described in the following two sections.

11-3 DEPENDENCE OF TARGET PHASE RELATIONSHIPS ON ARRAY ORIGIN

As mentioned in the introduction, incoherent transmitter location diversity (see Chapter 6) belongs to a class of techniques that can be used to decorrelate the radiation fields of coherent targets. Each of these techniques changes the location of the apparent array origin* from snapshot to snapshot. Changing the array origin changes the geometrical relationships between the target and array element locations. As a consequence, the phase relationships between the targets also change from snapshot to snapshot. The dependence of the phase relationships between the coherent targets on the location of the array origin can therefore be exploited to generate several independent snapshots of the coherent target scenario. This permits estimation of the data correlations and hence the AR parameters via Eq. 11-19. Thus array motion (array origin is physically changed) or target motion (array origin can be synthetically changed) can be exploited to decorrelate the echoes from coherent targets.

The dependence of the phase relationships between the targets is now quantified. These results are applied to periodic arrays in Section 11-4, where a technique called *spatial smoothing* is used to synthesize several snapshots of the data correlation from the echoes due to a single transmitted pulse.

The following terminology is defined for future use. A set of snapshots of a particular target scenario are called phase-independent if the phase relationships between the targets change from snapshot to snapshot. Thus a set of snapshots from incoherent targets are phase-independent (they are also statistically independent because of the incoherence of the targets). This and the following section develop methods for synthesizing phase-independent snapshots of coherent targets. These phase-independent snapshots are then averaged to form the data correlations $R(i, j)$.

Consider an N-element array whose elements are located at $x = x_c + x_n$, $0 \leqslant n \leqslant N - 1$, and $x_0 = 0$. The location of the array origin, that is, the location of the first element is therefore x_c. Consider the phase of the echo e_n at the nth element due to a far-field target at $u = u_t$ whose phase is ψ'_t. It is equal to

$$\operatorname{Arg}\{e_n\} = \psi'_t + ku_t(x_c + x_n) = (\psi'_t + ku_t x_c) + ku_t x_n = \psi_t + ku_t x_n \tag{11-21}$$

where $\operatorname{Arg}\{z\}$ is the argument of the complex number z. The phase of the echo can be separated into two parts: the first $\psi_t = (\psi'_t + ku_t x_c)$ is the sum of the target phase and a term that depends on the array origin x_c, and the second term depends on the array element locations relative to the array origin x_n. The first term ψ_t can be viewed as the phase of the target when the array origin is at x_c. The phase difference between two targets at $u = u_1$ and $u = u_2$ is

$$\psi_2 - \psi_1 = kx_c(u_2 - u_1) + \psi'_2 - \psi'_1 \tag{11-22}$$

The phase difference between the two targets is therefore a function of the array origin x_c. Changing the array origin is equivalent to moving the whole array along the X-axis. Thus phase-independent snapshots of the target scenario can be obtained by moving the array between transmitted pulses. Moving

* The array origin is conventionally defined as the location of the first array element.

arrays, for example, airborne phased arrays, can therefore generate several phase-independent snapshots of the target scenario in order to estimate the data correlation matrix. This result was anticipated in Chapter 5. Reference [36] describes an application where array motion is exploited to generate SR images of a coherent target scenario.

The dependence of the target phase difference on array origin location can be exploited in a subtler fashion when the array is periodic, as described in the following section.

11-4 DATA CORRELATION ESTIMATION FOR PERIODIC ARRAYS: SPATIAL SMOOTHING

This section discusses methods for estimating the elements of the correlation matrix **R** in Eq. 11-19 using the N data samples $\hat{e}_n$ collected by an N-element periodic array with interelement spacing $\lambda/2$. The datum at the nth element in the presence of additive noise is given by Eq. 11-11, which is repeated here for convenience:

$$\hat{e}_n = \sum_{t=1}^{T} A_t \, e^{j\psi_t'} \, e^{j\pi u_t n} + \text{noise} \tag{11-23}$$

where A_t, ψ_t', and u_t represent the amplitude, phase, and location of the tth target.

As mentioned in Section 11-2, the data samples $\hat{e}_n$ in Eq. 11-23 can be modeled as an AR process with model order $P \geqslant T$. The parameters of this model, a_1 through a_P, can be obtained by estimating the data correlations $R(i, j)$ and substituting these estimates into Eq. 11-19. We shall assume that a model order $P \geqslant T$ has been selected and estimate all the data correlations required for this model order. Selection of the model order is a nontrivial problem and is often the key to obtaining good SR images. Reference [66] discusses techniques for model order selection.

The data correlation matrix **R** in Eq. 11-19 is Hermitian, positive semidefinite, and Toeplitz. It thus appears that the best estimate of **R** should also possess the same three properties because the assumptions used in deriving Eq. 11-19 will be violated otherwise. However, sometimes very few samples are available to estimate $R_{i,j}$, leading to large errors in their estimated values. Under such circumstances, estimates of **R** that are not Toeplitz sometimes provide images with fewer spurious peaks and smaller errors in the target location estimates. Several methods for estimating **R** are described next.

An estimator that produces a Hermitian and Toeplitz **R** is [43]

$$\begin{aligned} R_1(i, i+k) &= \frac{1}{N-k} \sum_{p=0}^{N-1-k} \hat{e}_p e_{p+k}^*, \qquad && 0 \leqslant k \leqslant P \\ R_1(i, i+k) &= R_1^*(i, i-k); && k < 0 \end{aligned} \tag{11-24}$$

This estimator uses all the $(N - k)$ possible values of the product $\hat{e}_p\hat{e}^*_{p+k}$ in order to estimate $R(p, p + k)$, that is, it yields the best available estimate of $R(i, i + k)$ from N data samples. It is also unbiased, that is, the $R_1(i, i + k)$ asymptote to their correct values as $N \to \infty$. However, it can be shown that substitution of Eq. 11-24 into Eq. 11-19 leads to a correlation matrix that may not be positive semidefinite. This violates the positive semidefiniteness property of **R**. Experience has shown that the SR images obtained using the $R_1(i, i + k)$ have several problems associated with them, including peak splitting and biases in the target location estimates, especially when the number of data samples is small. One reason for this phenomenon is that very few data samples are available for estimating the correlation for large lags k, thereby increasing the error in those estimates. The problem can be partially alleviated by down-weighting the correlation estimates for larger lags. The following estimator for the data correlations, denoted $R_2(i, i + k)$, uses a triangular weighting function to give less importance to data correlations having large lags:

$$
\begin{aligned}
R_2(i, i + k) &= \frac{1}{N} \sum_{p=0}^{N-1-k} \hat{e}_{p+k}\hat{e}^*_p, \qquad 0 \leqslant k \leqslant P \\
R_2(i, i + k) &= R^*_2(i, i - k), \qquad k < 0
\end{aligned}
\tag{11-25}
$$

The only difference between Eqs. 11-24 and 11-25 is the normalizing factor $(= 1/(N - k)$ in Eq. 11-24 and $1/N$ in Eq. 11-25). This triangular weighting improves the noise properties of the estimates as compared to the $R_1(i, i + k)$ for nonzero lags. It can be shown that the triangular weighting also restores the positive semidefiniteness property of **R** [43]. Thus the correlation matrix obtained by substituting Eq. 11-25 into Eq. 11-19 has all the three properties of the ideal correlation matrix, that is, it is positive semidefinite, Hermitian, and Toeplitz. However, it also has some deficiencies, which are described in the following paragaph.

First, note that the triangular weighting creates a bias in the correlation estimates for $k > 0$ because only $(N - k)$ products $\hat{e}_i\hat{e}^*_{i+k}$ are averaged, whereas the normalizing factor is $1/N$. Further, the data correlations $R_2(i, i + k)$ can be considered to be obtained from a data sequence $\hat{e}_n$, $-P \leqslant n \leqslant N - 1 + P$, where the first and last P elements of the data sequence are zero. In other words, the actual data sequence of length N can be considered to be a windowed version of a larger data sequence of length $N + 2P$ while estimating the $R_2(i, i + k)$. This inherent windowing of the data decreases the resolution of the AR image and causes peak splitting [49].

Thus both $R_1(i, i + k)$ and $R_2(i, i + k)$ lead to spectra with peak splitting and biases in the target location estimates. Further, neither $R_1(i, i + k)$ nor $R_2(i, i + k)$ calculate the data correlation in the usual fashion, by averaging phase-independent snapshots of the data.

Ulrych and Clayton [122] and Nutall [71] independently suggested a method for mitigating the problems associated with the SR images obtained using the $R_1(i, i+k)$ and $R_2(i, i+k)$. This technique, called the *modified covariance method* or *forward–backward linear prediction* (FBLP) method in time series literature, finds the AR parameters a_1 through a_P by minimizing the sum of the squares of the forward and backward prediction errors, denoted $|b_0|^2$ and given by

$$|b_0|^2 = \sum_{n=p}^{N-1} |f_n|^2 + |b_n|^2 \tag{11-26a}$$

where f_n is the forward prediction error

$$f_n = \hat{e}_n + \sum_{p=1}^{P} a_p \hat{e}_{n-p} \tag{11-26b}$$

and b_n is the backward prediction error

$$b_n = \hat{e}^*_{n-P} + \sum_{p=1}^{P} a_p \hat{e}^*_{n+p-P} \tag{11-26c}$$

The minimization is performed with respect to the P AR parameters. The optimal a_p are obtained by equating the derivative of $|b_0|^2$ with respect to the real and imaginary parts of the a_p to zero and solving the resulting set of linear equations for the a_p.

Comparison of Eq. 11-26a with Eq. 11-10 shows that the forward prediction errors f_n are zero in the absence of noise. Equation 11-10 can be used to show that the b_n are also zero at infinite SNR. Therefore, determining the a_p by minimizing $|b_0|^2$ is theoretically sound because $e_n \approx -\sum a_p e_{n-p}$ and $e^*_{n-P} \approx -\sum a_p e^*_{n+p-P}$ within the limits allowed by noise.

Another approach is used to justify the FBLP method. This approach, given in the following paragraphs, is physically more meaningful to practitioners in the field of spatial signal processing. The approach is based on estimating the data correlations using phase-independent snapshots of the radiation field and uses the results developed in the previous section.

As mentioned in Section 11-3, phase-independent snapshots of the radiation field of coherent targets can be obtained by changing the location of the array origin, either synthetically or by moving the array, from pulse to pulse. Figure 11-4 shows a five-element array with interelement spacing d divided into two subarrays, each consisting of four elements. Both subarrays are identical in that they have four elements and are periodic with the same interelement spacings. However, the origin of the second subarray is displaced by d from the origin of the first subarray. The results of Section 11-3 therefore imply that the phase relationships between the targets as seen by the first subarray are different from those seen by the second subarray (see Eq. 11-22). In other words, the two data sets $\{\hat{e}_0, \ldots, \hat{e}_3\}$ and $\{\hat{e}_1, \ldots, \hat{e}_4\}$ can be considered to be two phase-independent snapshots of the target scenario. The technique of exploiting the periodic nature

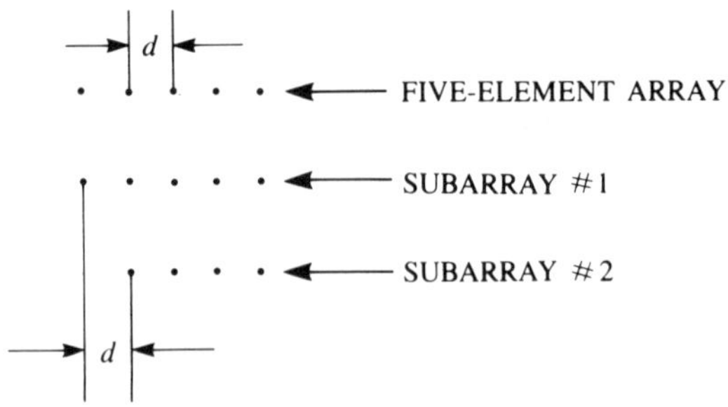

Figure 11-4 Subarray division for a periodic array.

of the array to synthesize phase-independent "snapshots" of a coherent target scenario is called *spatial* or *subarray smoothing* and was discovered in [26], [62], [87]. This principle is elucidated in the following paragraphs.

11-4-1 Spatial Smoothing

An N-element periodic array can be divided into $(N - P)$ subarrays, each consisting of $(P + 1)$ elements. The pth subarray consists of the elements at $x = x_p$ through $x = x_{p+P}$, $0 \leqslant p \leqslant N - P - 1$. Each of the $(N - P)$ subarrays can be treated as a $(P + 1)$-element periodic array, with the array origin at $x = x_p$. The phase relationships between the targets change from subarray to subarray because the origins of the subarrays are at different locations, as shown below.

The nth element of the pth subarray is at $x = (n + p)\lambda/2$, $0 \leqslant n \leqslant P$. The field at the nth element of the pth subarray, denoted $\hat{e}_n(p)$, is therefore

$$\hat{e}_n(p) = \hat{e}_{n+p} \approx \sum_{t=1}^{T} A_t\, e^{j(\psi'_t + \pi p u_t)}\, e^{j\pi n u_t}; \qquad 0 \leqslant n \leqslant P, \quad 0 \leqslant p \leqslant N - P - 1 \tag{11-27}$$

where the approximation holds to within the limits allowed by noise. The first exponential term in Eq. 11-27, $(\psi'_t + \pi p u_t)$, represents the phase of the tth target as seen by the pth subarray. Let this quantity be denoted $\phi_t(p)$; that is, let

$$\phi_t(p) = \psi'_t + \pi p u_t \tag{11-28}$$

The phase difference between the rth and tth targets, as seen by the pth subarray, is

$$\phi_r(p) - \phi_t(p) = \psi'_r - \psi'_t + \pi p(u_r - u_t) \tag{11-29}$$

Notice that the phase difference between the targets depends on the subarray index p, that is, it changes from subarray to subarray. Thus one could view the data at the $(N - P)$ subarrays as phase-independent snapshots of the data obtained using a $(P + 1)$-element array. The correlation between the kth and lth elements of an array consisting of $(P + 1)$ elements, denoted $R_3(k, l)$, can be estimated by averaging the $(N - P)$ snapshots as

$$R_3(k, l) = \frac{1}{N - P} \sum_{p=0}^{N-P-1} \hat{e}_{p+k} \hat{e}^*_{p+l}; \qquad 0 \leqslant k, l \leqslant P \tag{11-30}$$

The correlation estimates $R_3(k, l)$ can be substituted into Eq. 11-19 in order to compute the AR parameters. Minimizing the forward prediction error $\sum |f_n|^2$ in Eq. 11-26b leads to the correlation estimates that are identical to Eq. 11-30. The covariance method [63] forms the data correlations using Eq. 11-30.

Observe that $R_3(k, l)$ is an unbiased estimator of the $R(k, l)$, that is, the $R_3(k, l)$ asymptote to their correct values as $N \to \infty$. It can easily be shown that estimating the $R_3(k, l)$ using Eq. 11-30 leads to a positive semidefinite correlation matrix $\mathbf{R}$ in Eq. 11-19 as long as $(N - P) > (P + 1)$. The disadvantage of estimating $R(k, l)$ according to Eq. 11-30 is the loss in spatial stationarity of the correlation estimates, that is, $R_3(i, i - k) \neq R_3(j, j - k)$ for $i \neq j$. This leads to a correlation matrix estimate that is not Toeplitz, whereas the ideal correlation matrix $\mathbf{R}$ is Toeplitz.

The SR images obtained by substituting the $R_3(k, l)$ into Eq. 11-19 also exhibit spurious peaks, biases in the target location estimates, and peak splitting [49]. This is because the AR parameters are estimated by minimizing only the sum of the forward prediction errors. The image quality can be further improved by minimizing the sum of the forward and backward prediction errors with respect to the unknown AR parameters. It is shown below that backward prediction techniques can be used to generate $(N - P)$ more phase-independent snapshots of the target scenario. The additional snapshots decrease the errors in the estimated data correlations, thereby improving image quality. They are obtained by reversing the order of the old data samples and conjugating them. Let $\hat{e}'_n$, $0 \leqslant n \leqslant N - 1$, denote the new set of data samples formed:

$$\hat{e}'_n = \hat{e}^*_{N-n-1}; \qquad 0 \leqslant n \leqslant N - 1 \tag{11-31}$$

Use of Eq. 11-23 in Eq. 11-31 results in

$$\hat{e}'_n \approx \sum_{t=1}^{T} A_t \, e^{-j\psi'_t} \, e^{-j\pi u_t (N-n-1)} \tag{11-32}$$

where the approximation holds to within the limits allowed by noise. The new data samples can also be divided into $(N - P)$ subarrays, each consisting of $(P + 1)$ elements. The datum at the nth element of the pth subarray, denoted

$\hat{e}'_n(p)$, is

$$\hat{e}'_n(p) = \hat{e}'_{n+p} = \sum_{t=1}^{T} A_t \, e^{-j[\psi'_t + \pi(N-1-p)u_t]} \, e^{j\pi n u_t} \tag{11-33}$$

Comparison of Eq. 11-33 with Eq. 11-27 shows that the Fourier kernel $e^{j\pi nu}$ stays the same after the data samples are conjugated and reversed in order. However, the phase of the tth target, as seen by the nth element of the pth subarray, denoted $\phi'_t(p)$, does change for the data sample set $\hat{e}'_n$. It is equal to the sum of all the exponential terms in Eq. 11-33 that do not depend on n. That is,

$$\phi'_t(p) = -\psi'_t - \pi(N-1-p)u_t \tag{11-34}$$

The phase difference between the rth and tth targets is

$$\phi'_r(p) - \phi'_t(p) = -\psi'_r + \psi'_t - \pi(N-1-p)(u_r - u_t) \tag{11-35}$$

Thus the phase relationships between the targets depend on the subarray index p, that is, they change from subarray to subarray. Further, comparison of Eq. 11-35 with Eq. 11-29 shows that the phase relationships between the targets in the new set of subarrays are different from their phase relationships in the old set. In fact, the phase difference between the targets changes its sign for the new set of subarrays. Hence each of the new $(N-P)$ snapshots obtained using the $\hat{e}'_n$ are phase-independent of $(N-P)$ snapshots obtained using the $\hat{e}_n$. A total of $2(N-P)$ phase-independent snapshots are therefore obtained by using both the $\hat{e}_n$ and the $\hat{e}'_n$.

The correlation between the kth and lth elements of an array consisting of $(P+1)$ elements, denoted $R_4(k, l)$, can be estimated by averaging the $2(N-P)$ snapshots as

$$\begin{aligned} R_4(k,l) &= \frac{1}{2(N-P)} \sum_{p=0}^{N-P-1} (\hat{e}_{p+k}\hat{e}^*_{p+l}) + (\hat{e}'_{p+k}\hat{e}'^*_{p+l}) \\ &= \frac{1}{2(N-P)} \sum_{p=0}^{N-P-1} (\hat{e}_{p+k}\hat{e}^*_{p+l}) + (\hat{e}^*_{N-p-1-k}\hat{e}_{N-1-p-l}) \\ &= \frac{1}{2(N-P)} \sum_{p=0}^{N-P-1} (\hat{e}_{p+k}\hat{e}^*_{p+l}) + (\hat{e}_{P+p-l}\hat{e}^*_{P+p-k}); \qquad 0 \leqslant k, l \leqslant P \end{aligned} \tag{11-36}$$

where Eq. 11-31 was used for the $\hat{e}'_n$ in addition to a change in the summation index for the sum over the products $\hat{e}'_m\hat{e}'^*_n$.

An upper bound on the model order P required to generate a positive definite $\mathbf{R}$ is now established. The estimate of $\mathbf{R}$ is positive definite when the number of snapshots $2(N-P) > (P+1)$, that is, $P \leqslant (2N-1)/3$. Thus the model order should lie between the number of targets T and $(2N-1)/3$ for an N-element array.

The procedure for forming the data correlation matrix in this fashion and solving for the a_p and $|b_0|^2$ using Eq. 11-19 is called the *modified covariance method* or the *forward–backward linear prediction* (FBLP) method in time series literature [71], [122]. The superresolved image obtained using the FBLP method, denoted $I_1(u)$, is

$$I_1(u) = \frac{|b_0|^2}{|\mathbf{a}^T \mathbf{b}(u)|^2} \tag{11-37}$$

where the weights $\mathbf{a}$ and $|b_0|^2$ are obtained by substituting Eq. 11-36 into Eq. 11-19 and solving the resulting set of linear equations, and $\mathbf{b}(u)$ is the beamsteering vector defined by Eq. 11-18.

In [122], it is shown that image $I_1(u)$ is less sensitive to the phase differences $(\psi_r' - \psi_t')$ between the targets as compared to the SR images obtained when either of $R_3(k, l)$, $R_2(i, i+k)$, or $R_1(i, i+k)$ is used to form $\mathbf{R}$. Further, the variance of the error in the estimated target locations obtained using $R_4(k, l)$ is smaller than the variances obtained when either of R_3, R_2, or R_1 is used [53]. Thus the R_4 lead to the best SR images among the four correlation estimators discussed in this section.

Note that the $R_4(k, l)$ do not produce a Toeplitz estimate of $\mathbf{R}$, whereas the ideal $\mathbf{R}$ is Toeplitz. Thus the computationally efficient Levinson–Durbin algorithm cannot be used to solve for the AR parameters when $R_4(k, l)$ are used in Eq. 11-19. However, an alternative algorithm that also requires only $O(P^2)$ operations, is described in [65]. Unlike the Levinson algorithm, this algorithm does not automatically determine the model order. Hence other methods for selecting P should be used before forming the correlation matrix using the $R_4(k, l)$. Several criteria for selecting the model order, such as the final prediction error, the Akaike information criterion, and the criterion AR transfer function, are described in [66]. In [122], it is shown that none of these criteria work well when the data segment is short. Experience has shown that model order selection with experimental data requires more subjective judgment as compared to data generated using controlled computer simulations. The effect of model order on experimental SR images is shown in Section 11-8.

Finally, note that only the estimate $R_2(i, i+k)$ produces a correlation matrix that possess all the three properties of the ideal $\mathbf{R}$, that is, the estimate of $\mathbf{R}$ is Hermitian, Toeplitz, and positive definite. However, the image obtained using the $R_2(i, i+k)$ is inferior to the image obtained using the $R_4(k, l)$, which do not produce a Toeplitz $\mathbf{R}$. Much work has been conducted on producing estimates of $\mathbf{R}$ that possess all three properties of the ideal correlation matrix and also yield good SR images. The method of Burg et al. [16] provides the maximum likelihood estimate of $\mathbf{R}$ that is consistent with theory when the data are statistically independent samples of a multivariate Gaussian process. The spectra shown in [16] appear to be of very high quality.

This completes the discussion of spatial smoothing and the FBLP method. A few other superresolution techniques are briefly described in the following section.

11-5 OTHER SUPERRESOLUTION TECHNIQUES

Several other model-based methods for superresolution are described in [50]. Most of them are based on all-pole models. The maximum entropy method [15], which maximizes the entropy of the measured data set, also results in an all-pole estimator. Two techniques that are not model-based are the maximum likelihood method (MLM) [18] and MUSIC [84]. They are now briefly described.

MLM maximizes the signal-to-interference (noise + sidelobe energy) ratio for every look direction under the constraint that the array gain is unity at the look direction. The resulting spectral estimator for model order P, denoted $I_2(u)$, is

$$I_2(u) = \frac{c_0^2}{\mathbf{b}^\dagger(u)\mathbf{R}^{-1}\mathbf{b}(u)} \tag{11-38}$$

where c_0 is a constant, $\mathbf{b}(u)$ is defined by Eq. 11-18 and $\mathbf{R}$ is the correlation matrix in Eq. 11-19.

The MLM spectrum of order P is equal to the parallel resistor average of the FBLP spectra of orders zero through P [66]. This leads to the conclusion that the MLM spectra have poorer resolution than linear predictive spectra.

The MUSIC algorithm is a vector space approach to spectrum estimation. A data sequence consisting of N samples can be expressed as a weighted sum of N orthogonal vectors, each consisting of N elements. Let the vector $\mathbf{e}$ represent the echoes received from T targets, that is, let

$$\mathbf{e} = [e_0 \quad e_1 \quad \cdots \quad e_{N-1}]^T \tag{11-39}$$

where the e_n are defined by Eq. 11-9 in the absence of noise. From Eq. 11-9, it is easy to show that

$$\mathbf{e} = \sum_{t=1}^{T} A_t\, e^{j\psi_t} \mathbf{b}^*(u_t) \tag{11-40}$$

where A_t, ψ_t, and u_t are the target amplitudes, phases, and locations, respectively, $\mathbf{b}(u)$ is the beamsteering vector

$$\mathbf{b}(u) = [e^{-jkux_0} \quad e^{-jkux_1} \quad \cdots \quad e^{-jkux_{N-1}}]^T \tag{11-41}$$

and the x_n are the element locations. Thus the noiseless data set from T targets can be expressed as a weighted sum of the T linearly independent vectors $\mathbf{b}^*(u_t)$. In other words, the noiseless data set spans a subspace of dimension T in the N-dimensional vector space. Hence it can be expressed as a weighted sum of T orthonormal vectors.

The received data set spans the whole N-dimensional space in the presence of additive noise. It can therefore be represented as a weighted sum of N

orthogonal vectors, of which only T are required to span the signal subspace. The remaining $(N - T)$ vectors span the noise subspace and are orthogonal to the signal subspace. Let $\mathbf{q}_n$, $1 \leqslant n \leqslant N$, represent N orthonormal vectors that span the N-dimensional space occupied by the received data. Let the first T vectors $\mathbf{q}_t$, $1 \leqslant t \leqslant T$, span the signal subspace. Then the noiseless data can be represented as

$$\mathbf{e} = \sum_{t=1}^{T} w_t \mathbf{q}_t \tag{11-42}$$

where the w_t are the weights associated with the vectors $\mathbf{q}_t$. Thus the noiseless data vector $\mathbf{e}$ is a weighted sum of the signal subspace eigenvectors $\mathbf{q}_t$, $1 \leqslant t \leqslant T$. Each of the signal subspace eigenvectors is orthogonal to all the noise subspace eigenvectors $\mathbf{q}_n$, $T + 1 \leqslant n \leqslant N$. Therefore the noiseless data vector $\mathbf{e}$ is also orthogonal to all the noise subspace eigenvectors, that is,

$$\mathbf{e}^\dagger \mathbf{q}_n = \left(\sum_{t=1}^{T} w_t \mathbf{q}_t^\dagger \right) \mathbf{q}_n = 0 \qquad \text{for } n > T$$

Combining this result with Eq. 11-40 yields

$$\mathbf{e}^\dagger \mathbf{q}_n = \sum_{t=1}^{T} A_t e^{j\psi_t} \mathbf{b}^T(u_t) \mathbf{q}_n = 0; \qquad n > T \tag{11-43}$$

Clearly, Eq. 11-43 should hold for all values of the target phases ψ_t. Because both $\mathbf{b}(u_t)$ and $\mathbf{q}_n$ are independent of the target phases, $\mathbf{e}^\dagger \mathbf{q}_n = 0$ implies that $\mathbf{b}^T(u_t)\mathbf{q}_n = 0$, $1 \leqslant t \leqslant T$, $n > T$. In other words, the inner product of the noise subspace vectors with the beamsteering vectors steered to the target locations is zero. The reciprocal of this quantity is infinite. MUSIC forms a spectral estimator that has the quantities $|\mathbf{b}^T(u)\mathbf{q}_n|^2$, $T + 1 \leqslant n \leqslant N$ in its denominator, thereby creating poles at the target locations. The spectral estimator formed by the MUSIC algorithm, denoted $I_3(u)$, is

$$I_3(u) = \frac{c_0^2}{\sum_{n=T+1}^{N} |\mathbf{b}^T(u)\mathbf{q}_n|^2} \tag{11-44}$$

where $\mathbf{b}(u)$ is the beamsteering vector defined by Eq. 11-42, c_0 is a constant, and the $\mathbf{q}_n$, $T + 1 \leqslant n \leqslant N$ are the $(N - T)$ eigenvectors that span the noise subspace. Note that the denominator of Eq. 11-44 is zero for $u = u_t$, all t. The set of N orthonormal vectors $\mathbf{q}_n$ can be estimated by performing an eigenvalue–eigenvector decomposition of the correlation matrix $\mathbf{R}$. At high SNR, the eigenvectors corresponding to the large eigenvalues can be associated with the signal subspace vectors and the remaining eigenvectors with the noise subspace vectors [84]. Thus, the vectors that span the signal and noise subspaces can be distinguished according to the strengths of their eigenvalues. The vectors corresponding to the noise subspace are used to form the MUSIC spectral estimator $I_3(u)$.

Three classes of superresolution techniques, viz. the FBLP method, MLM, and MUSIC were described in the last two sections. The linear predictive techniques model the data samples as realizations of autoregressive processes. MLM maximizes the signal-to-interference ratio in each look direction under the constraint that the array gain is unity in the look direction. MUSIC uses a vector space approach to estimate the angular spectrum. All three techniques require estimates of the data correlation matrix $\mathbf{R}$. These estimates can be obtained using transmitter location diversity, spatial smoothing, or by exploiting array or target motion.

The superresolution techniques do not require the targets to be located in the far field of the array. However, the following section shows that spatial smoothing requires the targets to be located in the far field of the array. A method for transforming the measured near-field data to the far field is also discussed. This transformation will permit the use of spatial smoothing on near-field data also.

11-6 NEAR-FIELD CORRECTION USING ADAPTIVE BEAMFORMING

Spatial smoothing exploits the variation of the phase relationships between coherent targets with the array origin location in order to synthesize several phase-independent snapshots of the target scenario from the returns due to a single transmitted pulse. It requires the receiving array to be periodic. It is shown below that spatial smoothing also requires that the targets be located in the far field of the periodic receiving array. A near-field to far-field transformation can be effected using the adaptive beamforming techniques described in Chapters 8 and 9. Modifying the data using beamforming therefore makes it conform to the requirements of spatial smoothing. It also compensates for phase errors in the data and errors due to element-position uncertainty.

The following discussion shows that the quadratic curvature in the phasefront of near-field targets introduces errors in the correlation estimates obtained by averaging the data from the various subarrays. Consider a subarray having origin x_c. Let the targets lie in the near field of the whole array. The quadratic part of the phase variation across the whole array is approximately $kx^2/2R$ when the Fresnel approximation of Eq. 2-2 is used, that is, when the field of view is narrow. This translates to $k(x + x_c)^2/2R$ for the subarray, where x is measured relative to the subarray origin x_c. The cross-products between the mth and nth elements of the subarray will consist of the quadratic term $[(x_m + x_c)^2 - (x_n + x_c)^2]/2R$ in the exponential kernel. The quadratic variation in the kernel changes from subarray to subarray because it depends on the subarray origin x_c. Consequently, data cross-products from different subarrays cannot be averaged to estimate the data correlations because of the variation in the Fresnel kernel. In contrast, the Fourier kernel $ku(x_m - x_n)$ does not depend on subarray origin, thereby permitting the data cross-products from different subarrays to be averaged to form the correlation matrix. Adaptive beamforming (ABF) can remedy the problem in the following manner.

As described in Chapters 8 and 9, ABF transforms near-field target scenarios to the far field by focusing the array to the range bin being examined. It can be used to modify the data before estimating the data correlations. The ABF correction typically consists of subtracting a phase, corresponding to the distance between the array element and the center of the field of view, from the phase of the measured data.* The phase correction for the array element at $(x_n + x_c)$ is $k(x_n + x_c)^2/(2R)$ when the center of the field of view is at broadside $(u = 0)$. Removing this phase from the data modifies the exponential kernel at the nth element to kux_n, that is, to the Fourier kernel. The modified data set therefore corresponds to far-field targets. The methods of Section 11-4 can hence be applied to estimate the data correlations using the modified data set. Thus ABF has made the data compatible with the requirements of spatial smoothing.

The following section discusses practical applications of SR techniques to coherent microwave imaging.

11-7 PRACTICAL APPLICATIONS OF SUPERRESOLUTION TECHNIQUES IN COHERENT MICROWAVE IMAGING

The echoes from coherent targets need to be decorrelated before SR techniques can be used to enhance image resolution. The key to decorrelating the coherent target echoes is to obtain several snapshots of the coherent target scenario, each snapshot corresponding to a different array origin location. Four techniques for decorrelating the echoes from coherent targets were described in Sections 11-3 and 11-4. They use either transmitter-location diversity, array motion, target motion, or subarray smoothing to generate several snapshots of the target scenario, with the phase relationships between the targets changing from snapshot to snapshot. Consequently, data from these snapshots can be averaged to estimate the data correlations. These estimated data correlations can then be used to form an AR image of the target scenario. The types of systems on which the decorrelating techniques can be implemented are described in Table 11-1.

Note that the first three techniques require multiple snapshots of the target scenario. In contrast, the data for subarray smoothing techniques, which require periodic receiving arrays, can be collected from a single transmitted pulse. Thus subarray smoothing is the recommended technique for decorrelating echoes from coherent targets when the array is periodic. The advantage of the remaining techniques is that they work equally well with periodic or aperiodic arrays.

* ABF also compensates for phase and element position errors; however, the error-compensation property of ABF is not relevant to this discussion and hence is ignored.

TABLE 11-1 Techniques for decorrelating echoes from coherent targets

Decorrelation Technique	*Comment*
Transmitter location diversity	Works with either periodic or aperiodic arrays. Requires at least $P+1$ snapshots when array is aperiodic. Also requires multiple transmitters
Exploiting target motion	Same as above except there is no need for multiple transmitters
Exploiting array motion	Same as above
Subarray smoothing	Requires periodic arrays. Does not require multiple snapshots of target scenario

The improvement in resolution obtained using SR techniques can be quantified using the following improvement factor I:

$$I = \frac{\text{resolution in Fourier image}}{\text{resolution in SR image}} \tag{11-45}$$

In other words, the SR image can distinguish between targets separated by more than $1/I$ times the Fourier resolution.

The improvement factor can be contrasted for two extreme imaging conditions. In the first case, the lengths of the apertures used for forming both the Fourier and SR images are equal. Such a situation arises when the data from the various snapshots cannot be combined coherently in order to form a Fourier image from a larger aperture. For example, assume that transmitter location diversity or target motion are used for generating the various snapshots. Further, assume that the transmitter locations (or the target trajectory) are not known accurately. In this case, it may not be possible to coherently combine the data from the various snapshots to form a Fourier image. Thus the SR and Fourier images are compared on an equal data set basis in this imaging situation.

In the second imaging condition, the length of the aperture used for forming the SR image is smaller than the length of the Fourier aperture. Such a situation arises when subarray smoothing is used to decorrelate the coherent targets. The length of the SR aperture is proportional to the model order P, whereas the length of the Fourier aperture is proportional to the number of array elements $N > P$. Another example of this imaging situation arises when the snapshots obtained using either transmitter-location diversity or array or target motion can be combined coherently. Then, the length of the aperture used for synthesizing the Fourier image via coherent diversity will be larger than the length of the SR aperture.

The improvement factor is largest when the length of the SR and Fourier apertures are equal and is smallest when all the snapshots are combined coherently to form the Fourier image. In practice, the value of I will lie between these two extreme values depending on the manner in which the received data are processed to form the Fourier image.

The following section examines the resolution capabilities of SR techniques using experimental microwave data.

11-8 EXPERIMENTAL COMPARISON OF SUPERRESOLUTION AND FOURIER IMAGES

Experimental data obtained using ISAR techniques will be employed as a basis of comparison between SR and Fourier imaging. As mentioned in the previous section, the comparison will be made under two extreme imaging conditions: (1) comparison on an equal subaperture basis, and (2) comparison on an equal data set basis. In addition, image quality is also examined subjectively as a function of model order. The data are taken from [44].

11-8-1 Subaperture-Based Comparison

Subaperture-based comparison contrasts the SR and Fourier images obtained using apertures of the same size. Comparison on an equal aperture size basis is justified when the various snapshots cannot be combined coherently. This happens when the transmitter locations are not known accurately while using transmitter location diversity or moving arrays, or when the target location is not known accurately at the time data are received from moving targets.

Consider a periodic array with N elements. Let P denote the model order. There are $2(N - P)$ independent snapshots for a model order P. These snapshots can be used to form the data correlation matrix, and an SR image can be formed with the weights calculated using the data correlations. In addition, the squared magnitudes of the Fourier images obtained using each subaperture (or subapertures that are disjoint) can be averaged to form a noncoherent diversity image. Subaperture-based comparison contrasts these two images. Clearly, since both images are obtained using apertures of the same size, one would expect the SR images to exhibit higher resolution. This is verified in the following microwave experiment.

A record of 160 echoes from a Boeing 727 aircraft, obtained using ISAR techniques, served as the data base. The aircraft was traveling at a speed of approximately 270 mph (120 meters per second) and was 3 km away from the radar. The interpulse period was 2.5 msec and the aircraft traveled 48 m during the observation period. Thus the length of the synthesized aperture was 48 m. The synthetic aperture beamwidth is approximately $\lambda/2L$ ($\lambda = 3$ cm). Thus the cross-range resolution cell size is approximately 1 m at a distance of 3 km from the radar.

Figure 11-5 shows three images of a Boeing 727 aircraft. Figure 11-5a is a

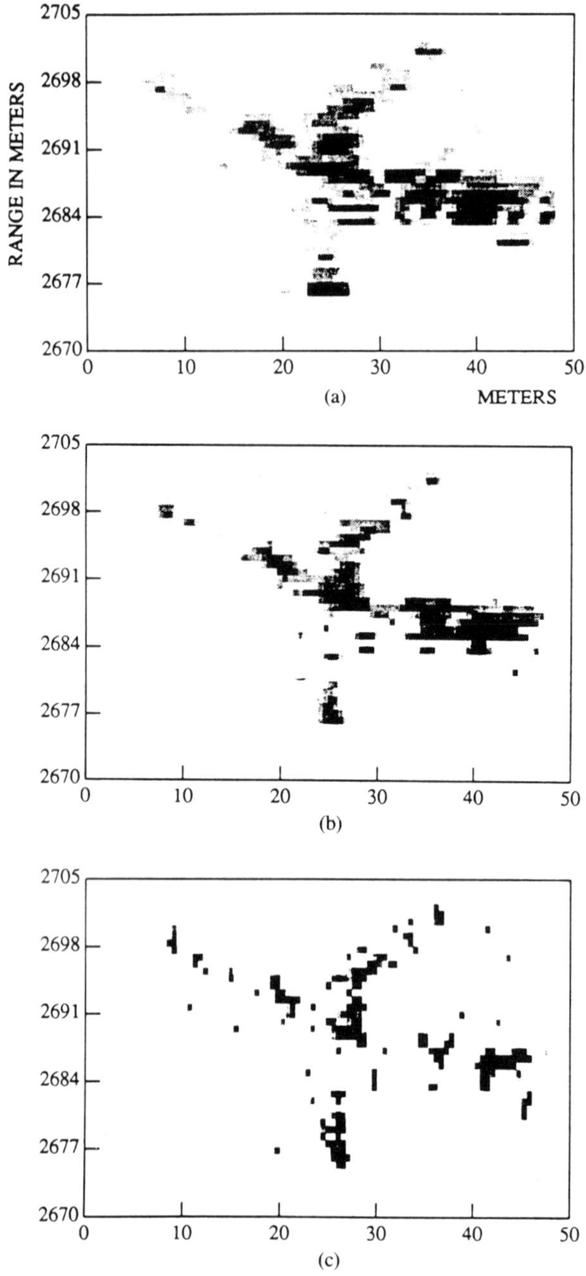

Figure 11-5 Comparison of 2-D Fourier and SR images on an equal aperture size basis. The data set consists of 160 samples obtained using ISAR techniques. (a) Quadruple-diversity image obtained by averaging four Fourier images. (b) MLM image (model order = 40). (c) FBLP image (model order = 40). (From ref. [44])

Fourier image; the others are SR images of model order 40. The aperture consisted of 160 equally spaced elements; consequently the subaperture is one-fourth the size of the full aperture. Figures 11-5b and 11-5c are the MLM and FBLP images. The Fourier image was obtained by first breaking the data set into four separate parts, each of 40 samples, forming four Fourier images and averaging their moduli. Thus Fig. 11-5a is a quadruple-diversity Fourier image. It is diffraction limited, the cross-range resolution being about 4 m at the target range of 3 km. The airplane is about 50 m long. The resolution cell is sufficiently small that the target is easily identified. The MLM image also shows the airplane shape: it is very similar to the Fourier image, the major difference being a reduced cross-range cell size by about a factor of 2. Thus the improvement factor I for the MLM (see Eq. 11-45) is approximately 2 when the Fourier and SR images use the same aperture size.

The FBLP image shows higher resolution than either of the two methods. The cross-range cell size of the FBLP image is approximately 4 times smaller than the cross-range cell size of the Fourier image, that is, $I = 4$ in. (11-45). Some separate targets that are indistinguishable in the Fourier and MLM images are resolved in the FBLP image. However, some false targets are introduced due to the large model order and the low SNR per element.

The differences between the images are more clearly discerned in the 1-dimensional slices of the 2-dimensional images. Figure 11-6 shows the Fourier, MLM, and FBLP images of two range bins: the first corresponds to the aircraft fuselage and the second to the wing. The superior resolving capability of the FBLP method compared to the other two is evident.

11-8-2 Comparison on an Equal Data Set Basis

This comparison contrasts the Fourier image obtained by coherently combining the data from the various independent snapshots with the SR images. It assumes that the data from the various snapshots can be combined coherently. This happens when the array is periodic (subarray smoothing is used to estimate the data correlations) or when the transmitter/target locations are known accurately for each snapshot.

The improvement in resolution obtained via SR is much less evident when this type of comparison is made because the aperture length for the Fourier image is N/P times larger than the aperture length for the SR images. Figure 11-7 shows the 2-dimensional Fourier image when the entire data set consisting of 160 samples is processed coherently. Comparison with Figs. 11-5b and 11-5c shows that this image exhibits higher quality than both the MLM and FBLP images obtained using a model order of 40. The Fourier image is not sensitive to noise and its resolution is inversely proportional to array length. The SR techniques are sensitive to noise, and use smaller aperture sizes in order to obtain stable estimates of the data correlation. The spurious peaks in the FBLP image of Fig. 11-5c arise due to the low SNR per element in the data. Comparing Fig. 11-5c with Fig. 11-7 shows that the cross-range resolution is roughly the

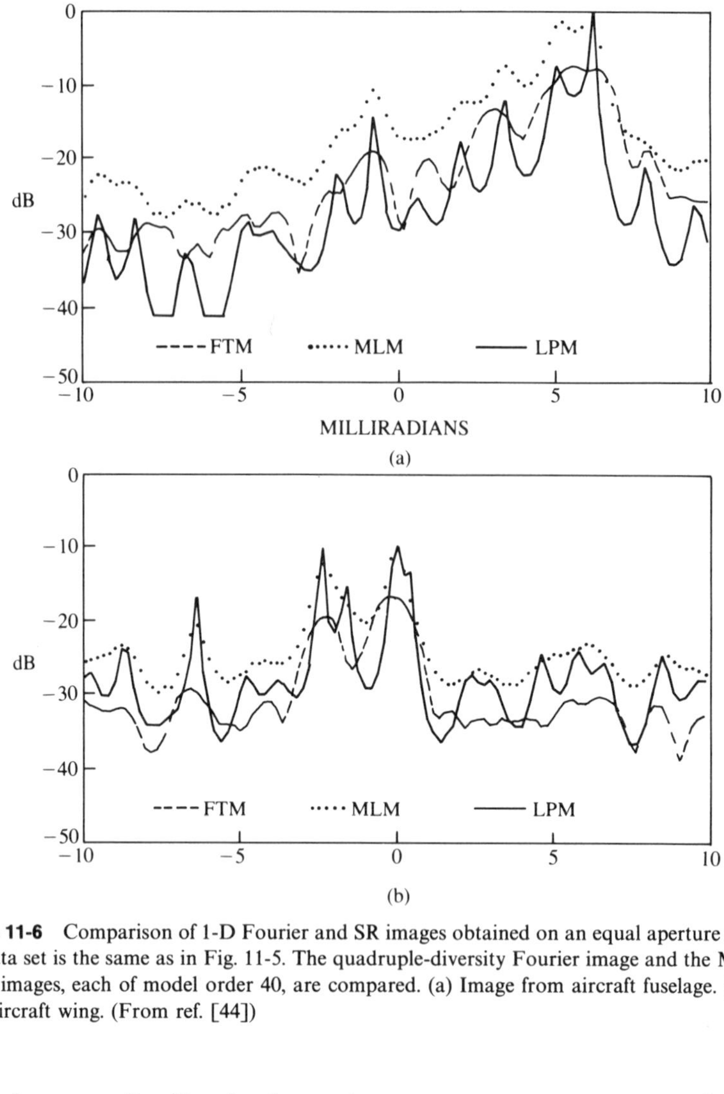

Figure 11-6 Comparison of 1-D Fourier and SR images obtained on an equal aperture size basis. The data set is the same as in Fig. 11-5. The quadruple-diversity Fourier image and the MLM and FBLP images, each of model order 40, are compared. (a) Image from aircraft fuselage. (b) Image from aircraft wing. (From ref. [44])

same; however, the Fourier image has greater target mass, superior shape definition, and less noise. Because the cross-range resolution of the FBLP image with model order 40 is approximately the same as the Fourier image with 160 data samples, the improvement factor I is unity when the Fourier and FBLP images are compared on an equal data set basis.

11-8-3 Other Results

The remaining figures document imaging results of the three methods for different model orders and aperture sizes. Figures 11-8a and b show MLM and FBLP images of model order 16. The images show lower resolution and missing

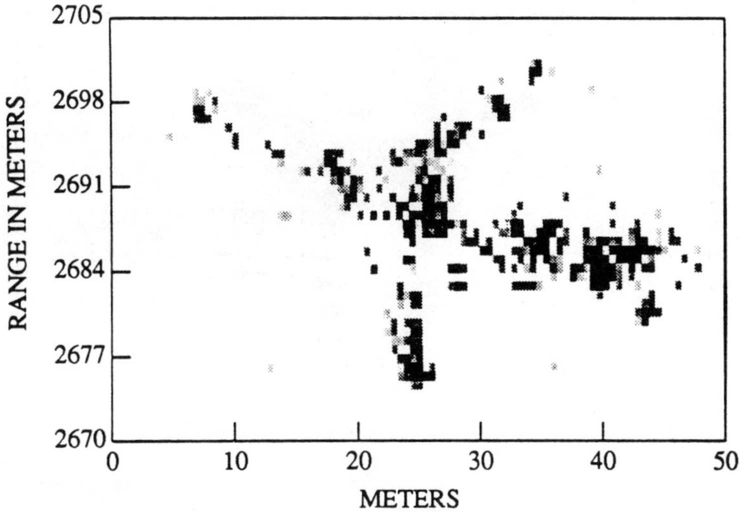

Figure 11-7 Fourier image obtained using 160 data samples. (From ref. [44])

targets and in general look inferior to the images of Figs. 11-5b and 11-5c. These characteristics result directly from the choice of lower model order.

The effect of reduced smoothing of the correlation coefficients is seen in Fig. 11-9. The model order is 40, as in Fig. 11-5, but the size of the data set is 80. The resolution cell size is unchanged because the model order is the same. The dual-diversity Fourier image (Fig. 11-9a) is little different from the quadruple-diversity Fourier image of Fig. 11-5a. However, the SR images (Figs. 11-9b and c) suffer badly from data truncation. Halving the data set from 160 to 80 while retaining the same model order reduces the spatial smoothing correspondingly. The result is increased variance in the data correlation estimates and poorer imaging. In neither image is an airplane recognizable.

The effect of reduced model order on SR imagery is further illustrated by the images of Fig. 11-10, which show 1-dimensional profiles of the aircraft fuselage. Figure 11-10a is a quadruple-diversity Fourier image with model order 40 and Fig. 11-10b is a single Fourier image (model order = 160). Figures 11-10c and d are MLM images of model orders 16 and 40, respectively. The effect of reducing the model order can be observed by comparing Figs. 11-10c and d. Figures 11-10e and f are FBLP images with model orders 16 and 40, respectively. Again the effect of the reduced model order is evident. FBLP produces superior images compared to MLM; however, the Fourier image (model order = 160) appears to be the best of the six images in Fig. 11-10. Figure 11-11 is a similar set of 1-dimensional images of the aircraft wing and the observations are similar.

The results show that the model order must be tailored to the specific data set being used. A low model order results in loss of targets and resolution.

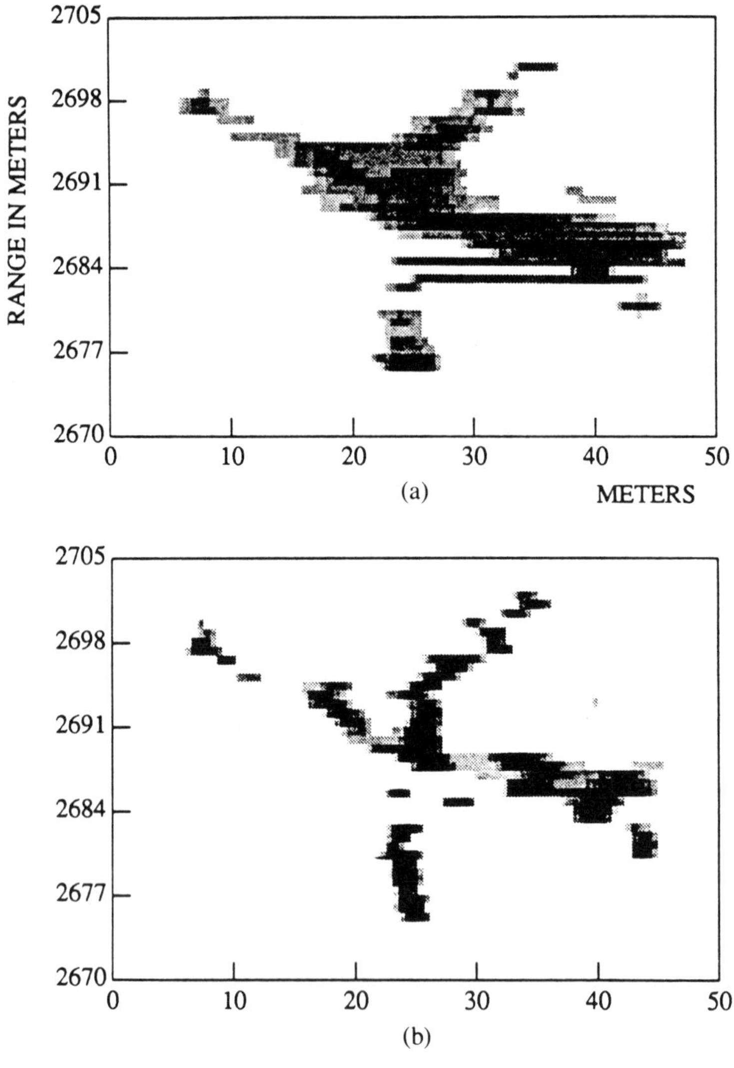

Figure 11-8 Effect of reduced model order on SR techniques. (a) MLM (model order = 16). (b) FBLP (model order = 16). (From ref. [44])

However, the model order cannot be raised arbitrarily because the effective SNR of the data correlation estimates (proportional to the amount of spatial smoothing) decreases with increasing model order for a fixed sample size. Thus the model order and the SNR are parameters that interact in their effects upon image quality. Furthermore, spurious peaks appear at very large model orders degrading image quality.

The most interesting overall observation is that the Fourier image obtained using the full data set (Fig. 11-7) produced the best image.

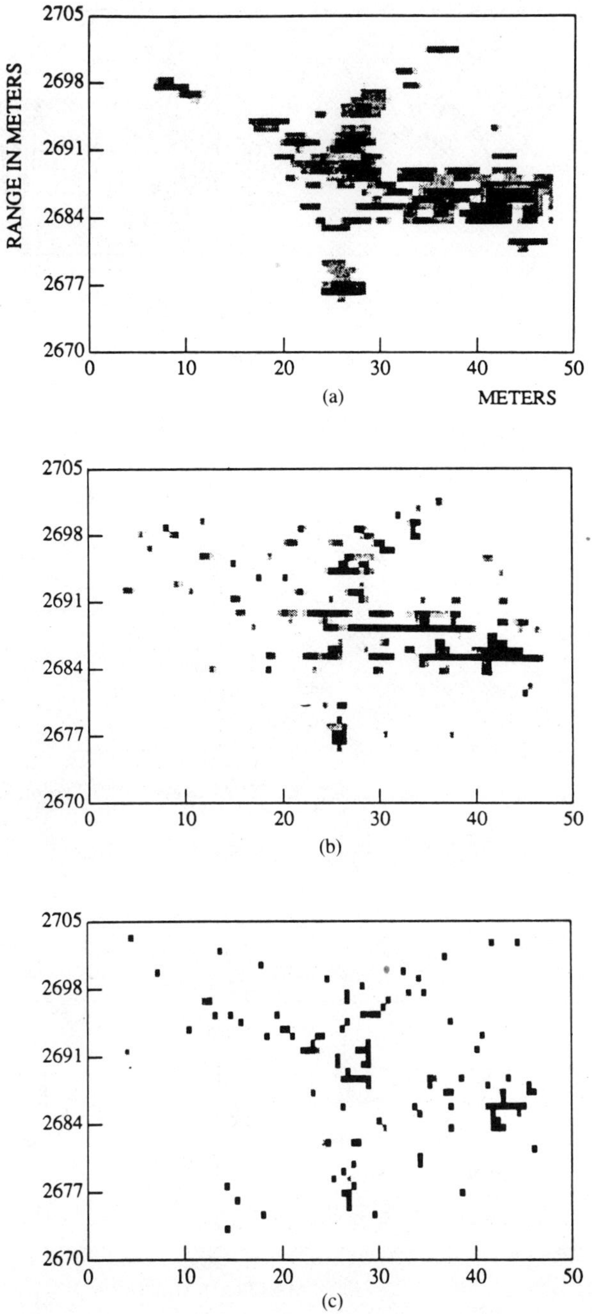

Figure 11-9 Effect of reduced number of data samples on Fourier and SR techniques. The data set of Fig. 11-5 is reduced from 160 to 80 and the same model order is retained. (a) Dual-diversity Fourier image (40 samples per image). (b) MLM image (model order = 40). (c) FBLP image (model order = 40). (From ref. [44])

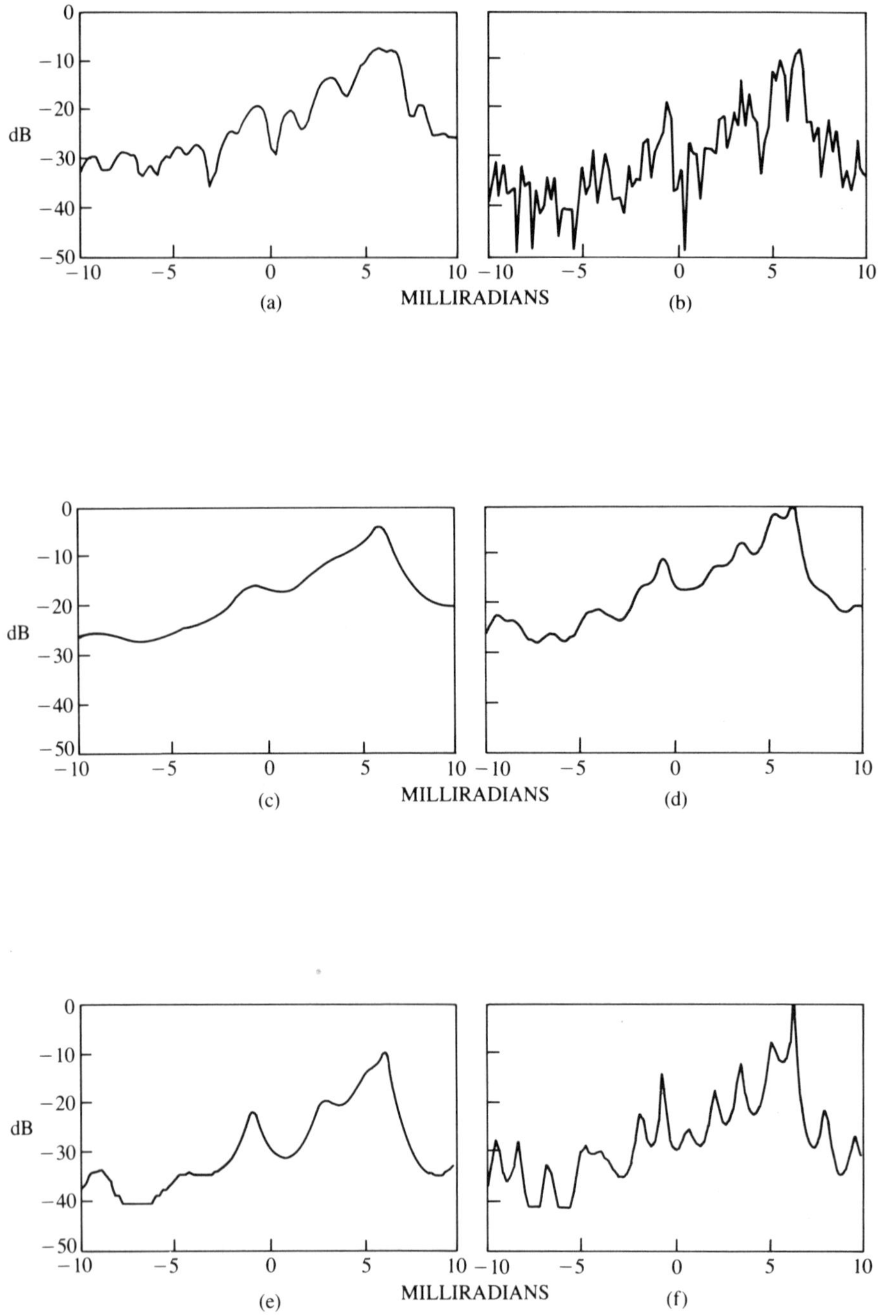

Figure 11-10 Effect of reduced model order on SR techniques. The data set consists of 160 samples of the echoes from the aircraft fuselage. (a) Quadruple-diversity Fourier image (model order = 40). (b) Fourier image with 160 data samples. (c) MLM (model order = 16). (d) MLM (model order = 40). (e) FBLP (model order = 16). (f) FBLP (model order = 40). (From ref. [44])

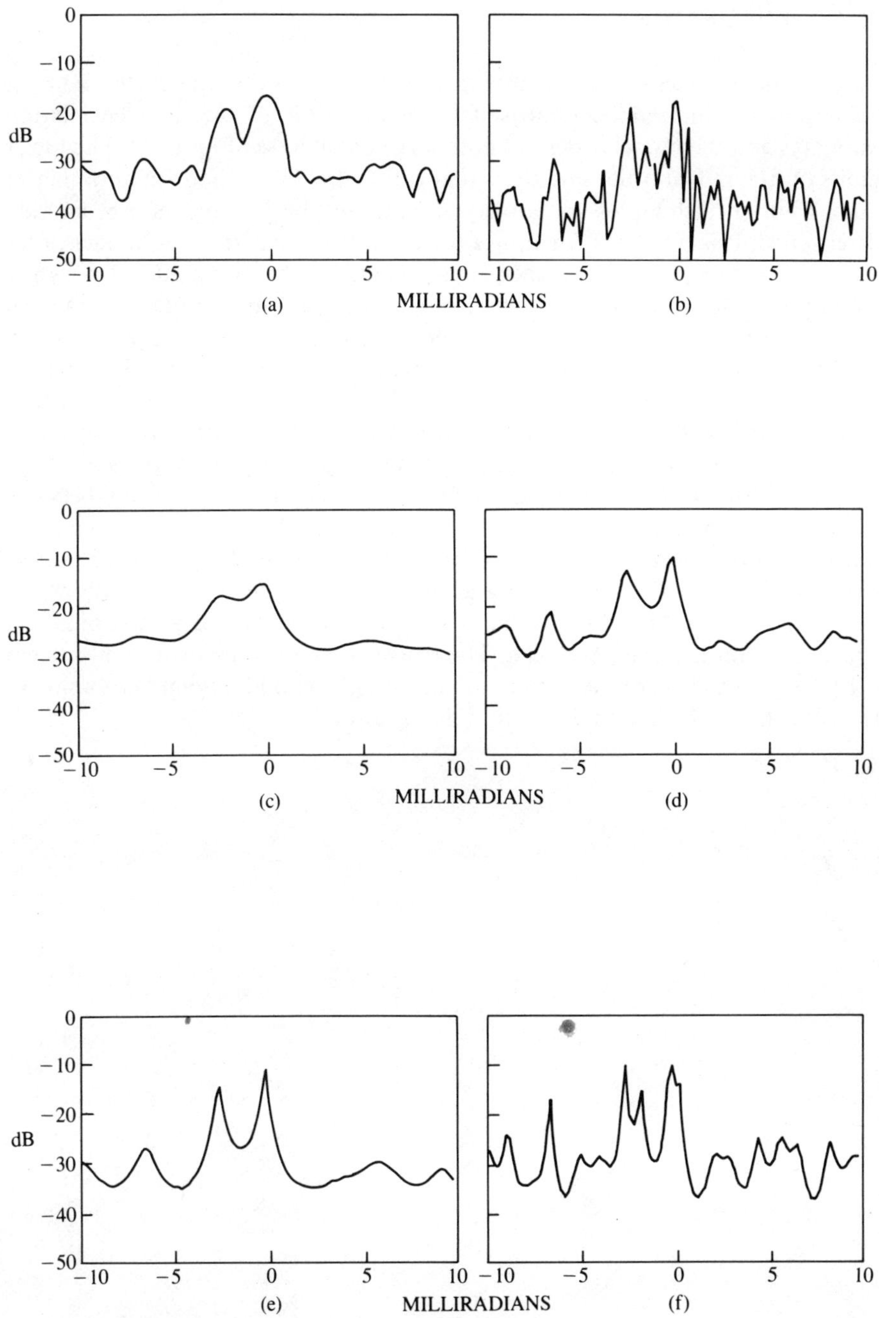

Figure 11-11 Effect of reduced model order on SR techniques. The data set consists of 160 samples of the echoes from the aircraft wing. (a) Quadruple-diversity Fourier image (model order = 40). (b) Fourier image with 160 data samples. (c) MLM (model order = 16). (d) MLM (model order = 40). (e) FBLP (model order = 16). (f) FBLP (model order = 40). (From ref. [44])

11-9 DISCUSSION

Testing the SR techniques using experimental microwave data showed that resolution can indeed be improved when the SNR of the data correlation estimates and the model order are both large enough (see Fig. 11-5). The lower limits on the model order and SNR that can make SR methods perform better than Fourier techniques on an equal data set basis have not been fully determined; it is a subject for future research. It is clear that the model order should be higher than the number of targets and probably significantly higher. Attempts to derive a direct measure of the optimal model order in practical applications have met with considerable difficulties [66]. The experiments in Section 11-8 show that the improvement in resolution obtained via SR is highly dependent on the number of snapshots available for estimating the correlation matrix. Further, the target structure, which invariably consists of distributed rather than point targets, significantly affects the performance of the algorithm. Much further work is required to fully understand the interplay between resolution and SNR, and model order selection and false alarm generation. The objective is to determine whether or not the modern SR techniques are better than conventional Fourier imaging techniques and under what conditions.

Alternately, an SR technique that is robust to noise and the process of model order selection is to be developed. The recent work of Pimbley and Silverstein [78], [90], inventors of the minimum free energy method, attempts to improve the robustness of SR techniques to noise and model order.

12

DISPLAY TECHNIQUES

Microwave images of manmade targets are difficult to recognize because their scattering characteristics differ so greatly from optical scattering. These differences were discussed in Chapter 4. The result is that images of targets consist largely of clusters of dots within confined regions that outline the target, exhibit large dynamic range, are highly sensitive to viewing angle, and have broken or discontinuous boundaries. The targets are nearly always immersed in backgrounds of clutter or noise or both. The display usually has a smaller dynamic range than the target image and some transformation, often nonlinear, between signal processor output and display input is required.

Certain display techniques that have been found useful at the Valley Forge Research Center (VFRC) are described in this chapter [55]. Their objectives are to increase target mass within a target's boundary, increase boundary continuity, and reduce noise and clutter, while preserving the resolution in the image.

VFRC's experience indicates that a high-resolution silhouette of a target is the most important asset leading to its recognition. Next, a modest gray scale is very helpful in accentuating or separating local regions on a target. An airplane engine is an example. The larger the dynamic range in the display the better, whether in gray scale or in false color, although the value diminishes with each increasing bit.

Discussed and illustrated below are three techniques: thresholding, spatial filtering, and the use of a target template or mask. The baseline image for comparison is the image of Fig. 12-1.⋆ The target is the Boeing 727 imaged by

⋆ All figures in this chapter are taken from [55]. Reprinted with permission.

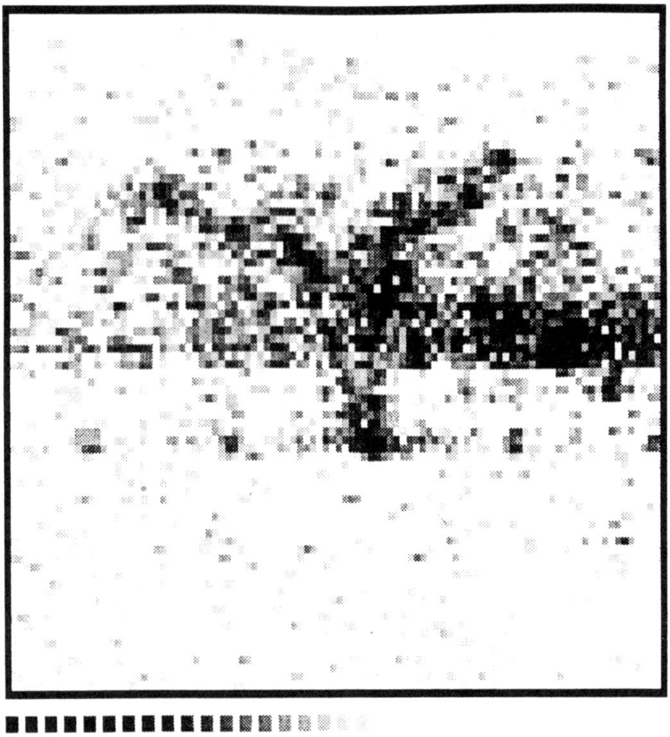

Figure 12-1 Original image. No display processing. No diversity enhancement. The target is the Boeing 727 aircraft shown earlier in Figs. 4-9 and 5-15. Its range is 2.7 km; the wavelength is 3 cm.

ISAR plus ABF and displayed earlier in Figs. 4-9 and 5-15. The raw image is shown. Neither diversity combining nor any other image enhancement technique is employed.

12-1 THRESHOLDS

Three thresholds are described in this section. They are called the *noise threshold*, the *line rms threshold*, and the *window threshold*. The noise threshold suppresses noise, the line rms threshold reduces image artifacts due to sidelobes, and the window threshold reduces the noise threshold in the target regions, thereby increasing target mass and boundary continuity.

12-1-1 Noise Threshold

The noise background is evident everywhere in Fig. 12-1. Outside the target region (that is, in the lower and upper portions of the image) the noise is statistically uniform and can easily be measured. Its rms value in Fig. 12-1 is 21.2 in arbitrary units. Because each noise sample is the amplitude of a bivariate Gaussian process, its distribution is Rayleigh, for which the probability density function (pdf) is

$$p(x) = \frac{2x}{\psi^2} e^{-x^2/\psi^2} \tag{12-1}$$

The variable x in Eq. 12-1 is the noise amplitude and ψ is its rms value.

The probability that noise exceeds a threshold x_0 is

$$\alpha = \int_{x_0}^{\infty} p(x)\, dx = e^{-x_0^2/\psi^2} \tag{12-2}$$

Equation 12-2 implies that if a threshold x_0, which is called the *noise threshold,* is established such that no values $x < x_0$ are displayed, $\alpha = \exp(-x_0^2/\psi^2)$ is the expected fraction of display pixels in which noise will appear. The noise background can be suppressed to any arbitrary degree by raising the threshold x_0. However, x_0 is constrained by the requirement that the target not be obliterated. Let y represent the target amplitude and $p(y)$ its pdf. Then

$$\beta = \int_{x_0}^{\infty} p(y)\, dy \tag{12-3}$$

is the fraction of the target distribution that survives when a threshold x_0 is imposed. Both α and β diminish with x_0. Because small α and large β are desired, a compromise must be established in choosing x_0. In practice x_0 is set to provide the highest noise level (largest α) that the user can tolerate.

Table 12-1 lists α for several values of x_0/ψ, the ratio of the noise threshold to the rms noise. This ratio is called the *rms noise multiplier.*

TABLE 12-1 **α is the fraction of the display occupied by noise, x_0 is the noise threshold, and ψ is the rms noise level.**

$\frac{x_0}{\psi}$	α
1	0.37
2	1.8×10^{-2}
2.5	1.9×10^{-3}
3	1.2×10^{-4}
4	1.1×10^{-7}

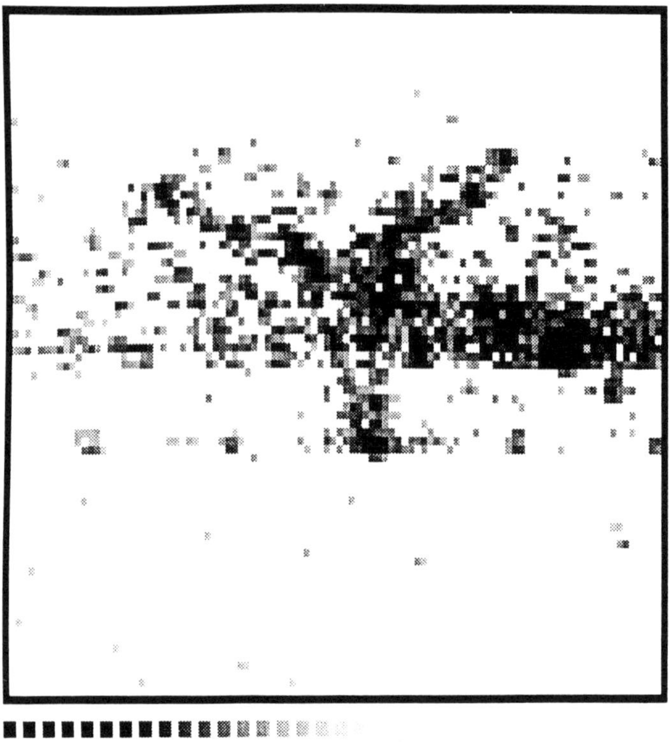

Figure 12-2 Noise threshold applied. Threshold level = 55 units.

Figure 12-2 shows the aircraft image when a noise threshold of 55 units is applied. The rms noise multiplier is 2.6 in this case. The residual noise is still excessive. A multiplier in excess of 3 is needed to remove the remaining noise. Such a value removes too much target, however. Instead, two threshold techniques are used in sequence. The first is a relatively low noise threshold as used in Fig. 12-2. This suppresses the bulk of the noise and leaves the remainder as relatively isolated dots in the image. Such isolated interference can be cleared by other techniques, such as window thresholding and median filtering, both of which are discussed later.

With the noise largely removed from Fig. 12-2, another problem becomes evident: smearing or redistribution of energy in the cross-range or azimuth dimension. It is a one-dimensional phenomenon. It is independent from range bin to range bin. It is most prominent in those range bins that have strong scatterers. The level of the smeared or redistributed energy is directly related to the strength of the large scatterers. Unlike the noise, this error energy is signal related. It is due to distant sidelobes or grating lobes resulting from residual phase errors in the array following the adaptive beamforming process. Its removal requires another threshold that varies from range bin to range bin and is a function of the target-echo strengths in each range bin.

12-1-2 Line rms Threshold

The line rms threshold x_1 performs this function. The rms value of the image data in each range bin is calculated and multiplied by a constant to form the second threshold x_1. In those range bins in which $x_1 < x_0$, x_1 is ignored, and the noise threshold is applied. When $x_1 > x_0$, however, the larger signal-related threshold is needed to reduce the artifact content.

Figure 12-3 shows the improvement in the quality of the image for a line rms multiplier of 0.3. The artifact content in the central range bins is largely removed. While a larger value of the multiplier would provide further suppression, it would also remove weak target echoes in this region. Thus this threshold limits the target dynamic range on a line-by-line basis. The compromise value of 0.3 has been found to be useful.

Examination of Fig. 12-3 discloses that the residual artifact content consists largely of relatively isolated false targets. The third thresholding technique, called *windowing* applies to this problem as well as to the residual noise.

12-1-3 Window Threshold

Window thresholding is a two-dimensional procedure. It specifically attempts to

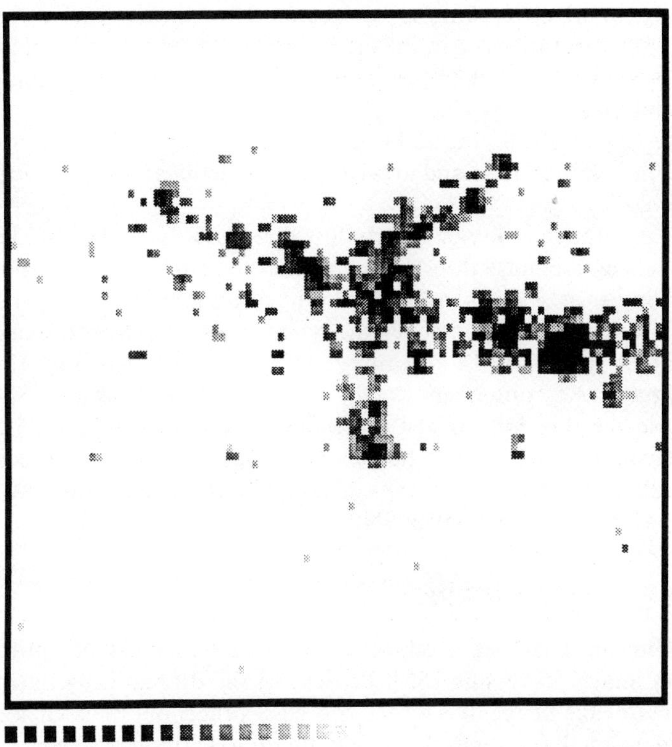

Figure 12-3 Line rms threshold added. Noise threshold = 55 units. Line rms multiplier = 0.3.

reduce strong, isolated noise and artifact values while maintaining the integrity of the target. The method capitalizes upon an important distinction between the residual noise and artifacts and the target: the residual noise is distributed randomly and uniformly throughout the image and the residual artifacts are distributed more or less uniformly in their range bins, while target echoes cluster within the target boundary. The objective of window thresholding is to remove the residual interference outside the target area while permitting a low threshold within the target.

The image is divided into windows a few pixels on each side. A window size of 4×5 is used in the example that follows. Define $x_0 = k_0 N_{rms}$ and $x_2 = k_2 W_{rms}$, where x_0 is the noise threshold, N_{rms} is the rms value of the noise in all range bins not containing target, and k_0 is the rms noise multiplier. Similarly, x_2 is the window threshold, W_{rms} is the rms value within the window and k_2 is the rms window multiplier. The rule is to replace x_0 with x_2 in every window for which $W_{rms} < x_0$.

The value of k_2 is between 4 and 5, while $k_0 \sim 2.5$. Thus $x_2 > x_0$ in most windows in the nontarget portions of the image. In other words, the effective noise threshold rises from x_0 to x_2 in the noise-alone regions, which reduces α in Eq. 12-1 to a suitably small value.

Within the target region, however, a different condition occurs. The rms target echoes in a window need exceed the noise by no more than 10 dB for $W_{rms} > x_0$, in which case x_0 is not replaced by x_2 and the lower threshold governs. The low threshold in the target region favors the preservation of small scatterers, while the higher threshold in the noise region rejects noise and isolated interference.

Figure 12-4 illustrates the addition of a window threshold with rms multiplier $k_2 = 5$. The residual noise and array-distortion artifacts are gone and the target is preserved.

The quantitative utility of the window threshold x_2 can be found experimentally by raising the noise threshold x_0 so that the background, without window thresholding, is as clean as in Fig. 12-4. The result is shown in Fig. 12-5. Raising x_0 to 105 units not only removes the noise; it also suppresses weak scatterers in the target. Thus Fig. 12-5 exhibits less target mass than Fig. 12-4 and is a poorer image as a consequence.

Because the threshold in the target region was raised from 55 to 105, the target strength would have to be 20 log(105/55) = 5.6 dB larger to achieve the same quality image. This is a typical example. In general, windowing permits imaging with 5 to 6 dB smaller SNR.

12-2 SPATIAL FILTERING

Spatial filtering has been used with considerable success in optical, IR, and television image processing [83]. Because of the dramatically different nature of the microwave image of a target, this class of techniques is far less valuable. Two-dimensional (nonlinear) median filtering and two-dimensional (linear) lowpass filtering are demonstrated in this section.

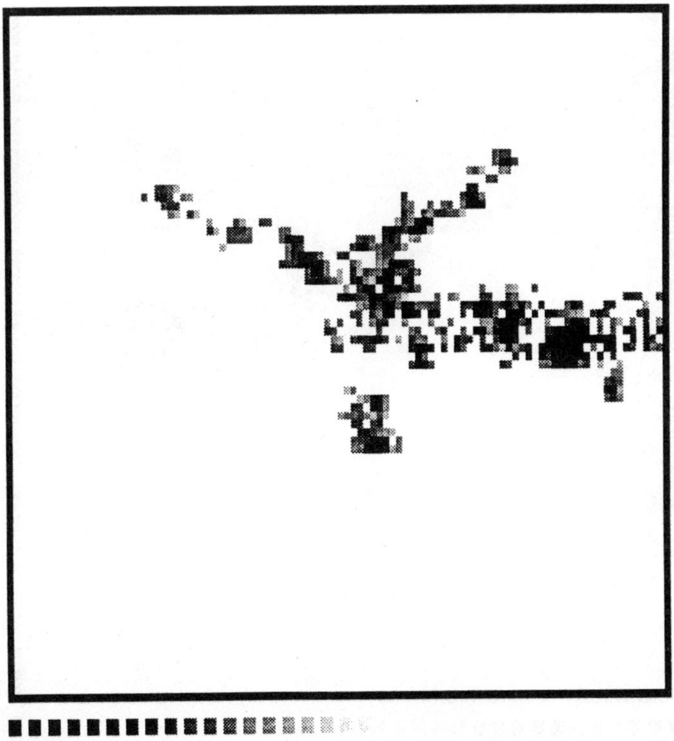

Figure 12-4 Window threshold added. Noise threshold = 55 units. Line rms multiplier = 0.3. Window rms multiplier = 5.0.

12-2-1 Median Filtering

In this method a small window (3 × 3 pixels in the illustrations) is placed about each pixel and the median value of the (9 in this case) samples in the window replaces the value in the center [56]. Thus large and small values are filtered out. No loss in resolution occurs because there is no spatial smoothing (as with linear lowpass filtering). It is an effective procedure in the noise regions but unfortunately it is equally effective in suppressing relatively isolated target pixels in the target regions.

Figure 12-6 shows median filtering applied to the airplane image. Use of this procedure allowed the noise threshold to be lowered to 37 units. However, the target image suffered.

12-2-2 Square Averaging

Square averaging is two-dimensional, linear, lowpass spatial filtering. A square mask or window of uniform amplitude is convolved with the image. The average noise power is lessened by the averaging process. However, the resolution in each dimension suffers by a factor equal to the size of the window.

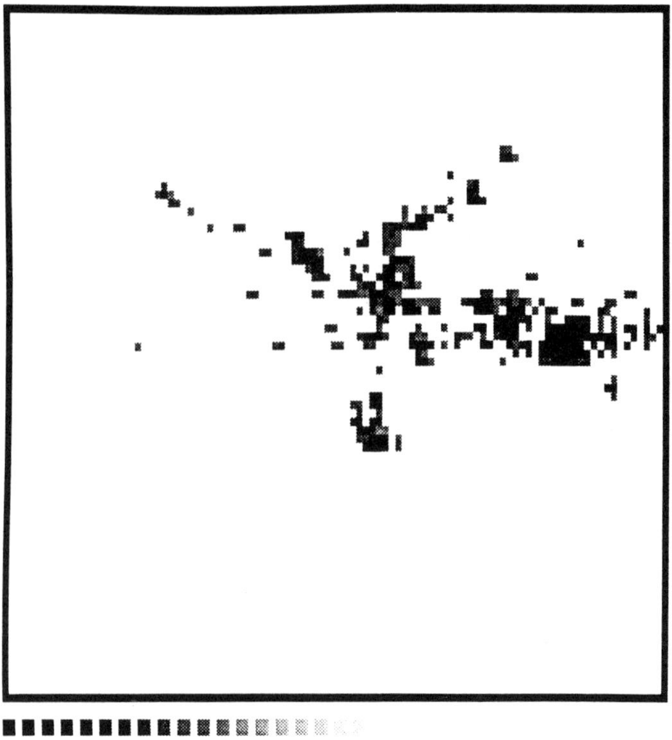

Figure 12-5 Without window threshold. Noise threshold adjusted for same noise background. Noise threshold = 105 units. Line rms multiplier = 0.3.

Figure 12-6 Median filtering applied to original image. Noise threshold = 37 units. Line rms multiplier = 0.3. Window rms multiplier = 5.0.

As before, we will assume that the noise has the Rayleigh distribution (12-1). The noise power $Ex^2 = \psi^2$ and its mean value $Ex = \sqrt{\pi}\,\psi/2$. The noise variance $Ex^2 - (Ex)^2 = \psi^2(1 - \pi/4)$ and its standard deviation $\sigma = \psi(1 - \pi/4)^{1/2}$. After averaging $M \times M$ independent noise samples (in an $M \times M$ window) the

TABLE 12.2 $x_0 = Ex + 4\sigma_M$ **is the threshold level which noise exceeds with probability** $\sim 10^{-4}$. ψ **is the rms noise level and** M **is the size of the square averaging window. The last two columns show the allowed reduction in the threshold resulting from the** $M \times M$ **averaging.**

M	$\frac{x_0}{\psi}$	*Reduction in* x_0/ψ	
1	2.74	1	0 dB
2	1.81	0.66	3.6
3	1.50	0.55	5.2
4	1.35	0.49	6.1
6	0.95	0.35	9.2
∞	0.89	0.32	9.8

standard deviation reduces to $\sigma_M = \psi(1 - \pi/4)^{1/2}/M$, while the mean remains unchanged. The noise threshold x_0 is approximately the mean Ex plus $4\sigma_M$, which is tabulated in Table 12-2. The table indicates that a 3×3 window will permit the noise threshold x_0 to be lowered by 5.2 dB.

Figure 12-7 is a 3×3 square average image using the same parameters as in Fig. 12-3. The threshold was adjusted to remove the noise. A reduction from 105 (Fig. 12-5) to 56, or 5.5 dB, was achieved, which is consistent with the predicted value from Table 12.2.

The loss in resolution resulting from the convolution of the image with the 3×3 window is evident.

When window thresholding is added (Fig. 12-8) the noise threshold is lowered an additional 4.3 dB for a total reduction of 9.8 dB. The result is greater target mass in the image.

12-2-3 Gaussian Averaging

Tapering the 3×3 mask results in a lower loss in resolution; the penalty is a slightly lower reduction in the noise threshold. Figures 12-9 and 12-10 apply a Gaussian window, for which the weights are

$$\begin{matrix} 0 & 1 & 0 \\ 1 & 4 & 1 \\ 0 & 1 & 0 \end{matrix}$$

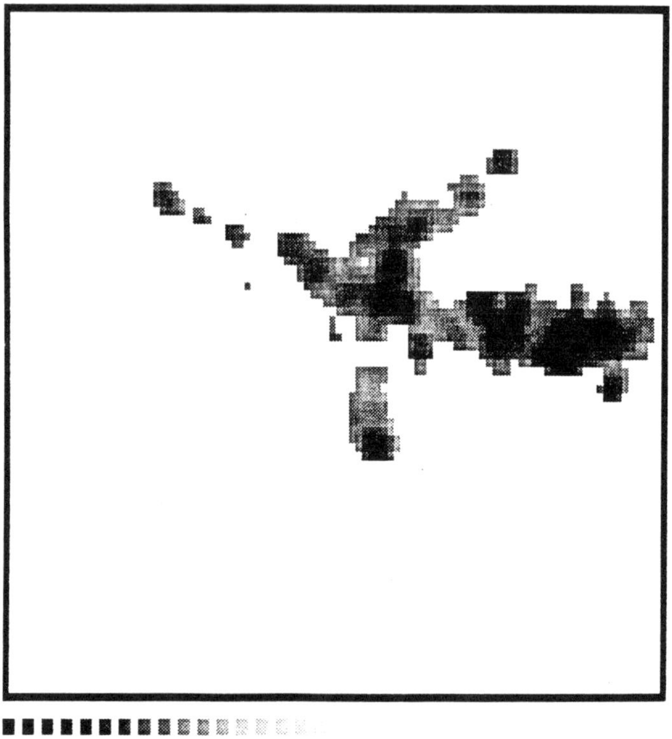

Figure 12-7 Square averaging without window thresholding. Noise threshold = 56 units. Line rms multipliers = 0.3.

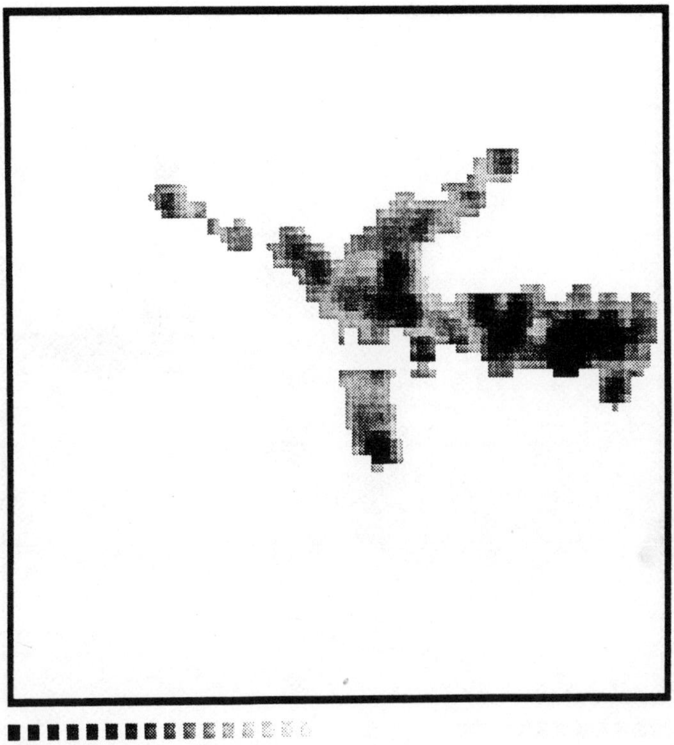

Figure 12-8 Square averaging with window thresholding. Noise threshold = 34 units. Line rms multiplier = 0.3. Window rms multiplier = 5.0.

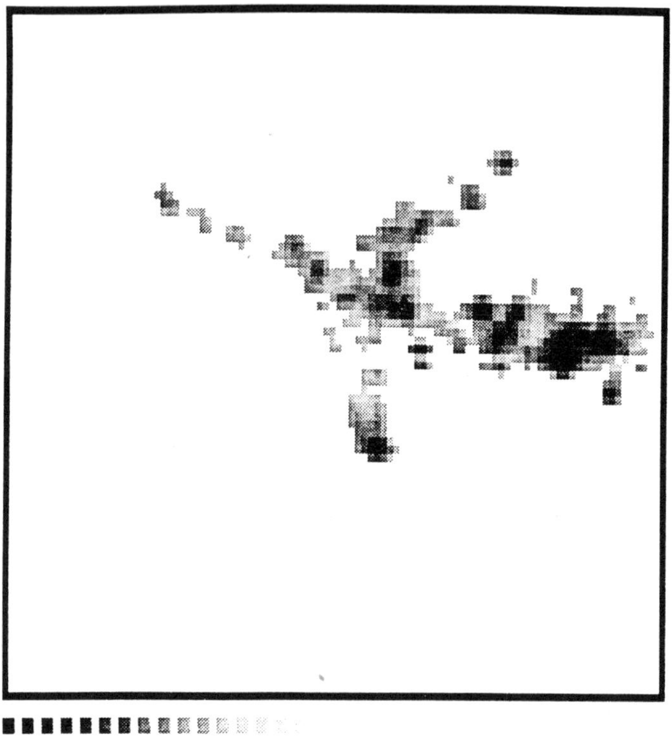

Figure 12-9 Gaussian averaging without window thresholding. Noise threshold = 58 units. Line rms multiplier = 0.3.

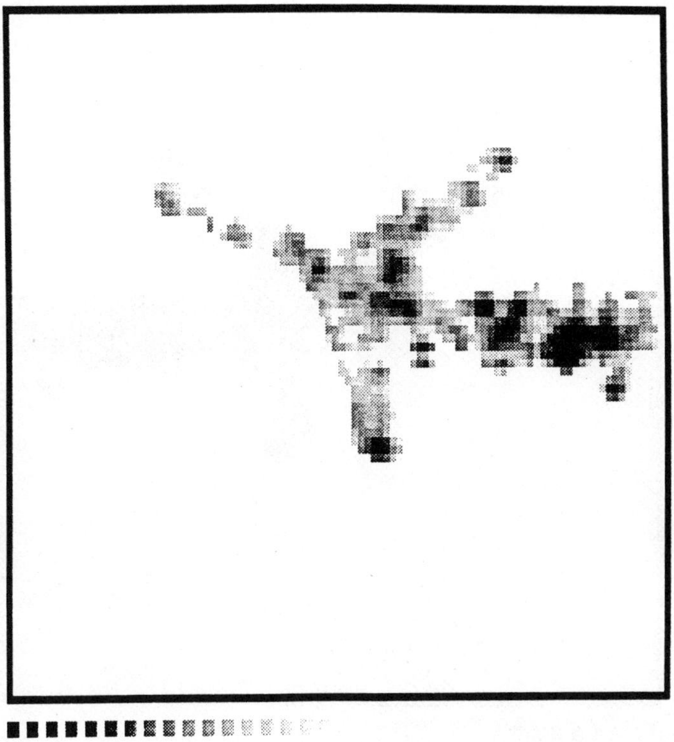

Figure 12-10 Gaussian averaging with window thresholding. Noise threshold = 36 units. Line rms multiplier = 0.3. Window rms multiplier = 5.0.

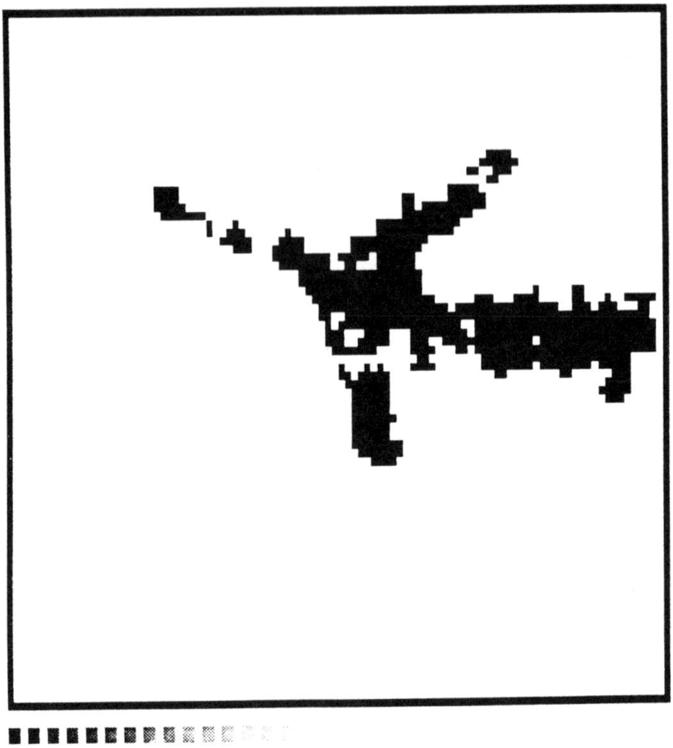

Figure 12-11 Target template made from Gaussian filtered image (Fig. 12-10).

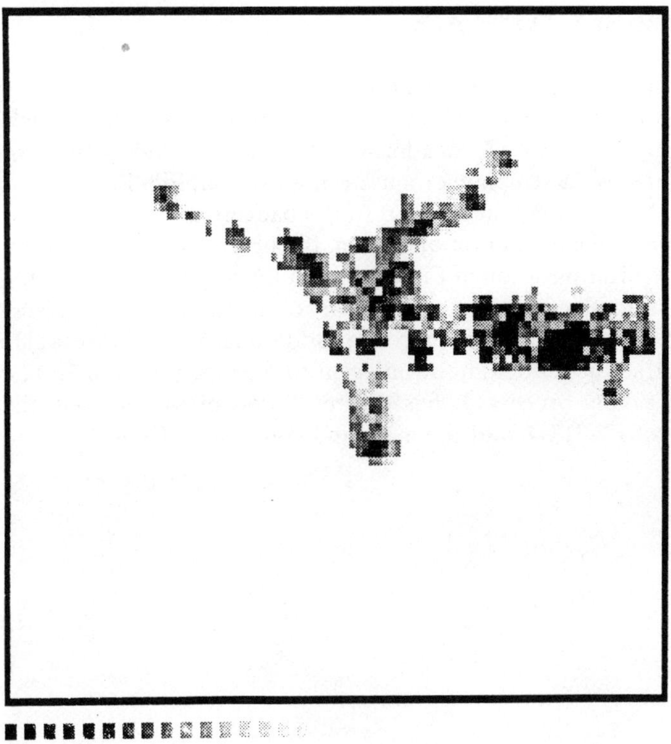

Figure 12-12 Original image (Fig. 12-1) multiplied by target mask (Fig. 12-11).

Without window thresholding (Fig. 12-9) the noise threshold reduced to 56 (from 105), or 5.5 dB. With window thresholding (Fig. 12-10) the reduction was 9.3 dB. In general 8 to 9 dB can be expected from 3 × 3 Gaussian filtering plus window thresholding.

Both images are very good. The noise is gone, the target mass is satisfactory and resolution is good. Comparison of Fig. 12-10 with Fig. 12-4 shows the target mass and boundary continuity to be better in the former, at a sacrifice of resolution.

12-3 TARGET TEMPLATE

Resolution can be preserved and target mass and boundary continuity increased when the target is in the clear, as in the airplane image. A black-and white target template is made from a linearly filtered and window thresholded image. Figure 12-11 is the template or mask obtained from Fig. 12-10. Assign the value 1 to the (black) target mass and 0 to the background field. Because Fig. 12-10 resulted from a convolution operation, the black area completely encompasses the target. Multiplication of Fig. 12-11 and the original image (Fig. 12-1) rejects all noise more than a pixel from any part of the target and preserves the highest resolution available in the image. An additional step of thresholding may be used to clean the residual noise adjacent to the target. Figure 12-12 is the result of this masking process. It has the resolution of the original, the noise-free property of Fig. 12-4, and the mass and continuity of Fig. 12-10.

REFERENCES

[1] Abend, K., Attia, E. H. and Kilpatrick, D. "Distributed array radar demonstration," Interspec Phase I SBIR Final Report to Rome Air Development Center, Griffis AFB, NY, 13441-5700, under Contract No. F19628-87-C-0136, Report No. RADC-TR-89-207, October 1989.

[2] Attia, E. H. "Self-cohering large antenna arrays using the spatial correlation properties of radar clutter," Ph.D. Dissertation, The Moore School of Electrical Eng., University of Pennsylvania, December 1984.

[3] Attia, E. H. "Self-cohering airborne distributed arrays," INTERSPEC Final Report on Contract No. F19628-84-C-0080 for Rome Air Development Center, Hanscom, MA, December 1985.

[4] Attia, E. H. "Self-cohering airborne distributed arrays using the robust minimum variance algorithm," *Proc. IEEE AP-S and URSI International Symposium*, Philadelphia, Symposium Digest, Vol. 2, pp. 603–606, June 1986.

[5] Attia, E. H. and Steinberg, B. D. "Self-cohering large antenna arrays using the spatial correlation properties of radar clutter," *IEEE Trans. Ant. and Prop.*, **AP-37**(1), pp. 30–38, January 1989.

[6] Ausherman, D. A., Kozma, A., Walker, J. L., Jones, H. M. and Poggio, E. C. "Developments in radar imaging," *IEEE Trans. Aerosp. Electron. Syst.*, **AES-20**(4), pp. 363–398, July 1984.

[7] Baghdady, E. J., Ed. (1961). *Lectures on Communication System Theory*, McGraw-Hill, New York.

[8] Berkowitz, R. S., Ed. (1965). *Modern Radar Systems*, Wiley, New York (see Chap. 3, Part VI, by B. D. Steinberg).

[9] Berran, M. J. and Parrent, G. R. (1964). *Theory of Partial Coherence*, Prentice-Hall, Englewood Cliffs, N.J.

[10] Boerner, W. M. (1982). "Basic concepts of radar polarimetry and its applications to target discrimination, classification, imaging, and identification," Technical Report EMID-CL-82-05-18-02, Communications Laboratory, University of Illinois, Chicago.

[11] Booker, H. G. and Clemmow, P. C. "The concept of an angular spectrum of plane waves, and its application to that of polar diagram and aperture distribution," *Proc. IEEE (London)*, **97**, January 1950.

[12] Born, M. and Wolf, E. (1965). *Principles of Optics*, 5th ed., Pergamon Press, New York.

[13] Box, G. E. P. and Jenkins, G. M. (1970). *Time Series Analysis: Forecasting and Control*, Holden-Day, San Francisco, CA.

[14] Brennan, D. G. "Linear diversity techniques," *Proc. IRE*, **47**, pp. 1075–1102, June 1959.

[15] Burg, J. P. "Maximum entropy spectral analysis," *Proc. 37th Annual International Meeting*, Society of Exploration Geophys., Oklahoma City, OK, October 1967.

[16] Burg, J. P., Luenberger, D. G. and Wegner, D. L. "Estimation of structured covariance matrices," *Proc. IEEE*, **70**(9), September 1982.

[17] Butler, J. and Lowe, R. "Beamforming matrix simplifies design of electronically scanned antennas," *Electron. Design*, **9**(7), pp. 170–173, April 1961.

[18] Capon, J. "High resolution frequency-wavenumber spectrum estimation," *Proc. IEEE*, **57**(8), pp. 1408–1418, August 1969.

[19] Chen, C. C. and Andrews, H. C. "Target-motion-induced radar imaging," *IEEE Trans. Aerosp. Electron. Syst.*, **AES-16**(1), pp. 2–14, January 1980.

[20] Clark, B. G. (1980). "An efficient implementation of the algorithm CLEAN," *Astron. Astrophys.*, **89**, pp. 377–378.

[21] Cornwell, T. J. "Self-calibration," Lecture Notes, National Radio Astronomy Observatory, Socorro, NM.

[22] Cornwell, T. J. "The application of closure phase to astronomical imaging," *Science*, **245**(4915), pp. 263–269, July 1989.

[23] Dallas, W. J. "Phase quantization—a compact derivation," *Appl. Opt.*, **10**(3), pp. 673–674, March 1971.

[24] Dudgeon, D. E. and Mersereau, R. M. (1984). *Multidimensional Digital Signal Processing*, Prentice Hall, Englewood Cliffs, N.J.

[25] Evans, D. L., Farr, T. G., Van Zyl, J. J. and Zebker, H. A. "Radar polarimetry: analysis tools and applications," *IEEE Trans. Geoscience and Remote Sensing*, **26**(6), pp. 774–789, November 1988.

[26] Evans, J. E., Johnson, J. R. and Sun, D. F. (1981). "High resolution angular spectrum estimation techniques for terrain scattering analysis and angle of arrival estimation," *Proc. 1st ASSP Workshop on Spectral Estimation*, Hamilton, Ontario, pp. 134–139.

[27] Farhat, N. H. and Guard, W. R. "Millimeter wave holographic images of concealed weapons," *Proc. IEEE*, **59**, pp. 1383–1384, September 1971.

[28] Farhat, N. H. Unpublished Notes, The Moore School of Electrical Eng., University of Pennsylvania, Philadelphia, PA.

[29] Feinup, J. R. "Space object imaging through a turbulent atmosphere," *Opt. Eng.*, **18**(5), pp. 529–534, September–October 1979.

[30] Feinup, J. R. "Iterative method applied to image reconstruction and to computer-generated holograms," *Opt. Eng.*, **19**(3), pp. 297–305, May–June 1980.

[31] Gerchberg, R. W. and Saxton, W. O. (1972). "A practical algorithm for the determination of phase from image and diffraction plane pictures," *Optik*, **35**, pp. 237–246.

[32] Goodman, J. W. and Silvesteri, A. M. "Some effects of Fourier domain phase quantization," *IBM J. Res. Develop.*, pp. 478–484, September 1970.

[33] Goodman, J. W. "Some fundamental properties of speckle," *J. Opt. Soc. Amer.*, **66**(11), November 1976.

[34] Goodman, J. W. *Statistical Optics*, Wiley, New York.

[35] Giuli, D. "Polarization diversity in radars," *Proc. IEEE*, **74**(2), pp. 245–269, February 1986.

[36] Haber, F. and Zoltowski, M. "Spatial spectrum estimation in a coherent signal environment using an array in motion," *IEEE Trans. Ant. and Prop.*, **AP-34** (3), March 1986.

[37] Halat, J. "A survey of image-embedded passive beamforming sources for a radio camera," Masters Thesis, The Moore School of Electrical Eng., University of Pennsylvania, Philadelphia, PA, December 1985.

[38] Hall, H. J. and Howell, H. K. (1965). *Photographic Considerations for Aerospace*, ITEK Corp., Boston, MA.

[39] Hamaker, J. P., O'Sullivan, J. D. and Noordam, J. E. (1977). "Image sharpness, Fourier optics and redundant-spacing interferometry," *J. Opt. Soc. Amer.*, **67**, pp. 1122–1123.

[40] Haykin, S. S. and Reilly, J. "Mixed autoregressive moving average modeling of the response of a linear array antenna to incident plane waves," *Proc. IEEE*, **69**, November 1981.

[41] Hoctor, R. T. and Kassam, S. A. "The unifying role of the coarray in aperture synthesis for coherent and incoherent imaging," *Proc. IEEE*, **78**(4), pp. 735–752, April 1990.

[42] Hogbom, J. (1974). "Aperture synthesis with a non-regular distribution of interferometer baselines," *Astrophysics Journal Supplement*, **15**, pp. 417–426.

[43] Jenkins, G. M. and Watts, D. G. (1968). *Spectral Analysis and its Applications*, Holden-Day, San Francisco, CA.

[44] Jin, C. and Steinberg, B. D. "Application of modern spectral estimation techniques to microwave imaging," Valley Forge Research Center Report UP-VFRC-15-87, Valley Forge Research Center, The Moore School of Electrical Eng., University of Pennsylvania, Philadelphia, PA, March 1988.

[45] Johnson, R. C. and Jasik, H. (1984). *Antenna Engineering Handbook*, 2nd ed., McGraw-Hill, New York.

[46] Kang, B. "Self-calibration technique [for] phased arrays using the multiple dominant scatterer algorithm," Masters Thesis, The Moore School of Electrical Eng., University of Pennsylvania, Philadelphia, PA, May 1987.

[47] Kang, B. "Research on self-calibrating algorithms for high resolution microwave imaging systems," Ph.D. Dissertation, Dept. of Electrical Eng., Drexel University, Philadelphia, PA, August 1990.

[48] Kang, B., Subbaram, H. M. and Steinberg, B. D. "Improved adaptive beamforming target for self-calibrating a distorted phased array," *IEEE Trans. Ant. and Prop.*, **AP-38**(2), pp. 186–194, February 1990.

[49] Kay, S. M. and Marple Jr., S. L. (1979). "Sources of and remedies for spectral line splitting in autoregressive spectrum analysis," *Proc. Int. Conf. Acoustics, Speech, and Sig. Proc.*, pp. 151–154.

[50] Kay, S. M. and Marple Jr., S. L. "Spectral analysis—a modern perspective," *Proc. IEEE*, **69**(4), April 1981.

[51] Kermisch, D. "Image reconstruction from phase information only," *J. Opt. Soc. Amer.* , **60**(1), January 1970.

[52] Lafrance, P. and Osbourne, J. J. "Effect of reduced SAR azimuth processing," *IEEE Trans. Aerosp. Electron. Syst.*, **AES-23**(2), pp. 152–157, March 1987.

[53] Lang, S. W. and McClellan, J. H. "Frequency estimation with maximum entropy spectral estimators," *IEEE Trans. Acoustics, Speech, and Sig. Proc.*, **ASSP-28**, pp. 716–724, December 1980.

[54] Lee, H. and Wade, G. "Evaluating quantization error in phase-only holograms," *IEEE Trans. on Ultrasonics*, **SU29**, pp. 251–254, September 1982.

[55] Lee, W. "Threshold and image enhancement techniques for microwave imagery," Valley Forge Research Center Quarterly Progress report No. 56, Valley Forge Research Center, The Moore School of Electrical Eng., University of Pennsylvania, Philadelphia, PA., pp. 14–26, December 1988.

[56] Lee, Y. H. and Kassam, S. A. "Generalized median filtering and related non-linear filtering techniques," *IEEE Trans. Acoustics, Speech, and Sig. Proc.*, **ASSP-33**, pp. 672–683, June 1985.

[57] Levinson, N. (1947). "The Weiner (root mean square) error criterion in filter design and prediction," *J. Math. Phys.*, **25**, pp. 261–278.

[58] Lewis, B. L., Kretschmer, Jr., F. F. and Shelton, W. W. (1986). *Aspects of Radar Signal Processing*, Artech House, Norwood, MA (see Chapter 2, Pulse Compression).

[59] Liang, Z. and Steinberg, B. D. "Optimum data quantization in microwave imagery and its effect upon image quality," *IEEE Trans. Acoustics, Speech, and Sig. Proc.*, **ASSP-36**(12), pp. 1889–1898, December 1988.

[60] Liang, Z., Steinberg, B. D. and Kesler, S. B. "An exact solution of the analytic equation of image quality from optimum quantization of microwave imaging data," *IEEE Trans. Acoustics, Speech, and Sig. Proc.*, **ASSP-37**(9), pp. 1442–1445, September 1989.

[61] Lohmann, A. W. (1970). "Comments about phase-only holograms," in *Acoustical Holography*, Vol. 2, A. F. Metterhill and L. Larimore (Eds), Plenum, New York, Ch. 14, pp. 203–210.

[62] Luthra, A. K. (1981). "Maximum entropy method in the space-angle domain and a new technique with superior performance," Ph.D. Dissertation, Systems Eng., University of Pennsylvania, PA.

[63] Makhoul, J. "Linear prediction: a tutorial review," *Proc. IEEE*, **63**(4), April 1975.

[64] Mangin, P. A., Von Ramm, O. T. and Thurstone, F. L. "Delay quantization error in phase array images," *IEEE Trans. Sonics and Ultrasound*, **SU-28**(5), pp. 305–310, September 1981.

[65] Marple Jr., S. L. "A new autoregressive spectrum analysis algorithm," *IEEE Trans. Acoustics, Speech, and Sig. Proc.*, **ASSP-28**, August 1980.

[66] Marple Jr., S. L. (1987). *Digital Spectral Analysis with Applications*, Prentice-Hall, Englewood Cliffs, N.J.

[67] Metterhill, A. F. (1969). "The relative importance of phase and amplitude in acoustical holography," in *Acoustical Holography*, Vol. 1, A. F. Metterhill, H. M. A. El-Sum and L. Larimore (Eds), Plenum, New York, Ch. 14, pp. 203–224.

[68] Mitchell, R. H. and Marder, S. (1981). "Synthetic aperture radar (SAR) image quality considerations," *Proc. SPIE, Image Quality*, **310**, pp. 58–68.

[69] Muller, R. A. and Buffington, A. "Real-time correction of atmospherically degraded telescope images through image sharpening," *J. Opt. Soc. Amer.*, **64**(9), pp. 1200–1210, September 1974.

[70] Munson, Jr., D. C. and Sanz, J. L. C. "Phase-only image reconstruction from offset Fourier data," *Opt. Eng.*, May 1986.

[71] Nutall, A. H. "Spectral analysis of a univariate process with bad data points," Tech. Document 5419, Naval Underwater Systems Center, New London, CT, May 1976.

[72] Oppenheim, A. V. and Lim, J. S. "The importance of phase in signals," *Proc. IEEE*, **29**(5), pp. 529–541, May 1981.

[73] Panter, P. (1972). *Communication System Design: Line-of-Sight and Troposcatter Systems*, McGraw-Hill, New York.

[74] Papoulis, A. (1962). *The Fourier Integral and Its Applications*, McGraw-Hill, New York.

[75] Papoulis, A. "A new algorithm in spectral analysis and bandlimited signal extrapolation," *IEEE Trans. Circuits and Systems*, **CAS-22**, pp. 735–742, September 1975.

[76] Patrick, S. "Validation of Tsao's energy conservation algorithm for self-calibration of distorted phased arrays," Masters Thesis, The Moore School of Electrical Eng., University of Pennsylvania, Philadelphia, PA, May 1990.

[77] Perley, R. A. and Dreher, J. W. Courtesy National Radio Astronomy Observatory, Socorro, New Mexico.

[78] Pimbley, J. M. and Silverstein, S. D. "Recursive minimum free energy spectral estimation," 22nd Asilomar Conference on Signals, Systems and Computers, Pacific Grove, CA, November 1988.

[79] Powers, E. N. (1975). "Adaptive arrays for microwave imaging," Ph.D. Dissertation, The Moore School of Electrical Eng., University of Pennsylvania, Philadelphia, PA.

[80] Powers, J., Landry, J. and Wade, G. (1970). "Computed reconstructions from phase-only and amplitude-only holograms," in *Acoustical Holography*, Vol. 2, A. F. Metterhill and L. Larimore (Eds), Plenum, New York, Ch. 13, pp. 185–202.

[81] Readhead, A. C. S. "Radio astronomy by very long baseline interferometry," *Sci. Am.*, pp. 53–61, June 1982.

[82] Rihaczek, A. W. (1969). *Principles of High-Resolution Radar Imaging*, McGraw-Hill, New York.

[83] Rosenfeld, A. and Kak, A. C. (1982). *Digital Picture Processing*, Academic Press, New York, N.Y.

[84] Schmidt, R. O. "Multiple emitter location and signal parameter estimation," *IEEE Trans. Ant. and Prop.*, **AP-34**, March 1986.

[85] Schwab, F. R. (1984). "Relaxing the isoplanatism assumption in self-calibration; applications to low-frequency radio interferometry," *Astron. Journal*, **89**, pp. 1076–1081.

[86] Schwartz, M., Bennett, W. R. and Stein, S. (1966). *Communications Systems and Techniques*, McGraw-Hill, New York.

[87] Shan, T. J., Wax, M. and Kailath, T. "On spatial smoothing for direction of arrival estimation of coherent signals," *IEEE Trans. Acoustics, Speech, and Sig. Proc.*, **ASSP-33**(5), pp. 806–811, August 1985.

[88] Silva, M. T. and Robinson, E. A. (1979). *Deconvolution of Geophysical Time Series in the Exploration of Oil and Natural Gas*, Elsevier, Amsterdam.

[89] Silver, S. (1949). *Microwave Antenna Theory and Design*, MIT Rad. Lab. Ser., McGraw-Hill, New York.

[90] Silverstein, S. D. and Pimbley, J. M. "Robust spectral estimation: autocorrelation based minimum free energy method," 22nd Asilomar Conference on Signals, Systems and Computers, Pacific Grove, CA, November 1988.

[91] Skolnik, M. I., Nemhauser, G. and Sherman, III, J. W. "Dynamic programming applied to unequally spaced arrays," *IEEE Trans. Ant. and Prop.*, **AP-12**, pp. 35–43, January 1964.

[92] Skolnik, M. I., Ed. (1970). *Radar Handbook*, McGraw-Hill, New York.

[93] Steinberg, B. D. "Design approach for a high resolution microwave imaging radio camera," *Journal of the Franklin Institute*, December 1973.

[94] Steinberg, B. D. (1976). *Principles of Aperture and Array System Design*, Wiley, New York.

[95] Steinberg, B. D. "Radar imaging from a distorted array: the radio camera algorithm and experiments," *IEEE Trans. Ant. and Prop.*, **AP-29**(5), pp. 740–748, September 1981.

[96] Steinberg, B. D. "Properties of phase synchronizing sources for a radio camera," *IEEE Trans. Ant. and Prop.*, **AP-30**(6), pp. 1086–1092, November 1982.

[97] Steinberg, B. D. (1983). *Microwave Imaging with Large Antenna Arrays: Radio Camera Principles and Techniques*, Wiley, New York.

[98] Steinberg, B. D. and Attia, E. H. "Sidelobe reduction of random arrays by element position and frequency diversity," *IEEE Trans. Ant. and Prop.*, **AP-31**, pp. 922–930, November 1983.

[99] Steinberg, B. D. and Yadin, E. "Effect of multipath and scattering on array gain of a large adaptive beam forming phased array," *IEEE Trans. Ant. and Prop.*, **AP-33**(5), pp. 481–487, May 1985.

[100] Steinberg, B. D. "Distortion correction by image feedback control," Valley Forge Research Center Quarterly Progress Report No. 49, Valley Forge Research Center, The Moore School of Electrical Eng., University of Pennsylvania, Philadelphia, PA, pp. 54–58, March 1986.

[101] Steinberg, B. D. "A theory on the effect of hard limiting and other distortions upon the quality of microwave images," *IEEE Trans. Acoustics, Speech, and Sig. Proc.*, **ASSP-35**(10), pp. 1462–1472, October 1987.

[102] Steinberg, B. D. "Microwave imaging of aircraft," *Proc. IEEE*, **76**(12), pp. 1578–1592, December 1988.

[103] Steinberg, B. D., Carlson, D. and Lee, W. "Localized RCS of targets," *Proc. IEEE*, **77**(5), May 1989.

[104] Stockburger, E. (1988). "CLEAN performance in near-field inverse synthetic aperture radar imagery," Master's Thesis, Dept. of Electrical Eng., University of Pennsylvania, Philadelphia, PA.

[105] Subbaram, H. M. and Steinberg, B. D. "Diversity techniques for sidelobe reduction in phased array antennas," *Proc. 1986 National Radio Science Meeting*, Philadelphia, PA, June 1986.

[106] Subbaram, H. M. and Steinberg, B. D. "Self-calibration of phased array antennas using the Muller Buffington theorem and transmitter location diversity," Valley Forge Research Center Quarterly Progress Report No. 50, Valley Forge Research Center, The Moore School of Electrical Eng., University of Pennsylvania, Philadelphia, PA, April–September 1986.

[107] Subbaram, H. M. and Steinberg, B. D. "A scene-independent technique for self-calibrating phased arrays in a coherent target environment," *Proc. 1987 URSI Radio Science Meeting*, Blacksburg, VA, June 1987.

[108] Subbaram, H. M. (1987). "Artifact suppression and superresolution in coherent microwave imaging," Ph.D. Dissertation, The Moore School of Electrical Eng., University of Pennsylvania, Philadelphia, PA.

[109] Subbaram, H. M. "Scene independent self-calibration of phased array antennas," Valley Forge Research Center Quarterly Progress Report No. 52, Valley Forge Research Center, The Moore School of Electrical Eng., University of Pennsylvania, Philadelphia, PA, pp. 17–26, June 1987.

[110] Subbaram, H. M., Stockburger, E. and Fell, B. "Data quantization for digital beamforming radar systems", Interspec Final Report on Contract No. DAALO2-87-C-0068 for the Department of the Army, Harry Diamond Laboratories, Adelphi, MD, January 1988.

[111] Subbaram, H. M., Abend, K., Martinson, L. and Attia, E. H. "Integrated design of 'smart' phased array systems," Interspec Final Report on Contract No. F19628-89-C-0179 for Rome Air Development Center (Hanscom), MA, March 1990.

[112] Subbaram, H. M. and Steinberg, B. D. (1990). "Decorrelation of coherent targets using transmitter location diversity," to be published in *IEEE Trans. Ant. and Prop.*

[113] Swenson, G. W. and Mathur, N. C. "The interferometer in radio astronomy," *Proc. IEEE*, **56**(12), December 1968.

[114] Taheri, S. H. "Coherent transmitter diversity," Unpublished Notes, Valley Forge Research Center, The Moore School of Electrical Eng., University of Pennsylvania, Philadelphia, PA.

[115] Thomas, J. B. (1969). *Statistical Communication Theory*, Wiley, New York.

[116] Thompson, A. R., Clark, B. G., Wade, C. M. and Napier, P. J. (1980). "The very large array," *Astrophys. Journal Supplement*, **45**, pp. 151–167.

[117] Thompson, A. R., Moran, J. M. and Swenson, Jr., G. W. (1986). *Interferometry and Synthesis in Radio Astronomy*, Wiley, New York.

[118] Tsao, J. (1985). "A position error correction technique and a revised CLEAN technique for thinned array imaging," Ph.D. Dissertation, Systems Eng., University of Pennsylvania, Philadelphia, PA.

[119] Tsao, J. "Phased array beamforming by the Parceval's theorem," *IEEE AP-S Symposium Digest*, IEEE AP-S International Symposium, Philadelphia, PA, Vol. 1, pp. 335–338, June 1986.

[120] Tsao, J. (1988). "Radar imaging with a partially coherent array," *Proc. IEEE AP-S International Symposium*, Vol. II, pp. 794–797.

[121] Tsao, J. and Steinberg, B. D. "Reduction of sidelobe and speckle artifacts in microwave imaging: the CLEAN technique," *IEEE Trans. Ant. and Prop.*, **AP-36**(4), pp. 543–556, April 1988.

[122] Ulrych, T. J. and Clayton, R. W. "Time series modelling and maximum entropy," *Phys. Earth Planetary Interiors*, **12**, August 1976.

[123] Van-Hove, P. L., Lim, J. S. and Oppenheim, A. V. (1983). "Signal reconstruction from Fourier transform amplitude," *Tech. Digest*, Topical Meeting on Signal Recovery and Synthesis with Incomplete Information and Partial Constraints, Lake Tahoe, NV.

[124] Van-Hove, P. L., Hayes, M. H. and Oppenheim, A. V. "Signal reconstruction from signed Fourier transform magnitude," *IEEE Trans. Acoustics, Speech, and Sig. Proc.*, **ASSP-31**, October 1983.

[125] Wehner, D. R. (1987). *High Resolution Radar*, Artech House, Norwood, MA.

[126] Wong, A. C. C. "Radar digital beamforming," *Military Microwaves*, **82**, London.

[127] Woodward, P. M. (1953). *Probability and Information Theory, with Application to Radar*, McGraw-Hill, New York.

[128] Yadin (Jadlovker), E. (1981). "Airborne radars with distributed antenna arrays," Ph.D. Dissertation, Systems Eng., University of Pennsylvania, Philadelphia, PA.

INDEX